T0271358

Construction Claims

Construction Claims
Prevention and Resolution

Third Edition

Robert A. Rubin
Virginia Fairweather
Sammie D. Guy

JOHN WILEY & SONS, INC.

New York – Chichester – Weinheim – Brisbane – Singapore – Toronto

Library of Congress Cataloging-in-Publication Data:

Rubin, Robert A.
    Construction claims : prevention and resolution / Robert Rubin, Virginia Fairweather, Sammie D. Guy. — 3rd ed.
        p. cm.
    Rev. ed. of: Construction claims / Robert Rubin ... [et al.]. 2nd ed. c1992.
    Includes index.
    ISBN 978-0-471-34863-4
    1. Construction contracts—United States. I. Fairweather, Virginia. II. Guy, Sammie D. III. Construction claims. IV. Title.
KF902.C596   1999
343.73'07869—dc21   99-22909

10 9 8 7 6 5 4 3 2

Contents

Preface

In the sixteen years since publication of the first edition of this book, much has happened in the field of construction claims, and paradoxically, much has also remained the same.

Perhaps the 1980s can be characterized as the era of hardball construction litigation, conducted by pitbull attorneys, and encouraged by more-than-willing clients. What followed in the early 1990's was a quiet revolution shifting the focus to alternative dispute resolution of construction disputes.

Dispute review boards have been used in well over 100 projects worth over $10 billion. The success rate of the DRBs has been impressive. The overwhelming number of recommendations for settlements have been made and accepted with no litigation. Minitrials and mediations, once textbook concepts, have become commonplace. Now ADR is firmly established but there is a discernable trend back to litigation.

Environmental laws proliferated rapidly in the last decade and a half and interpretations and amendments of those laws will continue. We concentrate on the legal implications of hazardous and toxic wastes as they relate to the construction site.

Delivery systems have also changed in the last sixteen years. Construction and program management, fast track construction, and design-build gained widespread acceptance. Other methods such as design-build-operate-transfer have surfaced in the attempt to streamline the process.

What is the same? The technology of the claims analysis has not progressed significantly in the past sixteen years. Perhaps there is a greater uniformity in approach. We believe the first edition of this book and our seminars with the American Society of Civil Engineers on this subject can be credited in some measure with this uniformity.

For the third edition, we have changed the book's title. The focus of the first was on "analysis, presentation, and defense." The focus of this book is on "prevention and resolution". This change is reflected in the substantial revisions to and expansions of Chapter 1, 2, 9, and 10, which deal with these subjects in great detail.

The authors thank Bruce Moore, U.S. Bureau of Reclamation, Curtis Wideman, The Parsons Corp., and Amy N. Fairweather, third year law student at Hastings, San Francisco and for their assistance in preparing this third edition.

We remain ever grateful to our co-author in the first two editions, Alfred C. Maevis. Al retired after a long and distinguished career, having held positions as the Chief Engineer of the New York City Transit Authority, New York City Commissioner of Public Works and Assistant Postmaster General for Real Estate and Buildings. Al's insight and wisdom pervades the text of this edition, as well.

Part I: Claims—Understanding and Avoiding Them

1. The Groundwork: The Contract

The scenario for construction claims is invariably written right into the contract documents. Long before men and machines reach the job site, conditions for claims and disputes have often been signed by both parties. This happens, for example, when plans and specifications are incomplete or defective. More commonly, it happens when construction contracts are ambiguous, overly restrictive, or unfairly allocate particularly burdensome risks to one party alone.

There is a growing trend in the construction industry toward realistic risk allocation and cooperative project management. To that end, the use of Geotechnical Baseline Reports and alternate dispute resolution approaches such as mediation and dispute review boards are agreed upon at the earliest planning stages. While claims and disputes remain inevitable, proactive dispute management can lead to quick resolution and more rewarding use of resources.

RISK MANAGEMENT

The construction industry is notoriously risky. Much of the preparatory paperwork that precedes construction projects can be viewed as the formulation of risk allocation between the owner, the contractor, and the designer.

DISPUTE ASSESSMENT

Realistic assessment of potential disputes can lead to equitable risk allocation and a reduction in disputes and delays. Some tools have been developed to assist parties with such assessments. For example, the Construction Industry Institute's (CII) Dispute Prevention and Resolution Task Force has developed the Dispute Potential Index (DPI) to help predict the likelihood of specific types of disputes. The software breaks down factors that contribute to disputes into three broad groups: people, process, and project. Each grouping is further divided into specific issues such as "success with past projects," "environmental issues," and "financial planning."

The software transposes detailed input into dispute scores, which can be used to develop precontract changes in design and implementation, and ultimately reduce disputes. (See Appendix 5 for dispute potential questionnaire.)

The Owner

The owner is taking the risk that his project will not get built on schedule, that it will not get built for what he has budgeted, and that it will not be of the quality he expected. The owner naturally seeks to insure that these three factors will be satisfied, and he often thinks he can accomplish this through the contract language. In some cases, the owner has other risks beyond these; for example, building a water treatment plant or some other project subject to public protest or environmental and regulatory delays. Typically, however, an owner will seek to control whatever risks he can through his contract documents.

The Contractor

The contractor faces a multitude of risks. Among them are inflation, bad weather, strikes and other labor problems, shortages of materials, accidents, and unforeseen conditions at the construction site. Many of the risks posed by archeological finds, environmental, or other citizen opposition are shared by the contractor. Work stoppage can affect his allocation of men and materials, or prevent him from bidding other work. Ultimately, the contractor faces the possibility of losing a great deal of money or of being forced out of the business. Naturally, he too would like the contract wording to be protective of his interests.

The owner usually has the upper hand here. He is the party who generates the contract documents. Contractors, and to a lesser extent designers, merely react to these. One way to react, of course, is not to bid on the job. But that is not the route most contractors take. By and large, contractors tend to be inveterate optimists, believing that either the risk contractually foisted upon them will not materialize or, if it does, that the contract clause by which the risk is being imposed will not be enforced.

In this, and subsequent chapters, we hope to convince all parties that allocating or sharing risk fairly can ultimately save time and money for everyone, and get the job done on schedule for the least part overall cost. The more enlightened people in the industry are convinced that equitable sharing of risks can get the work done for less money and diminish the likelihood of claims and litigation. However, these people seem to be in the minority. Many owners still view the harsh contract as consummate protection. In reality, such contracts often turn out to be bad business, for the following reasons:

1. Harsh contracts discourage responsible bidders. The only way a responsible contractor can protect himself from a high-risk situation is to include a high contingency in his bid. If the risk and the contingency are sufficiently high, his bid will not be competitive, and probably not be within the owner's budget.

2. Ambiguous language or exculpatory clauses through which the owner hopes to escape responsibility almost inevitably result in conflict. When the conflict escalates to the courtroom, judges almost always rule against the party who drew up the ambiguous contract—the owner.

3. Such contracts attract those bidders willing to take any kind of chance, or those who expect from the outset to make up their dollars via claims. This is particularly true in the public sector, where, in most cases, by law, the lowest responsive, responsible bidder must be awarded the work.

PUBLIC SECTOR PROBLEMS

As we have just observed, in the public sector the lowest bidder usually gets the job. Public officials are under special pressure; they must answer to the taxpayers. Even if the lowest bidder happens to be a firm that has a reputation for being contentious, it would still be difficult to prove to the public that there is a legitimate reason not to award a contract to that firm. Suspected or conjectured irregularities can result in adverse publicity and charges of favoritism or corruption. It is a lot easier to suggest to the public that an official has improperly awarded a contract than it is to show to the public that an award to other than the low bidder was proper.

In an alternate design competition for the Dame Point bridge in Florida, a prestigious review board evaluated the various proposals and rejected the low bid on technical grounds. In fact, the board awarded the contract to the seventh lowest bidder. The outcry was sufficient to result in a million-dollar settlement to protesting contractors. The money was paid by the successful contractor and the public authority owner. With the vision of hindsight, observers note that this contract was extremely complex, including separate design alternatives for approaches and other project components. The experience did not diminish Florida's efforts toward innovation. The state has developed alternative design/build bidding procedures, which are examined in greater detail in the design/build portion of this chapter.

Though the lion's share of public sector contracts is still awarded through competitive bidding, state and federal use of alternative contract methods is here to stay. In 1991, the Federal Government General Services Administration announced a plan to use a wider variety of contractual arrangements and methods of procurement of consultants and contractors in its construction program. The General Services Administration has adopted the use of partnering and value engineering. More recently, in accordance with the Clinger-Cohen Act of 1996, a two-step design/build procurement procedure has been added to the Federal Acquisition Regulations (FAR). First, a limited pool of capable design/build contractors are chosen, then the contract is awarded based on design approach and price. The FAR provisions offer little in the way of design/build authorization or selection criteria. It remains to be seen how these provisions will be implemented.

Many public officials continue to consider the harsh contract their best protection against the contractor, who is often viewed as the enemy. A fair allocation of risk set out in the contract would alleviate most of the problems we described. The result would be fewer lawsuits and delays and, possibly, better quality work.

Sometimes public agencies retain the protective language of their contracts, but administer them in an enlightened manner. Damages have been awarded to contractors where the contract expressly forbids relief. Contract administrators have judged that the contractor's request was equitable, and bypassed the harsh language. The thinking seems to be that this language can be invoked when necessary, thus protecting the public entity, while risk-sharing is de facto agency philosophy.

This practice is questionable at best. The ironclad contract language discourages bidders, restricting response to those contractors familiar with the agency's contract administration practices. Also, the "lenient" contract interpretation is under public scrutiny, and an action, however just, can be misconstrued. Furthermore, the new FAR provisions are open to the interpretation of contracting officers. Personnel change, and with some administrators, the harsh language may stand.

Public entities are also burdened with inherent bureaucracy. Procedures and standard contract forms are typically mandated, and change is a slow and unwieldy process. Many parties have to approve any modification and the bureaucratic prejudice is usually on the side of conservatism.

TAILOR-MADE CONTRACTS

Were it not for public-bidding requirements, all parties—the engineer, the designer, the owner, and the contractor—might sit down and work out the most equitable terms, with the lowest possible cost and in the most expeditious time frame.

In the private sector, of course, there is more leeway. Parties may tailor the contract to the job and to the parties' desires. But in both sectors, parties need to be educated. Contractors need to understand contract terms so that they don't assume great risks without clearly realizing the potential consequences. Contracts should be read with care and with skepticism. Owners need to understand that the "toughest" contract is not necessarily the best one.

ANYTHING GOES

Notwithstanding this prefatory warning about inequitable risk allocation, it must be remembered that anything can be written into a contract as long as it is not illegal. For example, an owner could require that workers wear yellow socks on the site; if the contractor signs the contract agreeing to the condition, yellow socks it must be.

In a more costly instance of an arbitrary requirement, the documents for a dam financed by an international lending agency called for grouting a cutoff wall. During construction, the contractor found conditions that meant a great deal more grout than estimated would be needed and that a different kind of cutoff wall would have been as effective and far less costly. But the contract said grout, and grout it was, at tremendous expense to the contractor.

In the following pages, we will analyze more commonly used contract terms and devote special attention to those that often cause difficulties and give rise to claims. We will analyze standard contract terms from several perspectives, with the caveat that these analyses are conceptual tools, aimed at a better understanding of contracts and of the risks involved. The analyses are perhaps somewhat arbitrary and over-simplified, but each should help shed light on contracts and their inherent risks.

INTERNATIONAL ISSUES

International projects can be complicated by unexpected risks. As American, European, and Asian parties team up, cultural, organizational, and legal differences can hamper contractual relationships. Issues as seemingly benign as differing management styles can result in serious miscommunication. In the era of globalization, it is important to acknowledge and respect these cultural differences. (See Appendix 10 tables concerning cultural priorities.)

GETTING ALONG

"Partnering" has become a popular buzzword in the construction industry, andis being touted as the means for building cooperation and preventing disputes.

The Corps of Engineers is credited with having developed the concept of Partnering in government contracting as a logical outgrowth of its goal of achieving timely, quality, dispute-free construction. Since government projects must, in most instances, go to the lowest bidder(s), it is not always possible to set up a team that suits the needs and personalities of all involved. Because of this, communication, respect, and cooperation must be actively cultivated if the work is to proceed with a minimum of conflict. To this end, the Corps set up Partnering workshops to promote the development of positive relationships between the parties to the project, and to reduce the potential for friction on the job. The number of disputes reported has dropped dramatically and the concept of Partnering has taken hold. Partnering seminars have become a new and productive trend in both the public and private sectors, saving the construction industry two important commodities: time and money.

The Partnering Process

The stage is set for an atmosphere of cooperation and fair play even before the construction project is bid. Requests for Proposals (RFPs) indicate the intent to follow the principles of Partnering during the course of the project and give early indication of the type of trusting relationship that will characterize the job. This not only puts bidders into the appropriate frame of mind, it also may serve to lower the bid price since contractors anticipate fewer contingencies to prepare for.

Once the parties to the project have been selected, a seminar is planned where key personnel are taught to work as a team. Senior management representatives, parties with decision-making authority, and project members who will be regularly involved in the construction process attend these seminars, which can be as short as a day or as long as a week, depending on the scope of the project and the personalities involved. An independent and objective facilitator is selected to run the seminar and teach participants group dynamics and team building. This facilitator may also serve as a friendly neutral during the course of the Partnering project, available to help the team build and maintain consensus. Once the parties complete the seminar, a mutual "win-win" charter is developed which articulates the common expectations and objectives in a positive manner. The participants draft a mission statement to pinpoint goals and pledge to work toward achieving them.

To preserve the ideals of trust and cooperation inherent in any Partnering agreement, certain common business practices should be re-evaluated and alternative approaches more in keeping with the Partnering philosophy should be adopted. Parties are encouraged not to create defensive paper trails, but to correspond principally to clarify intentions, transmit submittals, or comply with contractual notice requirements.

Synergistic bonds between the contracting parties must be created and protected, when necessary by novel and creative approaches, if the benefits of Partnering are to be fully realized.

PARTIES TO THE CONTRACTOR/OWNER RELATIONSHIPS

The first analysis is from the point of view of the parties to a contract. The four general types of contractual relationships reflective of current practice are those between an owner and:

1.   a general contractor;
2.   a design/build;
3.   a construction manager;
4.   separate independent contractors.

There are scores of risks inherent in any construction undertaking regardless of the contractual arrangement. Among these are availability of labor, materials, and equipment; defective design; supplier failure; mistakes; accidents; traffic maintenance; inflation; and differing site conditions, to name just a few. The allocation of these risks between owner and contractor is central to how people often think of, and define, contractual relationships. But communications is a pivotal problem. If an owner says that he intends to engage a general contractor for project A, whereas he feels a construction manager is better suited to the circumstances of project B, that on project C he wants a turnkey contractor, and that on project D he feels independent prime contractors would be preferable—the owner has his own concepts of each of these relationships and an understanding of how the relationships differ one from another.

If a dictionary were consulted—even a construction dictionary—it is likely that the definitions would not satisfactorily differentiate the several types of relationships. Unfortunately, people often use the same terms while having in mind a different meaning. This leads to confusion, and can, in the construction context, lead to claims and disputes when the expectations of one party are not fulfilled by the performance of the other party. Thus, instead of resorting to dictionary definitions of those relationships, one might better define them in terms of how various risks are allocated between contractor and owner in each one. The following is our attempt at such "risk" definitions:

General Contractor/Owner Contracts

This is probably the most widely used type of contractual relationship, in spite of a recent shift to design/build and other means to bypass rigid constraints inherent in the standard GC/owner contract. The time-honored GC arrangement is a three-step process: design-bid-build. The owner engages an independent architect or engineer to prepare detailed plans and specifications for the work to be done. In some cases, the owner has the in-house capability to do this himself. Bids are then solicited from contractors on the basis of those plans and specifications. Here the design responsibility is the owner's; he is deemed to have adopted the design of that engineer or architect as his own. If that design turns out to be inadequate or defective, insofar as the contractor is concerned, the owner is stuck. The owner, of course, can turn to the designer for recovery. If the design was generated in-house, the owner naturally bears total responsibility.

The owner can shift some of that risk to the contractor in those instances where performance specifications are included. The contract documents may not set forth a particular design but instead simply specify, for example, a sprinkler system that complies with applicable building codes. Performance specifications aside (they will be discussed in another chapter), the design responsibility is the owner's.

Where the contract is for a lump sum price, responsibility for the cost falls on the contractor. He bears all the risk of bringing the job in at the price bid. While there are a number of clauses that can mitigate this risk, it is basically up to the contractor to perform within cost limits and to earn his profit within that limit as well.

The general contractor also assumes liability to subcontractors and typically agrees to indemnify the owner for projected-related damage and liability to third parties.

The contractor is responsible for coordinating the work of the various trades. The owner makes progress payments, typically on a monthly basis and according to an agreed upon engineer's estimate of the percentage of the work completed, less the amount agreed upon to be retained until completion of the work.

Usage.   This type of contractual relationship is the one most widely used for buildings and for heavy construction—dams, tunnels, sewers, and highways.

Design/Build (D/B) Contractors

In a design/build operation, the owner gives performance criteria to either a single firm that does both design and construction work or, in some instances, to a co-venture of a designer and a contractor. Design/build is a variation of turnkey contracts, which are frequently used in complex manufacturing projects. Thus, the responsibility for the design is shifted from the owner's side to the design/build contractor side.

Design/build contracts have been heralded as a return to the concept of the master builder because one party has single-source responsibility for design and construction. A perceived benefit to the owner is that by choosing design/build, it will not be subjected to the designer's and contractor's competing claims of "unworkmanlike construction" and "design error." The design/build contractors' greatest benefit, the ability to control and coordinate project design and construction, naturally leads to its greatest risk, the expanded responsibility for design and construction.

Standard documents have been developed by the AIA, EJCDC, and FIDIC. In design/build contracts, critical areas of risk allocation include scope of work, budget and schedule, statutory or regulatory restrictions, quality of drawings and specifications, independent review and dispute resolution, limitation of liability, and bonds and insurance. Design/build contracts are not yet the norm in the United States, and traditional insurance policies do not cover the risks associated with design/build projects. Specifically a traditional contractor need not, and in many instances cannot, be covered for professional services. In some cases, risks and responsibilities can be contractually allocated along traditional lines between co-venturers.

The construction insurance industry has responded to the growing need for design/build coverage by developing new insurance instruments. However, the design/build contractor is well advised to guard against gaps in coverage.

Usage.   Design/build is frequently used in the private sector, where an owner may want to turn over all responsibility for a job for various reasons. Complex manufacturing or processing plants are common instances where performance specifications are given; for example, the design/build contractor is to produce a facility capable of turning so much tonnage of iron ore or refined crude per day into a finished product, that is, a certain grade of oil or a certain level of ore derivative.

Design/build is increasingly being used in public sector. Many states began to use design/build contracts in the early 1990s, and their use has increased. As complicated design/build projects are not well suited to competitive bidding, states have developed alternative methods for design/build procurement.

Washington's Department of General Administration provides one example. It built correctional facilities using design/build contracts, arguing that the method provided space for inmates sooner. State officials modified existing contract forms to deal with these projects and expected to continue refining the method and the documents.

In another case, Washington built a new Department of Ecology office building using a design/build variation dictated by an elaborate financial arrangement. The state had a construction management contract with a trustee (a bank), which in turn used a design/build contract to construct the office building. The "rent" from the lessees was the revenue stream for the work, since the courts had ruled that the state could incur no further financial obligations through bonds, and did not want to raise its debt limit. Other ingenious methods will no doubt develop to get funds for public projects when federal budgets funds are curtailed.

In 1996, the Florida Legislature authorized the use of design/build for all Department of Transportation projects. In complex projects, such as bridges, where alternative designs are sought, the agency uses an Adjusted Score Design-Build bidding procedure. Letters of interest are solicited, from which a short list of prequalified candidates is invited to submit designs and price proposals. The bids are reviewed by qualified experts and receive a technical proposal score. The winning bid is determined by dividing the price proposal by the technical proposal score. When appropriate, unsuccessful firms may be compensated for their bid efforts.

BOTS, BOOTS, and International Design/Build

Design/build is used in more than 50 percent of nonresidential construction in Europe and more than 70 percent of nonresidential construction in Japan. Worldwide privatization has also fostered the increased use of delivery systems that outlive the design and construction phase.

Project delivery systems that include ownership or operations of the project add more cooks to the construction broth. The list of acronyms is expanding as fast as new parties—BOT (build-operate-transfer), BOOT (build-own-operate-transfer), DBOM(design-build-operate-maintain)— and new risks, are added to the mix. Once these systems were reserved for the most sophisticated international projects, whose owners were quasi-governmental agencies.

Originally, these systems were used in the poorest of developing nations, which could not afford to finance their own infrastructure projects. For example, a power plant might be built and financed by a BOT consortium, and after a period of operation, title would be transferred to its owner, usually a public entity. These delivery systems can require many contracts with many different parties. While parties to a large project will be familiar with international contract agreements such as those issued by FIDIC (Federation Internationale des Ingenieurs-Conseils) or ENAA (Engineering Advancement Association of Japan), it is often necessary to tailor and combine them to meet the stakeholders' needs.

Now, BOT is being used in smaller ventures, and the risks associated with this new way of doing business have yet to be settled. For example, BOT was recently used in traffic-bound Orange County California to build a toll road. Where "operate" includes user fees, who bears the risk of insufficient trips per day on a toll road, or insufficient daily tonnage of waste meeting defined disposal criteria?

Additionally, BOT and BOOT projects require a significant long-term capital investment. A/Es, who traditionally got paid for a complete design, regardless of whether a project got built, risk being viewed as project equity holders, whose return is dependent on project completion and profitability. Subcontractors and suppliers, who traditionally were protected by mechanics' lien laws, may no longer be able to rely on these statutes if they become "owners."

New contract forms being issued for BOT-type contracts lack the track record of established form contracts. At present, it is impossible to predict how these contracts will be interpreted when disputes develop.

Construction Management Contracts

An owner may engage a construction manager, or CM, during the preliminary study, the planning phase, the design phase, or in other instances during construction, to provide advice and to effect economies. The CM's role is purely organizational, quintessential project delivery. In contrast to the linear design/bid/build system, a CM project typically compartmentalizes the overall project into trade contracts which are competitively bid to prequalified firms. CM eliminates some of the risks of fast-tracking a conventional project.

Two principal kinds of construction management have evolved: "agency" and "at risk." In agency CM, the CM is the owner's advisor, and the owner contracts directly with the A/Es and trade contractors. The agency CM does have risks, but they are generally limited to making a good-faith effort to perform its contractual duties and to exercise reasonable skill and judgment in performing those duties.

In at risk construction management, the CM holds the trade contracts. The at risk CM typically offers a guaranteed maximum price, which includes the trade contracts and the CM's fee. Where the CM contracts directly with the trade contractors, but is reimbursed on a cost-plus basis, the relationship is closer to that of agency CM.

Because the CM concept is so flexible, standard contracts may inadequately address the actual relationship between the owner and CM on a given project. For example, the CM's risks related to cost, schedule, and quality may vary with different CM arrangements. With pure agency CM, all three are the owner's primary legal responsibility. When the CM is at risk and committed to a GMP, cost, schedule, and quality are the responsibility of the CM. As yet there is no consensus within the industry on CM contracts. Nearly every CM project has a different contractual arrangement. The lack of consistency is a notable characteristic of the method.

The benefits most often ascribed to agency CM are that it can provide independent oversight of the A/E's work, early constructibility reviews, value engineering, and early completion due to the phased construction. The CM is considered to be a neutral party, whose prenegotiated fee limits financial incentives for the CM to cut corners on design or construction. Even so, the CM's tendency, and financial incentive, to "move the project along" may tip the balance from the triangular "owner, A/E, contractor" relationship to a lopsided one where the CM sides with the contractor regardless of the merits of the A/E's position. The principal drawback to an owner with an at risk CM with a GMP arises when the CM seeks to protect its fee by accepting marginal work from the trade contractors.

Typically, the CM assists the owner in contractual agreements with the subtrades, improving time scheduling, reducing claims, improving plans, and coordinating the work. Those subtrade contracts, however, are generally entered into under the name of the owner or by the CM as agent for the owner. Therefore, the subcontractors are entitled to look directly to the owner for payment, unlike the general contractor relationship just described.

The owner may also indemnify the CM from claims arising from the work. Owners are becoming increasingly reluctant to do this, however, because the law regarding CM liability has changed somewhat since the inception of the method. Recent court decisions have modified the concept of privity of contract and allowed for third-party liability. This makes the owner's potential exposure much greater. As a result, it is becoming rare for CM contracts to include indemnification clauses.

In addition, the owner shoulders the liability for the subcontractors and finances the work. Cost of the project also falls to the owner, in many cases. Sometimes a CM is engaged on the basis of a guaranteed maximum project cost, where the owner and the CM share in savings realized below that figure.

As is the case in each of these relationships, various contractual arrangements are possible. But the overriding idea is that the CM is to save the owner money through his coordination of the work. The method is less clearly defined than other time-honored approaches. However, there is a growing body of legal precedent that can serve as guide when disputes arise. The spectrum of contractual arrangements is still very wide, and contracts are likely to be custom-made.

Usage. Like the design/build arrangement, CMs are used where the owner lacks in-house capability; and typically, they are used on large complex projects where an objective coordinator can effect savings in time and money.

The state of Oregon has pioneered a combination of CM and general contractor, in which the CM's fee is based on a percentage of a guaranteed maximum project cost, or "GMC." This contrasts to the traditional CM arrangement in which there is a consulting fee. That GMC is determined by the public entity, in cooperation with the architect retained by that body. The CM becomes the "CM/GC" by virtue of the fact that he is contractually responsible for completion of the work at the agreed-upon GMC. He oversees the subs, and insures that the final cost does not exceed the maximum cost. The method includes provision for savings incentives, the split between the owner and the CM/GC to be set forth in a given contract. Another version of this approach has been used in Washington, where any savings must accrue to the state. Other variations may appear in the effort to attach contractual responsibility to a "traditional" CM.

Separate Prime Contracts

Here the owner bears the responsibility for financing the work, for coordination of the work, and for the design. The separate contractors bear the risk of the cost of the construction under lump-sum contracts, for liability to their subcontractors, and for indemnification for casualties relating to their portions of the work.

Usage. This method can only be used, or should only be used, when the owner has the capability to oversee and coordinate these separate components of the work. For example, an owner with skilled in-house construction management personnel may want to maintain complete control over the project. He may have in-house capabilities for design and for construction management. The separate primes would be in special areas, with the owner managing the project.

SUMMARY

These broad brush-stroke definitions by allocation of risk will in each case be altered by the actual contract language. The point is that the name given the relationship means nothing. The only important factor is how the contract allocates risks between the owner and the contractor.

Under the definition given earlier, the owner potentially bears the most risk and responsibility in a construction management arrangement. The CM is essentially immune from contractual responsibility, but this immunity does not extend to tort liability. In the past ten years, tort liability has expanded greatly. The contractor bears the most responsibility in the turnkey relationship, as we have defined it. However, with appropriate contract language, allocation of responsibility could be completely reversed.

Time is also a factor in the choice of the parties to a contract. The general contractor method takes longer than some other arrangements, because all plans and specs must be completed before work can go to bid. However, since they are complete, the final cost should be more predictable. In the CM or turnkey types of arrangements, plans and specs need not be complete before the work is awarded. Fast-tracking, phased construction, or other methods of acceleration may be incorporated. If designs do not need to be complete, final cost is also more nebulous. But an owner may be willing to sacrifice a better handle on final cost for swift completion—or at least for the promise of saving time.

There has been criticism of fast-tracking, especially from designers and contractors who suggest the quality of work suffers and that proper inspection is sacrificed for speed. The owner who wants a swift return on his money may spend part of that money defending and settling claims. The final consideration is the capability of the owner. A large public agency, such as the Army Corps of Engineers, is clearly able to provide design work and to perform construction management functions in-house. However, many public agencies are under some pressure to use private-sector designers for their work, and use outside construction contractors as well. In special assignments such as the air force bases constructed in Israel under an international agreement of the U.S. government, the Corps also contracted out CM work. The Environmental Protection Agency administers Superfund cleanup work through four private sector firms that oversee work in each of the agency's regions. The Department of

Energy awarded CM contracts for its Super Conducting Collider project to a joint venture of two private sector firms. Thus, there may be special time and "diplomatic" constraints on the contractual relationship, even in the public sector.

In the private sector, many large corporations and utilities have continuing and sizable construction programs, and their choice of contractual arrangement will be dictated by the current capabilities and needs of the organization.

MONEY—THE BIGGEST RISK

Nothing has quite the potential for putting all parties at odds as does money and how it is paid. In the first analysis of contracts and the allocation of risk and responsibility commonly assigned, we dwelt on the overall contractual arrangements between the parties. We noted that, in many instances, risk of the cost of construction is allocated to the contractor. This risk can be quite finely tuned according to contractual variations that are described in the following pages.

FIXED-PRICE TO COST-REIMBURSABLE CONTRACTS

All contracts fall into a range between a firm fixed price at one end of the spectrum (where the contractor bears all responsibility for project costs) and actual cost plus fixed fee, or percentage of cost fee, at the other end of the spectrum (where the contractor seemingly need only present his bill to the owner).

There are variations within this range, and reasons for choosing one type or another. An owner may choose a cost-plus or a fixed-price contract with a general contractor or with a turnkey, or with a CM or with independent primes. The permutations are easy to see. The most common choice is, however, a fixed-price contract with a general contractor. This is in part because the method is the oldest one in use and the one that seems the simplest.

Fixed-Price (Hard Money)

The fixed-price contract is most frequently used and is commonly called a "hard money" contract; it can be either a lump-sum or unit-price contract, or a combination of both. For example, in a dam contract, the pumping station portions could be let on a unit-price basis. Whichever is used, it is the type that results in more construction claims than any other type.

It does not follow, however, that the widespread use of fixed-price contracts directly relates to the incidence of claims. The likelihood of claims is more related to the high contractor risk in the fixed-price contract.

Before we begin a more detailed analysis, let's look at three variables that govern most jobs and that can be affected by the type of contract. These are time, cost, and quality of the work. Everyone involved wants to get the job done fast. The owner wants to get the highest-quality work he can for the money he is willing to pay. The contractor wants to do high-quality work out of pride and a desire to maintain his reputation, but is reluctant to risk losing a profit to do so. If the terms of the contract make it unprofitable for him to deliver a high-quality job at the time specified and for the contract price, the most likely outcome is a project impeded by claims and costly delays. It's much better to begin with an equitable contract, and one suited to the type of project.

There is no ideal type of contract. Claims can always arise, from circumstances beyond the control of the parties or from litigious contractors. Owners can make claims against contractors as well. But owners generate the contract, which is the starting point for conflict, or for a mutually profitable relationship.

COST-REIMBURSABLE CONTRACTS

1. **Time and Materials.** At the very end of the spectrum, an owner might engage a contractor to provide men and materials at a rate fixed in the contract. This arrangement is close to rental service.

   Application. An owner might want to use this method when he is in a position to maintain complete control of the work, or in order to complement his own in-house capabilities.

   Disadvantages. It is difficult for the owner to control his costs; he bears the entire risk of project cost.

2. **Cost Plus Percentage of Cost.** This type of contract, it should be noted, is prohibited by the federal government. This suggests that there is something wrong with it. There are obvious disadvantages, the principal one being that there is almost a built-in disincentive for the contractor to watch the cost of the project since his fee increases in direct proportion to the cost of the work.

   Application. Why then, would an owner want to use a cost-plus percentage of cost contract? One application would be in an emergency, such as rehabilitation immediately after an earthquake. In such a case, there would be none but the most rudimentary plans and specs, and work would have to begin at once.

Few contractors would undertake work in such conditions without the protection this type of reimbursement affords.

Another application would be where an owner has had a long and harmonious working relationship with a contractor that induces mutual trust. The owner knows that the contractor will not take advantage of him; the contractor knows that the owner is fair.

Disadvantages. The main drawback of this contract type is that the owner cannot control his costs. He doesn't have a good estimate to begin with, and the contractor has no incentive to save money. It is easy to see why these contracts are disallowed in the public sector. Allegations of misuse of funds or of "sweetheart" deals are easily made and difficult to disprove. The contractor actually stands to make more profit if he spends more of the owner's (translate to taxpayer's) money. That state of affairs can be easily misinterpreted and, of course, abused. An owner must define allowable costs in the most detailed manner possible to maintain some control over the project. Vagueness here could lead to claims as to which costs are reimbursable and which are not, which are direct costs and which are included in the overhead allowance.

The owner bears most, if not all, of the risk in this arrangement.

3. **Cost Plus a Fixed Fee.** This arrangement is similar to the last, except that the fee is fixed. Thus the risk to the owner is somewhat diminished. Most contracts of this type provide that the fee can be adjusted if there is measurable change in the scope of work to be performed.

    Application. This type of contract might be used for similar situations such as those for which a cost plus percentage of cost contract is used; for example, in an emergency or where the parties have had a long working relationship.

    Disadvantages. The owner again must define allowable costs carefully. He still doesn't know beforehand exactly how much his project will cost. He does know what his contractor's fee will be, but there could be disputes or claims if the contractor perceives a change in scope when the owner does not. The owner still bears the majority of the risk in this contractual arrangement.

4. **Cost Plus Award Fee.** With this type of cost arrangement, the owner reimburses all allowable costs, agrees to pay a base fee, and sets up criteria on the basis of which the contractor may be paid more.

The criteria for "bonus" pay are set out in the contract documents. There should be no doubt about interpretation. It is also (typically) understood that the owner is the sole judge of the contractor's performance in these areas. The contractor, in signing on, must accept this fact and the idea that there might be a certain subjectivity in these judgments where money is involved. Clearly, parties have to know and trust each other in this arrangement. Early completion in certain areas of the work is an example of an opportunity for an award payment.

Application. This might be used for projects in which there is some urgency about time of completion and where the owner wants to provide some incentive for the contractor to excel.

The Army Corps of Engineers, the Department of Defense, the Navy, and other branches of the federal government have used such contracts in Vietnam or elsewhere overseas. The circumstances were deemed special—diplomacy, urgency of completion, or other reasons—so that an award arrangement might be considered a tool to attain swift and proper completion.

Disadvantages. The method has not been widely used, but where it has been, the problems have reportedly been minimal.

Potential drawbacks are these: a sophisticated organization is needed to administer such a contract, and there could be problems in judging whether a bonus payment is due. But as stated, the system does offer incentive for the contractor to work quickly and efficiently.

5. **Cost Plus Incentive Fee.** This is a more complicated approach to setting monetary incentives for the contractor's performance. A target cost is estimated; a target fee, a minimum fee, and a maximum fee are also set. A fee adjustment formula is set forth.

Typically, that formula might provide that the contractor be penalized a percentage of his fee if the final project costs exceed the target estimate. He might be rewarded at a specific percentage if final costs run under that estimate. Incentive is limited to the maximum fee set out; the fee cannot be less than the minimum, however. The maximum fee gives the owner some degree of control over the final cost figure. Ideally, the owner would happily pay the higher contractor fee, since by definition that would mean the total project cost is less than estimated.

The advantages of the method are that the contractor has an incentive to save money and the owner has parameters on the final cost. The potential seems great for satisfying all parties to the contract.

Application. This method has not been widely used. Obviously, the approach requires a highly sophisticated organization to monitor costs; both the contractor and the owner must be sizable to do this properly, or even adequately. This method probably would be used on large, complex projects where the parties are amenable to working out these contractual arrangements and can oversee their implementation adequately.

Disadvantages. The incentive for both parties seems great. However, the cost plus incentive fee contract has not been applied in many instances. The greatest drawback would seem to be the difficulty in establishing and documenting cost savings that would result in the higher (or lower) contractor fee, and in determining responsibility for cost overruns that would affect the contractor's fee. Clearly, if the overruns are owner-caused, the contractor will not want to be penalized by a lower fee. Thus, while this arrangement in theory holds great promise, in practice it can become a nightmare.

6. **Fixed-Price Incentive Contracts.** The fixed-price incentive is a fairly innovative attempt at sharing risk, and has not been widely used. The fixed-price incentive contract sets a fixed project cost. However, the contractor's risk is mitigated by setting a target cost estimate and a price ceiling. A target profit is also set. A formula is agreed upon for dealing with the amount that the final cost falls below or above the target. The owner is protected by the ceiling price. The contractor has the opportunity to make unlimited profit; that is, there is no ceiling on his profit—if he lowers cost sufficiently to earn that extra money. His profit diminishes as the final cost approaches the target. If the final cost is above the target estimate, he loses. This contract resembles the cost plus incentive arrangement described earlier in the cost-reimbursable portion of this chapter. However, in the other version, there is a ceiling and a floor on profit, and all costs are reimbursable; here there is a ceiling on costs, none on profit.

Application. Use of this contract might be appropriate where the owner has plans and specs sufficient to allow a target cost, yet with enough leeway to allow a contractor to maneuver for that extra profit. Obviously, if plans and specs are complete, the contractor has less opportunity for innovations that might increase his profit.

Disadvantages. As in some of the other contractual arrangements, it takes a fairly sophisticated owner to attempt such a contract. There could be difficulties in setting up the sharing formula and in administering it.

7. **Guaranteed Maximum Price.** The guaranteed maximum price is a less complicated variation. For example, a maximum price might be set for the entire work, for an agreed-upon changed condition or for a portion of the work. Costs would be carefully audited. The contractor's fee could be a fixed sum or a percentage of cost. A percentage-sharing formula could be set up for the amount that the audited cost is below the guaranteed maximum price. In other words, this is a more limited and controlled application of the same idea as the fixed-price incentive arrangement. In the latter, presumably, the contractor would suggest changes in the work to save money and increase his profit.

Today, the fixed-fee, cost-reimbursable, guaranteed maximum price contract with shared cost savings, is most widely used in fast-track private construction based on less-than-complete specifications.

Contractors should be aware of a caveat in fast-tracked work, where the guaranteed price is based on drawings that may be only 50 percent or 80 percent complete: There are increasing claims based on substantial differences in the completed drawings.

The cost guarantee is frequently based on what is shown in the existing drawings and, in addition, what may reasonably be inferred to be required for the proper execution and completion of a fully functional work. Increases in the guaranteed maximum price are permitted only for changes in scope or increases in quality; for example, a change in concrete strength from 3,000 psi to 4,000 psi.

Disputes often arise as to whether changes from the initial drawings to the final drawings constitute scope or quality changes or really represent only the full depiction of things that could have been reasonably inferred from the initial drawings.

SUMMARY

Cost-Reimbursable Contracts

The contract types we have just described are most likely to be used in the private sector. The one overriding advantage of these contractual arrangements is that work can begin before all plans and specifications are complete. The great disadvantage is that the owner cannot have a firm idea of what his final costs will be.

Some of the more promising contract types, in which incentive is provided by the opportunity to earn a larger fee, require sizable organizations to administer them. For these reasons, there will likely be more use of these contract types by large firms in the private sector; for example, those with in-house construction capabilities and continuing construction programs. In the public sector, such methods might be used in extremely complex projects or special cases, such as the Israeli air bases mentioned previously that were part of an international agreement, or in urgent defense programs. The space program or other such special projects where there are many unknowns might be another application.

Fixed-Price Contracts

Most construction business in this country is conducted with fixed-price contracts. Anyone involved in the industry should understand this form thoroughly and be aware of the risks that exist and of clauses that can mitigate those risks.

In a bare-bones, fixed-price arrangement, the project risks fall almost entirely on the contractor. He either brings the work for the bid he proposed, or he takes a loss. This is true whether he is a general contractor, an independent prime, or a turnkey. It can be true with a construction manager as well if the contract is set up that way.

From the owner's point of view, this seems like a protective device. But the simplicity is deceiving. The contractor's bid must be high enough to cover those risks—the unknowns that will surface in the course of construction—and to make a profit as well. It must also be low enough to be competitive. The result of this guesswork is that one side inevitably loses. If the final project cost comes in lower than expected, the contractor makes more money, but the owner could have spent less.

If the final project cost comes in higher than bid, the contractor will not only lose money, he could go out of business. The owner, whether he wants to or not, can pass off the higher cost to someone else—the consumer, or in the public sector, the taxpayer. The solution could be as damaging in the long run as the contractor's loss is in the short run. An owner could be subjected to harsh criticism or lose sales.

Why Are Fixed-Price Contracts Used So Frequently?

"It's always been done that way." The method has gained respectability with age. Owners, particularly public owners, don't want to take a chance with an "innovative" contractual arrangement. Many managers in large corporations don't want to risk convincing their boss that another way might be better. Finally, most owners don't even think about other possibilities.

Since the risk in a fixed-price contract sits mainly with the contractor, many owners feel this is the only safe way to go. To a public official, particularly on the local level, the fixed-price contract can be a shield, a way of appearing to avoid favoritism and of saving the taxpayer from venal contractors.

Despite the reluctance of public sector officials to experiment in contracting, the use of two-step and competitive negotiated contracts for construction projects are beginning to be used. The contract remains firm-fixed price, but the major difference is the submittal of proposals for review by a team of public sector professionals (occasionally, some highly regarded private sector experts are included). The owner's reason for using these contracts is so that both parties have a better understanding of the work requirements with subsequent reductions in claims and overall project cost. The disadvantage to the contractor is that if he is an unsuccessful bidder, the cost and time spent to make proposals can sometimes be as much as three times the normal bid cost. The cost to the public agency of administering these contracts can also be considerable.

The use of these contracts is typically limited to specialty type construction projects such as rehabilitation of existing structures or those requiring innovative solutions to engineering problems.

The owner has the best handle on the final cost, an important consideration in the public sector where projects are funded through bond issues or are legislatively limited.

We have noted that the fixed-price contract results in many claims and that they are also the type most often used. One result of this situation has been the development of contract clauses that tend to mitigate the severity of the fixed-price contract. Such clauses shift from the contractor to the owner, some of the major risks over which the contractor has little or no control.

The principal clauses in use are escalation, differing site, suspension of work, and variation in quantity provisions. Owners have been slow to accept these, but their use is growing. They can, of course, be incorporated in a cost-reimbursable contract, although the necessity there is marginal, except where a guaranteed maximum price is included. These and other clauses will be dealt with in detail in the next chapter.

INTERNATIONAL CONTRACTS

The Conditions of Contract for Works of Civil Engineering Construction issued by The Federation Internationale des Ingenieurs-Conseil (FIDIC) are the standard in the international forum. The conditions are not a fixed-price contract and include more than 30 clauses that may entitle the contractor to request additional time or money. The result is lower initial bids, which are inevitably augmented with project specific cost increases.

The conditions bestow considerable power on the engineer, to whom claims are submitted for approval. As the engineer also acts as the owner's agent, this dual role can invoke skepticism. The newest FIDIC documents require impartiality and preclude the owner from influencing the engineer. However, when problems do arise, international arbitration is the standard recourse.

READING THE CONTRACT

Once the contract type has been chosen, parties often skim over what they consider to be boilerplate and leap into the plans and specifications. This is a major mistake. Much contractual risk is bound into the terms of the general conditions. Since the contractor typically has little voice in the contract choice, it is his responsibility—if he wants to stay in business—to assess that risk correctly. He may decide, on the basis of that assessment, not to bid at all. More often, if he perceives a high-risk situation, he will bid a higher price. If he gets the job, his awareness of that risk should result in defensive management of the project. He will, or should, take care not to go out on a limb in a risky area of the work. The experienced contractor will measure his risk, too, in terms of the owner's reputation. Some owners are known to stick to the letter of the law, no matter how harsh that interpretation might be.

Owners, as we have pointed out, have a perfectly legitimate concern with self-protection through the contract. Many contracts are formulated by a cut-and-paste procedure (or the computerized equivalent thereof), are inadequately reviewed by owners, or are antiquated documents that have achieved sanctity through age. The owner may not have considered the content for years, especially in a bureaucratic public owner case. Most often, however, owners tend to include harsh protective clauses that amount to an ill-advised defense and are the breeding ground for claims.

Both sides should, in other words, think about the contract carefully. In the following chapters, we will address some of the more problematic construction contract clauses.

2. Red Flag Clauses

The following are clauses, some of them meant to mitigate risk, that should be scrutinized thoroughly.

DRUG-FREE WORKPLACE

A new clause was added to federal contracts requiring the contractor to certify a drug-free workplace. In response to the concerns for controlling the use of drugs on federal construction sites, contractors are required to submit certification to establish a drug-free awareness program. Failure to submit this certification will mean the contractor will not be awarded the contract.

PROMPT PAYMENT

The federal government has adopted clauses designed to "force" prompt payment by all parties. (Many states have adopted similar clauses.) A detailed clause in construction contracts defines a proper invoice, interest penalties, and financing. A new addition to the clause is the requirement of the contractor to pay subcontractors within seven days after payment by the government to the contractor. If the contractor does not comply, he may be charged a late payment interest penalty. It is somewhat a paradox that the government seeks to insert itself into the contractor-subcontractor relationship while removing itself from involvement in any disputes between these parties.

TIME EXTENSION

Is the contractor entitled to an extension of contract time for acts of God and other causes beyond his control? What, specifically, are the cases in which delay is excused? What about delay caused by problems with the manufacture of materials or equipment? If the contractor is not entitled to time extensions, is he liable for liquidated damages or actual damages for delays? If he falls behind schedule because of a delay that does not entitle him to extra time, is he required to accelerate by working over-time?

To illustrate, two commonly used contract clauses differ somewhat in setting forth time extension conditions.

The Engineers Joint Contract Documents Committee's (EJCDC), Document 1910-8 (1996) (see Appendix 1 of this book for selected EJCDC articles), section 12.02 states that contract time may only be changed by a change order based on written notice delivered by the party making the claim within the time constraints provided in section 10.05. Notice must be submitted *within* thirty days of the start of the event giving rise to the claim. Notice of the extent of the claim with supporting data is due within sixty days. The American Institute of Architects' (AIA) clause, Document A201 (1997), para. 8.3.2, when read with para. 4.3.2, states that any claim must be initiated by written notice to the architect within twenty-one days of the occurrence or discovery of the condition that gives rise to the claim. Otherwise, the claim is waived.

These are minor variations in wording, but when the request arises, the correct procedure under the contract must be followed. Also, one clause might be more liberally interpreted than the other. The AIA sets out very explicit reasons for delay judged to be beyond the contractor's control, as well as causes that the architect determines may justify delay. The EJCDC prefaces its list with "shall include, but not be limited to." This leaves the contractual door open a little wider. The AIA version says the architect will decide what a reasonable time extension will be. The EJCDC clause says that the engineer will "render a formal written decision within thirty days," and that the engineer's decision is final and binding unless appealed through dispute resolution procedures set forth in the contract. If the contract does not specify such procedures, the parties may appeal to "a forum of competent jurisdiction within 60 days after the date of [the engineer's] decision or 60 days after substantial completion, whichever is later."

There are also some contracts in which it is stated that the contractor cannot claim any relief other than a time extension. For example, the City of New York's contract documents contain a clause stating that "the contractor agrees to make no claim for damages for delay in the performance of this contract occasioned by any act or omission to act of the City or any of its representatives, and agrees that any such claim shall be fully compensated by an extension of time to complete the work...."

Other municipalities or government bodies or a private owner may include similar clauses, known as "no damage for delay," which certainly put a burden on a contractor (see Chapter 6 on delays).

In fact—and we stress this throughout this book—an owner is free to put any kind of clause he pleases in any kind of contract. We would like the owner to think about the consequences and the contractor to be aware of high-risk contract provisions.

ESCALATION CLAUSES

These are clauses that allow reimbursement to the contractor for steep increases in labor or material costs (the latter more rarely) over a long period of time. Sometimes these clauses mention a time period after which the escalation applies; such clauses are most often used in contracts for work that will take at least a year to complete. Inflation, labor agreements, and the like are sometimes foreseeable, but the extent, as in the case of inflation, is not. It is reasonable to assume that a contractor cannot account for these in his bid when the project will be of long duration. If he does, he is not likely to get the job.

Neither the AIA nor the EJCDC contract documents include escalation clauses. Lawsuits and defaults are rarely attributed directly to losses from escalation risks. It's easy to see, however, that a contractor who takes a loss because of enormous increases in these areas could default, or in some cases, be looking to make up the money elsewhere, by cutting corners, or through litigation.

Basically, the escalation clause seems fair treatment of the contractor, and protects an owner from excessive contingency in bids. One federal agency has provided for escalation in contracts that will run for over two years. In part, the provision reads:

"Commencing one year after the date of award of contract, the payments earned thereafter under the contract will be adjusted by the amount of 75 percent of the difference between the amount of wages actually paid and the amount of such wages if computed at the hourly rates stated (elsewhere in the documents)."

Notice that the clause protects the owner for a year. If the labor rate changes considerably within that time, the owner doesn't pay before the year is out. It is assumed that the contractor might reasonably have foreseen a short-time cost hike. The shared cost after that time is intended to motivate the contractor to exercise what control he can have over such increases.

Industry surveys have shown that all parties—owners, contractors, designers, and attorneys—endorse the escalation clause. Under the risk section at the end of this chapter is a discussion of the results of such a survey on construction contract practices.

CHANGED CONDITIONS

This important provision, sometimes referred to as a "differing site conditions" clause, is now part of the federal government's standard documents, as well as the AIA and EJCDC documents.

Historically, unforeseen subsurface conditions have caused many construction claims and have driven more than a few contractors into default. A changed conditions clause allows reimbursement for unforeseen site conditions that affect the cost of work. For example, such unforeseen ground conditions can necessitate the driving of longer piles or the use of more support steel than was first planned. Both operations might disrupt and delay work, throw schedules out of kilter, and cost significantly more than a contractor could reasonably have expected. However, it has been the almost universal practice for the owner to include contract clauses disclaiming the borings and other data furnished about the site.

When disputes about such unforeseen conditions have reached the courts, enforcement has not been uniform. Many judges find excuses to avoid enforcement. On the other hand, some judges have strictly enforced the clauses and ruled that the contractor's signing of the documents denotes acceptance of the risk. There are better ways to deal with the risks of subsurface construction, and the changed conditions clause is the principal means used today.

These clauses are included in most standard contract documents. Owners should consider these as a means to reduce the incidence of lawsuits and to obtain more realistic bids from contractors who otherwise would include large contingency amounts. That said, there are major cities that refuse to include such terms. Contractors should not take the inclusion of changed conditions clauses for granted. Owners should consider their inclusion as insurance against subsequent lawsuits and against high bids.

The AIA and the EJCDC standard documents include provisions for changed conditions in existing structures as well as for subsurface conditions. The AIA documents allow for conditions that "differ materially from" conditions indicated by the contract documents or those generally inherent in the type of construction provided for in the contract. The EJCDC documents refer to conditions that "establish that any 'technical data' on which Contractor is entitled to rely" is "materially inaccurate," "differs materially from that shown or indicated in the Contract Documents" or that is "of an unusual nature, and differs materially from conditions ordinarily encountered" in the work provided for in the contract. The federal government clause on differing site conditions refers to differing subsurface conditions and also to "conditions at the site of an unusual nature differing materially from those ordinarily encountered and generally recognized as inhering the work of the character provided for in the contract."

Additionally, the 1996 edition of EJCDC 1910-8 and the 1997 revisions to AIA A201 contain provisions that specifically address hazardous conditions at the site. The EJCDC provision states that the differing subsurface and physical conditions clause is not intended to apply to hazardous conditions. (The provision is examined in greater detail in Chapter 5.)

The wording in all three cases is open for some interpretation, but inclusion of both subsurface and site conditions is prudent. This is especially true with the increased amount of rehabilitation and restoration work conducted in recent years. Retrofitting for seismic codes or energy-use reduction can result in structural changes for buildings where there are no existing as-built drawings or plans and specs. Nevertheless, the majority of unforeseen and differing conditions are found in subsurface work, tunneling, and the like. As to wording on notification, the AIA version (A201, para. 4.3.4) says that either party must give notice of the change to the other party no later than twenty-one days of observance of the condition. The architect will "promptly" investigate the condition and determine whether it calls for a change in contract time or amount. If either party opposes the architect's determination, they must initiate a claim within twenty-one days of the decision. The other two standard clauses, FAR, para. 52.236-2 and EJCDC 1910-8 para. 4.03, require that the contractor notify the owner "promptly," wording that is open to a range of interpretation.

These clauses shift responsibility for unknown conditions to the owner where it rightfully belongs. No contractor can be expected to have investigated the owner's site exhaustively before bidding (see Chapter 3 on changed conditions).

VARIATION IN QUANTITY

Many fixed-price contracts incorporate unit pricing. This entails the owner's setting forth estimated quantities of materials needed to accomplish the work, or portions of work, into units. For example, the owner might estimate quantities of cubic yards of concrete poured, the quantities being a combination of materials and work. The entire contract could be unit priced; if the design is complete, the key is the degree to which quantities can be reasonably estimated.

In complex projects, unit pricing becomes more difficult. The contractor generally takes more of a risk if unit pricing is used in such cases.

Wrong estimates can result in substantial losses or in windfall profits—for either party. If design changes are made, of course, a change order can allow for the difference in the originally estimated quantities. A different price for those unit-priced items might also be negotiated. But the owner might be hard-nosed and insist on application of the contract unit price.

Where quantities cannot reasonably be estimated, there is a substantial risk that wide variation from estimated to actual quantity will not be adequately compensated by the unit price. Even if the actual quantity is significantly lower than estimated, the contractor may not be able to recover his fixed cost for the work (say, for example, in a grouting operation where there is an expensive equipment setup, regardless of the amount of grout used). In other instances, he may be overly compensated.

Both the contractor and the owner are at risk here: The contractor can lose his shirt, or come into a windfall profit; the owner may have to pay considerably more than necessary or fair. If the actual quantity is significantly higher than estimated, the contractor may have to go to a more costly method. He may have to use a larger piece of equipment in an excavation operation, or steel in lieu of timber sheeting for earth support, or he may have a longer haul distance for the disposal of excavated material.

However, there are more fundamental problems with unit pricing. When the idea was conceived, labor costs were a much lower percentage of the cost of construction work than now. Over time, labor has assumed a greater percentage of the cost of construction and is probably about equal to that of materials in many cases. This change is not always reflected in unit pricing. Sometimes a design change without a quantity change might affect labor costs significantly. For example, a design change could require a special finish on concrete that does not change the actual amount used, but does require much higher labor costs.

Indirect costs are another area of contention in unit-priced contracts. These costs have to be spread over all unit-priced items in some rational fashion for the contractor to fully recover such costs that do not vary with the quantity of work performed, such as supervision, salaries of clerks, and so on, and which generally do not have separate unit-priced items themselves. When quantities change, as they often do, the result is frustration and argument. It is unlikely that the cost allowance included in the unit-priced item will be even close to the cost of the change. In fact, in many cases, the greater the magnitude of the change, the greater is the disparity between the cost effect and the compensation provided for in the contract.

Owners prefer to have the indirect costs spread pro rata over the unit-priced items. This is what they are used to and what makes sense to them. From the contractor's perspective, the method creates problems. What if a contractor expects that there will be a substantial underrun on one item? If he spreads his overhead pro rata, then he will not recover that overhead in the item or elsewhere, unless some other item similarly and conveniently increases. If it does not, he has no way to recover those fixed indirect expenses.

This is why contractors legitimately look through contracts for unit-priced items that may have an underrun in order not to spread fixed overhead to those items. Somewhat less legitimately, a contractor may look through the unit-priced items for one where he thinks an overrun may occur. If he attaches a heavier unit price on these, he may recover not only indirect costs on another item, but make a windfall profit. If he can correctly guess which items are likely to decrease substantially, he can apply a lower unit price to these and prevent a loss.

It is common for a contractor to perform work one month and not receive money for it until several months later. This means the contractor is always financing the cost of the work. If he can put more money in the items for work to be done at the beginning of the job—"front-end loading"—then he will be paid early enough in the job cycle to finance work at a lower cost. This can also mean that he can submit a lower and more competitive bid and be more likely to get the contract.

In all these instances—front-end loading, projecting overruns and underruns, and adjusting prices to cover indirect costs—the contractor is preparing an unbalanced bid. From his point of view, he is distributing the indirect costs in a manner more equitable than the pro rata distribution that owners prefer. This is one of the principal problems with unit-priced contracts.

An approach used in the United Kingdom is to establish bid items for all significant costs, not just for materials. These items can be selected so that the relationship between the quantity of the item and its total cost is a better one. There might be separate items covering direct cost elements and indirect elements. For example, there could be unit prices for a square foot of formed concrete surface and for special finishes.

The variation in quantity clause is another way to deal with the risk involved in a unit-price element of a contract. This clause distributes some risk by setting a limit on how much that estimated quantity can vary before making some adjustment in the price. For example, on the San Francisco Bay Area Rapid Transit (BART) project, this was the wording (paraphrased) of the variation in quantity clause. If the total price quantity of any contract item that amounts to 5 percent of the total contract bid price varies by 25 percent or less from the engineer's estimate, payment will remain as bid. If the variation is beyond 25 percent, a change order will be issued.

Under the BART arrangement, it would be possible to credit the district with any reduction in cost or to compensate the contractor for any increase in cost resulting from the changed quantity.

The U.S. Bureau of Reclamation in the past has used a split-quantity clause to deal with problems of variation in quantity. The provision was used for contracts where principal unit prices were susceptible to large variations. In the split-quantity provision, a 65 percent to 35 percent split was typically used.

The two ranges were provided so that bidders would include in the unit-price bid for the first 65 percent of the total estimated quantity for that item "that part of the contractor's cost for construction facilities mobilization and demobilization, plant, fixed overhead, etc." The price for the 65 percent portion would naturally be higher in most instances since it includes both fixed and variable cost. The 35 percent portion should only include the variable cost of the operation associated with completing this bid item of work. Therefore, no renegotiation of the unit price is appropriate, regardless of the final quantity, unless a change is made or the final quantity is less than 65 percent of the estimated quantity. If a change is made that affects the cost of the item of work, then an appropriate adjustment would be made pursuant to the changes clause in the contract. If the 65 percent threshold is not reached, the contractor is entitled to an adjustment for an unrecovered fixed cost.

The EJCDC and AIA standard forms offer somewhat similar clauses, but are less specifically worded than the BART provision quoted here.

The AIA clause (A201, para. 4.3.9) states that if unit prices are agreed on, and if the quantities originally contemplated are materially changed in a proposed change order, so that application of the unit price will cause "substantial inequity" to the owner or the contractor, then the unit price will be equitably adjusted.

The EJCDC clause (1910-8, para. 11.03) states that where the quantity of work with respect to any item covered by a unit price differs "materially and significantly" from the quantity of such work indicated in the contract documents, an appropriate change order shall be issued, on recommendation of the engineer to adjust the unit price. The EJCDC clause is applicable in a wider range of cases than is the AIA clause.

There is no doubt that variation in quantity clauses, if properly administered, spread construction risks more equitably and protect both parties to a contract. A notable instance illustrates the point. New York City's Third Water Tunnel involved tremendous overruns for unit-priced support steel for which the city had specified a lower-than-cost fixed-unit price. There was no changed conditions clause. To oversimplify the events, the contractor sued the city on the grounds that as owner, the city "knew or should have known" that much more steel would be needed.

The estimated quantities were vastly off the mark. Work stopped, the city countersued, inflation took its toll, and work remained unfinished for years, during which time taxpayers had substantial maintenance costs to keep the partially driven tunnel from regressing.

A good part of this scenario might have been avoided if a variation in quantity clause had been included in the contract and observed in good faith.

EXTRAS AND CHANGES IN THE WORK

Contracts generally include a clause that gives the owner the right to order extras or changes. For the contractor's protection, these clauses should include the extent of allowable change (for example, not to exceed 5 percent of the contract price) and a time extension for performing the change. The contractor should also be compensated for the impact of the change on other work, if any. A time limit should be included for payments relating to owner-requested changes. Owners sometimes take inordinate amounts of time to process change orders. This can tie up contractor dollars and erode cash flow. Most contracts do include time provisions for payments to the contractor for base contract work but not for extra work.

It should be apparent that some changes will have a ripple effect; that is, changes that could bring work into a winter season, call for different equipment, or tie up equipment needed for another portion of the work. That ripple, or impact, from the change should be compensated for in addition to the direct costs of the changed work.

EXCULPATORY LANGUAGE

Contractual disclaimers as to cost responsibility for incomplete plans or specifications, faulty or incomplete site information, or other matters should be viewed with caution. From the owner's point of view, inclusion simply guarantees that most contractors will come in with higher bids as insurance against those risks. When disputes over these clauses have reached the courts, judges have often ruled against the owner, citing "reasonableness." The courts have demonstrated a pendulum effect, with judges swinging from one extreme to the other in strictness of interpretation. Furthermore, there are variations from jurisdiction to jurisdiction. The courts have increasingly taken the position that the owner has the ultimate responsibility for the preparation of a reasonable contract, particularly competitively bid public contracts. Nevertheless, when such exculpatory language is present in a contract, it can be invoked. Parties cannot be sure how the court will rule.

Inclusion of exculpatory language is often a sign of a lack of professionalism on the part of an owner or designer. It represents an evasion of responsibility, an improper allocation of risk.

DISPUTE MECHANISMS

The omission of a provision for the resolution of disputes can force parties to the courts if negotiation is not successful. Standard contracts now almost universally include provisions for the use of private dispute resolution procedures such as mediation and arbitration. These forums are often advocated as being less costly and time-consuming than legal recourse. The advantages and disadvantages of each of these are set forth in the chapter on dispute resolution.

Article 16 of the EJCDC Supplementary Conditions "endorses mediation, followed by mandatory and binding arbitration" and notes the 1996 improvements in the American Arbitration Association's Construction Rules. It provides suggested language for mandatory mediation and arbitration clauses but "recognizes that this is a matter for Owners to decide after advice of legal counsel."

The 1997 revisions of the AIA general conditions provide an hierarchy of dispute resolution techniques. A201 section 4.4 requires that all claims initially be referred to the architect for resolution. The architect may request additional information, approve the claim, suggest a compromise, or conclude that it would be inappropriate for the architect to resolve the claim. The architect's decision is final and binding, but subject to mediation and arbitration if such a demand is made within thirty days of the architect's written decision. Section 4.5 requires the parties to "endeavor to resolve their claims by mediation." If the claim is not resolved, it may then move on to arbitration as set forth in section 4.6. Both mediation and arbitration are to be carried out under the construction industry rules of the American Arbitration Association, unless the parties mutually agree otherwise. The AIA provision also prohibits the joinder of the architect in arbitration proceedings between the owner and the contractor unless written consent is given by the architect.

The Metropolitan Atlanta Rapid Transit Authority (MARTA) specifications are an early example of a major project in which arbitration was mandated for disputes arising in connection with the contract. This provision also stated that the "arbitrators award and their decision of all questions of law and fact in connection therewith, shall be final and conclusive, and judgment upon their awards may be entered by the U.S. District Court for the Northern District of Georgia...."

With the Eisenhower Tunnel, part of a major highway project in Colorado, the contract provided for a preselected panel of industry experts to hear and deal with disputes. This dispute review board considered disputes directly as they arose on the job, before they developed into formal adversary proceedings. The method has been successful; however, it requires sophisticated parties who are willing to submit to the judgment of respected individuals in the industry. (Dispute review boards are addressed with more specificity in Chapter 12, Formal Dispute Resolution.)

PLANS AND SPECIFICATIONS/OTHER CONTRACT CAUTIONS

A number of other portions of the contract documents may lead to conflicts. To begin with, there are detailed specifications and there are performance specifications. In one case, the contractor might be asked to build a roof according to the detailed instructions in those specs; in the other case, he might be asked to build a "watertight" roof.

The contractor and the owner should know the difference. If the roof is built exactly according to detailed specs and still leaks, the contractor's obligation has been satisfied, and the owner will have to pay for repairs. If the contractor is working with a performance spec, the risk is his if the roof leaks. However, he has the opportunity to make more money by deciding himself how to construct that roof.

Other contract items that merit special scrutiny are provisions for the location of utilities, guarantees of right-of-way and, conversely, extension of time if the right-of-way is not available.

Some other provisions are more complex; the difficulty with these rests on whether certain decisions should be made by the designer or the contractor. These might include temporary support for deep excavations, responsibility for the consequences of dewatering, temporary tunnel linings, rock support-steel, use of timber or gunite (for support), the length of foundation piles, and so on. All are potential disaster areas and are often dealt with on a unit-price basis.

Cost is not the only consideration in these areas. The integrity of the structure and the safety of the workers are involved. Claims and delays can be averted by clearly delineating design requirements.

SUMMARY

Throughout this chapter, we have referred repeatedly to the owner and the contractor. Clearly, most of these references are also applicable to designers as parties to construction contracts. No bias was intended in this presentation. The point is, owners generate contracts. In turn designers, that is, architects and engineers, prepare the actual plans, specifications, and contract documents as "independent agents." They are actually part of the owner's team. Contractors are in a position only to react to these documents.

Most claims, not surprisingly, come from contractors. Nevertheless, all parties need to be educated. Construction contracts should not be cut-and-paste assemblages (or computer patchworks) of clauses picked up from here and there, nor should they be ancient documents used by city X or agency Y since time immemorial. Parties should be thinking about the contract documents and their content; that content should not be taken for granted.

Standard Documents

We have alluded to several sets of standard documents; these are reconsidered and revised regularly. These standard contract documents are published by the American Institute of Architects, by the Engineers Joint Contract Documents Committee (composed of the American Society of Civil Engineers, the American Council of Consulting Engineers, and the National Society of Professional Engineers), and by the federal government. The Federation Internationale des Ingenieurs-Conseils (FIDIC) issues standard contracts for use in international projects. These can be adopted and adapted. The important point, which cannot be stressed too often, is that all parties should think about the process and know what is in the contract and which risks are being borne by whom.

Footnote on Risk

Two documents in the appendixes of this book might be used as guidelines for equitable risk allocation. One is the result of a survey of construction industry parties taken at a 1979 American Society of Civil Engineers conference on risk and liability sharing. The respondents were a fair cross-section of industry practitioners: owners; designers; and contractors with a sprinkling of attorneys, academics, and insurance company representatives.

The results are notable because the majority of those who responded favored enlightened contract practice (particularly escalation clauses), variation in quantities clauses, and changed conditions clauses, none of which are in standard use today. The respondents also were opposed to exculpatory language in contracts; and they thought there should be provision for reimbursement for delay, in addition to time extensions. The group was also polled on risk allocation, as between the owner and the contractor.

The second item that indicates some changing attitudes is a report to the Board of Public Works of Los Angeles, entitled "Proper Allocation of Construction Risks Between Owners and Contractors as a Cost-Saving Concept." The author of the report was at the time the commissioner of Public Works in that city, and the document contains some unusually enlightened recommendations for sharing risks. Some of the proposals follow.

On the subject of delays, the report suggests that when causes of those delays are out of the control of the contractor, the owner and the contractor share extra costs. An escalation clause for labor is recommended; the owner and the contractor should work out the terms for sharing this increase on a case-by-case basis. The report also suggests that performance bond requirements and retainage be reduced as work progresses in order to reflect the owner's reduced exposure to damage.

The report concludes by stating that the city could save money by removing some risk from the contractor, and that all ambiguous, evasive, catch-all, and exculpatory clauses be eliminated from such contracts.

3. Differing Site Conditions/Changed Conditions/Geotechnical Baseline Reports

Differing site conditions were briefly discussed in Chapter 2. Such conditions are so often the cause of conflict between parties to the construction contract that we feel the subject merits further attention.

The terms "differing site conditions" and "changed conditions" are used interchangeably. In fact, the site does not change. The terms refer to situations in which construction conditions turn out to be different from those represented in the contract documents or from what parties to the contract could reasonably have expected from the information available. A common example occurs when true subsurface conditions vary from what typical site investigation borings have revealed.

The federal government added a clause to its standard construction contract, FAR, or Federal Acquisition Regulations, some years ago to deal with such cases. The alternative was, and still is, for a contractor to add a contingency factor to his bid to cover unforeseen situations. If they do occur, that contingency partially reduces the cost impact of those unexpected conditions. If unforeseen conditions don't occur, the contractor profits. There are cases, of course, in which a contractor might assert a claim even though he had included contingency money in his bid. That contingency is more often than not insufficient to cover the cost of unforeseen conditions.

In the first and second chapters, we discussed theories of risk allocation. The federal government's differing site conditions clause allocates the risk of unforeseen conditions to the owner. The owner doesn't have to pay for the contingency in cases where no changed conditions are discovered, nor does the owner have to risk lengthy and costly lawsuits if changed conditions are encountered. In other words, that risk is mitigated somewhat. The terms of the contract require the contracting officer to authorize payment by the government for the change when justified and substantiated. (It should be noted that the government is in a position to spread its risk over a large number of construction projects.)

Owners other than the federal government have adopted the differing site conditions clause, but it is not universally used. New York State and the Port Authority of New York and New Jersey, for example, do not incorporate such language in all of their contracts. Los Angeles and New York City, by and large, do. In the private sector, the practice also varies. However, changed conditions clauses have gained favor. The 1996 edition of the EJCDC general conditions includes the revised and renamed "differing subsurface or physical conditions." Owing to the cumbersome nature of these provisions, it is suggested that this language be deleted from the contract and replaced with the FAR wording.

Owners who plan to include a changed conditions clause in their contracts would be advised to use the Federal Acquisition Regulations language. Sufficient cases have been decided by the federal courts and boards of contract appeals to constitute a guide to interpretation and usage. The FAR language, found at 48 C.F.R. 52.236-2, reads as follows:

(a) The Contractor shall promptly, and before the conditions are disturbed, give a written notice to the Contracting Officer of (1) subsurface or latent physical conditions at the site which differ materially from those indicated in this contract, or (2) unknown physical conditions at the site, of an unusual nature, which differ materially from those ordinarily encountered and generally recognized as inhering in work of the character provided for in the contract.

(b) The Contracting Officer shall investigate the site conditions promptly after receiving the notice. If the conditions do materially so differ and cause an increase or decrease in the Contractor's cost of, or the time required for, performing any part of the work under this contract, whether or not changed as a result of the conditions, an equitable adjustment shall be made under this clause and the contract modified in writing accordingly.

(c) No request by the Contractor for an equitable adjustment to the contract under this clause shall be allowed, unless the Contractor has given the written notice required: provided, that the time prescribed in (a) above for giving written notice may be extended by the Contracting Officer.

(d) No request by the Contractor for an equitable adjustment to the contract for differing site conditions shall be allowed if made after final payment under this contract.

The distinction between the two types of differing site conditions allowed in such a clause is this: one kind refers to misrepresentation of site conditions, intentional or innocent, the other deals with conditions that could not have been reasonably foreseen by either the contractor or the owner. These are generally known in the industry as types I and II.

In general, most rulings have found that the contractor was not required to verify site representation as shown in the contract documentation. This is in spite of the fact that most contracts state that the contractor is "charged with the knowledge" that the owner is not responsible for those representations, and that the contractor shall examine the site prior to bidding and ascertain the prevailing site conditions. This language is exculpatory, and purports to disavow owner responsibility for the boring logs, geological data, and other information furnished with or referred to in the bid documents. As we mentioned in Chapter 1, exculpatory language is often disregarded when cases reach the courts or the contract appeals boards. However, this wording also speaks to the first type of changed condition, the one in which misrepresentation is alleged. Site investigations are a sore point in the contract-owner relationship.

As we have pointed out elsewhere in this book, it would be preferable for the owner to assume this risk. Owners have more time to conduct thorough investigations than do contractors. However, owners often investigate just as far as they need to for design purposes. The contractor may be given the right to make exploratory borings or to conduct other investigations, but in most cases it will not be practical to do so. For example, if an owner has been designing a tunnel project for several years, and performing geotechnical surveys during that period, it does not seem logical to expect each of several contractors to conduct another survey in the six-to-eight-week bid preparation period. Imagine the chaos if the bidders preparing a proposal for a tunnel project in New York City (or any large city) were to individually set up drill rigs and make test borings to verify the owner's data!

While the contractor should certainly visit the site and thoroughly examine it, court rulings have held that he does not have to verify to the extent that most contract documents require. Even if the owner states in the contract that he does not guarantee all of the representations, the contractor may recover his damages under the changed conditions clause. However, if the contractor alleges that the owner intentionally misrepresented the facts relating to the site or withheld relevant information, then both parties had better see their attorneys. Intentional misrepresentation constitutes a serious offense; such a claim is not taken lightly and is not easily settled. At the very least, a changed conditions clause gives parties to the contract the opportunity to adjust time and money if the owner's investigation turns out to be inadequate. The owner is also given an incentive to conduct a more extensive investigation than if the risk of changed conditions were placed on the contractor.

Examples of the second sort of differing site conditions, those that "could not have been reasonably expected," follow.

There might be unusual hydrostatic pressure or subterranean water conditions that might require different and more costly construction techniques. Another example might be the inability to obtain an adequate water supply where that water is essential to completion of the operation and where the water would ordinarily be expected to be available. Rehabilitation and restoration of older buildings for which plans are no longer available are also obvious candidates for unexpected conditions that could impact the work.

In most cases, the changed conditions clause would deal with site or subsurface conditions. Changes due to unusual weather would not come under a changed conditions clause; however, unexpected soil reaction to such weather may constitute a differing site condition. One such case dealt with unusual capillary action of soils that caused underground ice formations during freezing weather. A subsequent thaw resulted in a collapse.

The standard federal clause also says that the contracting officer shall promptly investigate the conditions. If it is found that such conditions differ materially and cause an increase or decrease in the contractor's cost of, or in the time required for, performance of any part of the work under this contract, "whether or not changed as a result of such conditions," an equitable adjustment shall be made and the contract modified in writing, accordingly. The phrase "any part of the work" in conjunction with "whether or not changed" allows for recovery of ripple, or impact, costs resulting from the differing site conditions.

NOTICE

The standard federal clause also states that the contractor cannot request an equitable adjustment after final payment has been made. Of course, an owner who is not under federal jurisdiction may write his own changed conditions clause and incorporate what he wishes and what a contractor will agree to. The federal clause states that some notice must be given in writing and that this must be done before conditions are disturbed. Failure to comply with these provisions can jeopardize recovery. The Los Angeles version of the changed conditions clause states that failure to give timely notice constitutes a waiver of the claim. The federal government clause states this and that the government may extend the time required to report changed conditions. In other words, "promptly" can have different meanings.

The purpose of the time limitation is to permit an owner to make any field investigations necessary to determine whether or not a changed condition does exist. The owner then also has the opportunity to do something about the changed condition; he may devise an alternate construction scheme to overcome difficulties presented by the changed condition. He also has time to keep records of work performance in dealing with the changed condition. Those records will enable him to judge the contractor's claim in a rational manner.

If timely notice is not given and it is possible to reconstruct the nature and extent of the changed condition, the federal government contract appeals boards and the courts have been known to be lenient. In one case, a contractor alleged sizable change in rock volume, but did not submit this claim until after the site had been backfilled and the structure built. Recovery was allowed, however. The contractor produced the records of an independent survey from which it was possible to estimate the quantity of rock, and it was established that the owner's representative was on site and knew of the changed condition. This does not mean that the notice provision can be taken lightly. It is always advisable to follow these provisions to the letter and thus eliminate the need to defend the failure to notify the owner. The lenient example given here is not always the case. It does seem only fair to grant a contractor recovery if there is little doubt that there were indeed changed conditions that impacted the work, assuming the owner has not been seriously prejudiced by the failure of notice.

If the contractor does not realize that he has a differing site condition until after he has disturbed the site, he is still protected as long as he has given notice immediately on becoming aware of the changed condition.

What happens if the owner declares that there is no changed condition? The contractor must file a protest and state that he expects to be compensated for the change. If he fails to do this, he may waive rights to cash compensation on the grounds that he appears to agree with the owner's decision. He must continue with the work. He rarely has the right to stop work; and if he does stop work, he makes himself vulnerable under the default and termination clauses of the contract. (As explained in Chapter 5 on environmental issues, the contractor should, of course, have the right to stop work where hazardous materials are concerned.)

When the contractor objects to the decision, his next step is to follow the procedures set forth in the contract. There is usually a disputes clause that deals with those procedures (see Chapters 2 and 11). In the federal government, once a final decision has been issued by the contracting officer, the contractor has the right to appeal to the appeals board or to the U.S. Claims Court. In all of these cases, the final decision may take some time. It is in the best interest of both parties to attempt some agreement on the claim, rather than to continue through the time-consuming formal routes. Even under the best of circumstances, these all take time and money. Chapters 10 and 11 discuss these boards, various resolution methods applicable at other levels of government, the court, and arbitration.

To sum up, the changed conditions clause can be the best contractual means to deal with such conditions in the work. Exculpatory language can obscure the intent of these clauses, but is largely useless. The courts more often than not dismiss such language, and properly so.

Timely notice is called for; and while resolution bodies, including the courts, have been liberal in their interpretation of these claims, it is best to be timely. The need for changed conditions clauses has been argued for years within the construction industry. Enlightened owners have steadily increased usage of such clauses, but inclusion in contracts is not to be taken for granted. Considering that the federal government incorporated such a provision in its standard construction contract quite some time ago, and that there are many arguments in their favor, it seems surprising that many owners remain either ignorant or suspicious of these clauses.

But, there are also owners who simply cannot afford a changed conditions clause. For example, a local school board with a strictly limited amount of money would, in most instances, rather use a fixed-price contract, even if it does attract bids with contingency money built in. The school board wants to know exactly how much the job is going to cost. In other cases, owners are suspicious because they believe that contractors will look for changed conditions where they don't exist and then attempt to allocate costs improperly to those changed conditions.

Even contractors are not in full agreement concerning the use of changed conditions clauses. Some are still gamblers at heart and prefer to take the risks, and along with that risk the potential for a windfall profit. Unfortunately, those risks usually end in a lawsuit when the contractor loses that gamble. In summary: In the absence of a changed conditions clause, competition may be restricted to the so-called highrollers, thereby limiting competition. An owner may find this acceptable, or even desirable but he should be aware of that possibility. We recommend inclusion of the changed conditions clause to provide an equitable means of paying the contractor for overcoming conditions that neither he nor the owner could have expected from the information available at the time the contract was prepared.

GEOTECHNICAL BASELINE REPORTS

In projects involving complex underground construction, the inclusion of geotechnical baseline reports (GBRs) in the contract documents can facilitate risk management and dispute resolution. In a dispute over differing site conditions the GBR is used to establish just what the site conditions differ from.

GBRs have long been used to illustrate underground conditions and facilitate the bidding process. However, the documents were for "informational purposes" only and provided limited guidance in assessing risk allocation and controlling contingency costs. The contractor could assume the worst and include unknown costs in the bid, or more likely, hope for the best, bid low, and take on potentially substantial risk.

It is impossible, or at least economically infeasible to determine exact underground conditions. GBRs are not technical fact; they are realistic predictions of subsurface conditions based on technical data that the parties agree to use in developing design and cost proposals. The Technical Committee on Geotechnical Reports offered the following example in "Geotechnical Baseline Reports for Underground Construction" published by the American Society of Civil Engineers.

At a given tunnel site, investigations predict that between 100 and 300 boulders will be encountered. If the owner sets the baseline at 300, contractors will factor related costs into their bid. The owner may choose to pay the higher price in order to reduce (or eliminate) change orders and differing sites conditions claims. If the baseline is set at 100 boulders, and 200 boulders are actually encountered, the contractor will have a valid claim for the extra cost. In this scenario though, the owner will pay only the costs related to the boulders actually encountered during construction.

Contractors are not required to use the baseline in their bids. They are still free to hope for the best and bid low, but in doing so will openly assume the risk of any costs associated with conditions up to the baseline. Contractors are entitled to rely on the general accuracy of technical representations, but the GBR is not a warranty. The GBR does indicate the intentions and expectations of the parties at the time of contracting and the voluntary assumption of risk.

Uncertain subsurface conditions related to ground water, fault lines, bedrock, gas, contaminants, and the like should be addressed with specificity in the GBR. If the GBR is to be an effective tool in dispute resolution, the baselines must be definitive. Ambiguous language should be addressed and clarified before the contract is entered into.

ESCROW BID DOCUMENTS

Conditional and controlled disclosure of the contractor's escrow bid documents can also be a useful tool in preventing and resolving contract disputes. Under this scenario, the documents used to develop the bid are sealed and submitted at the time of the bid. When and if the parties think the documents may help resolve differences, the documents may be consulted.

Contractors understandably have misgivings about the practice. Such documents contain proprietary business information that should not fall into the hands of competitors. Measures can be taken to prevent this outcome. First, there should be written confirmation that the documents remain the property of the contractor. This will prevent them from becoming available in public projects under the Freedom of Information Act. Two, the documents should be held by a neutral and trustworthy third party, who should be present if and when the seal is broken. And, three, the documents should revert to the contractor when the project is complete.

In some cases, the mere presence of the documents can avoid disputes. If the parties know that the documents are available, they will be less likely to make bad faith claims. If it does become necessary to consult the documents, they may clear up honest misunderstandings and avoid protracted claims. The risk remains, however, that the contractor will prepare a separate set of bid documents for the escrow, anticipating certain areas of claim. Owner-held doubts related to this risk can limit the usefulness of escrow bid documents.

4. Changes

What are changes? For one thing, they are inevitable. The sooner all parties to a contract recognize that fact, the better. Changes can be complicated, but if handled in a cooperative working climate, they should not have to result in disputes, threats to make recourse to the law, threats to shut down the job, or threats to stop payments. For example, when Partnering or Dispute Review Boards are used, stakeholders may be more willing to accept changes because an experienced facilitator or board of neutrals has deemed it necessary.

The EJCDC defines change as "an addition, deletion, or revision in the work or an adjustment in the contract price or contract times, issued on or after the effective date of the agreement." As such, owners can require that work be accomplished under conditions provided by the change order as long as these are within the scope of the contract. For example, substitute types of cement can be required for a concrete wall due to a discovery of adverse soils. However, a concrete dam cannot be substituted for an earthfill dam under a change order. Such action would constitute a cardinal change and could invalidate the contract.

CARDINAL CHANGES

There are no hard and fast rules as to what constitutes a cardinal change, other than that such a change is beyond the scope of the contract. One ruling stated that a "cardinal change has been found to exist when the essential identity of the thing contracted for is altered or when the method or manner of anticipated performance is so drastically and unforeseeably changed that essentially a new agreement is created." This is still not very precise, and if a cardinal change is suspected, an attorney familiar with construction law should probably be consulted.

Of course, a contractor may accept the change, unless it is barred by applicable law such as a competitive bidding statute for public works. Most public contracts are governed by this sort of statute. In one public sector example, the contractor accepted a change order to build a tunnel connecting his project to another public building. That portion of the work was later deemed to have required a separate contract competitively bid. The contractor, even though he accepted the change in good faith, risked recovery of his expenses due to the laws governing public contracts.

If an owner requires a contractor to perform changes that are found to be cardinal, he might risk a breach of contract suit. So both owners and contractors should be careful about the extent of any changes—such changes should be within the scope of the contract.

The fact that the definition of a cardinal change is imprecise makes caution all the more important.

CLAIMS

Note that in the change order definition the word "claim" is used to mean a legitimate request for that adjustment in the case of a change. In usage, however, the word often connotes a dispute or a demand. "Claims" should be a neutral term; the expression "a claims-conscious" contractor certainly is not such a term. In this book we attempt to achieve an understanding of claims and changes so that claims need not be a fighting word. A change doesn't necessarily lead to a claim; a claim doesn't have to lead to a lawsuit. In the true sense of the word, a claim is a request for an equitable adjustment due to a change under the contract. However, those requests too often end up in less than neutral situations.

CHANGE ORDERS

Most contracts have a clause stating that a change order must be in writing. Even in cases where there is no such clause, written orders make for better communication on a job. A verbal change order or variation in procedures at the job site, no matter how insignificant at the time or how friendly parties may be at that time, can lead to disputes and claims later on. Recovery for extra costs or time can be difficult if those change orders were not in writing. Even if the owner's representative on the job knows that the changes have been ordered (he may have given the order) and knows that extra time and costs may be involved, some jurisdictions have held that no recovery can be made without that written notice where it is stipulated in the contract terms.

However, some decisions have gone the other way and held that the order need not have been in writing. Those exceptions have been based on prior actions of an owner. For example, if an owner acknowledged other claims related to extra work orders not given in writing, those prior actions might be held against him. Clearly, it is safest to insist on written change orders. Otherwise, the contractor could have difficulty proving later that a change was ever ordered or that the person who ordered that change had the authority to do so. From the owner's point of view, he should abide by the provisions of the contract or risk surrendering some of the protection afforded by the contract requirement that the changes be ordered in writing.

The 1996 edition of EJCDC 1910-8, section 10.01, which authorizes such changes, adds two new terms to the claims lexicon: Work Change Directive and Written Amendment.

> Without invalidating the agreement, the owner may anytime or from time to time, order additions, deletions or revisions in the work. These will be authorized by change orders. Upon receipt of the change order, the contractor shall proceed with the work involved. All such work shall be executed under the applicable conditions of the contract documents. If the change order causes an increase or decrease in the contract price or an extension or shortening of the contract time, an equitable adjustment will be made as provided in...(other portions of the document)...on the basis of a claim made by either party."

The Work Change Directive is a written statement to the contractor signed by the owner and recommended by the engineer that orders an addition, deletion, or revision, or responds to unforeseen subsurface or physical conditions, or emergencies. However, it cannot change the price or times and is only evidence that the parties expect it to be incorporated into a change order if they cannot agree to time and price adjustments themselves.

A Written Amendment is a statement modifying the contract documents that is signed by the owner and the contractor that normally deals with nontechnical matters rather than construction-related modifications.

These changes codify the practice of allowing parties to make necessary changes without stopping work while they hammer out a formal change order.

A Word about Timing

Change orders ought not to be issued between the bidding and the award stages of the proceedings. This has been held illegal on public contracts on the ground that the practice might favor one bidder. In other words, other bidders did not have the opportunity to price the changed or additional work. Changes also must ordinarily be ordered before final acceptance of the work. If a change is ordered late in the sequence of the work, extra cost allowances might have to be made to the contractor. Most of the labor force may have left the site, or the necessary equipment may be needed elsewhere by the contractor. Compensation for the change would have to take such factors into account. Sometimes parties negotiate a separate contract for last-minute changes, although this can be even more costly. For example, a separate contract let for the last-minute addition of a roof air-conditioning unit could, in effect, nullify the roof warranty. It is preferable, if at all possible, to deal with changes within the original contract.

THE OWNER'S RESPONSIBILITY

The owner sets the tone for claims and changes when he selects the contract type and when he decides, if he does, to put the contract out to competitive bidding. A cost-reimbursable contract arrangement is going to result in fewer disputes over changes, but a lump-sum, competitively bid contract virtually invites such disputes or at least questioning. (See Chapters 1 and 2 for discussions of allocation of risk.) However, the frequency and intensity of disputes can be mitigated by the elimination of exculpatory clauses and by clear contract provisions spelling out allowable costs of changes. A harsh contract with exculpatory language that puts most of the risk on the contractor is clearly going to set the scene for trouble.

Competition based solely on price can also lead to an undesirable job climate. If the job goes to a low bidder who bid too low and intends to improve on that low bid through "equitable adjustments," the job can be beset by claims. Most public bodies are required to accept the lowest responsible bid, so project management must be prepared to deal with the possibility of such a situation.

One might liken the parties in a construction contract to the principals in an arranged marriage. Each puts his best foot forward during negotiations. Once the contract has been signed, the honeymoon period is sometimes brief. We have mentioned the tone set by the contract type selected and the language used. But there are other reasons why that honeymoon period may deteriorate into an adversarial scenario. Some of them are described next.

The Environment for Claims and Changes

Here are several situations, often avoidable, that can lead to disputes and claims over changes:

- *Misunderstandings.* There can be perfectly honest failures in communication. The field office may misunderstand or misinterpret what the home office has done, or what they think about a certain aspect of the work. It's important to make sure everyone understands the same version of what has been agreed to. Minutes, memos, and other written records that are circulated can help.

- *Pride.* By this, we particularly mean pride on the part of the architect or engineer who maintains that he has designed the "perfect" project, and who thinks that any requests for changes or claims pertaining to those changes are a threat to his competence.

- *Greedy Owners.* There are owners who think that "as long as the contractor is here, why can't he do a little extra paving or painting here and there?" More frequently, what the owner might seek to extract from the contractor is to utilize his men, plant, or equipment for longer periods of time at no extra cost. These amount to unacknowledged changes and can only lead to resentment and a job atmosphere ripe for claims.

- *Avaricious Contractors.* There are contractors who knowingly enter agreements with the idea that they will file claims. No matter how enlightened an owner is, he could be dealing with such a contractor.

- *"Catch-up Profit."* There are also contractors who realize late in the day that they are losing money on a job, and scurry to see where they can make it up. One way to do this is to put in for claims for changes, even if there is only a shred of justification for doing so. This impulse should be repressed. The time and expense necessary to prepare a claim will not be offset if the likelihood of succeeding with the claim is remote. Few truly unjustified claims are ever settled or recovered in court.

- *Rigid Contract Interpretation.* Problems with interpretation are likely to occur when contract specifications are not absolutely clear. Sometimes this happens when an unreasonable owner or inspector goes beyond what is normal practice in the industry. The owner sets the tone for an adversarial situation in this case.

- *Vindictiveness.* Once an owner is presented with a claim, he can deal with it on a reasonable basis or he can retaliate. As we have said, "claims" need not be a dirty word. Some owners adopt an "I'll show him" attitude which can lead to rapid deterioration of the project climate and which is nothing but counterproductive. This attitude can be manifested by tougher inspection, rigid contract interpretations, and other unpleasant actions that only set the scene for more claims.

GROUND RULE MEETING

One way to establish a cooperative working climate is to have a meeting early in the project in which the rules for claims and changes are discussed. This serves to clear the air and prevent misunderstandings later on. The idea is to communicate during that honeymoon stage and before construction begins. Depending on the scope of the project, parties may choose to use Partnering, with or without an outside facilitator, or engage a Dispute Review Board. These practices are explained in Chapter 12 on formal dispute resolution.

First of all, the meaning and significance of contract provisions can be clearly presented and questions can be asked. Of course, both parties should always read their contracts before signing, but this doesn't always happen. Thus, the meeting allows the contract administrator to walk through the documents, pointing out salient features that might have been missed or skimmed over. A lot of contract provisions about notice and allowable costs are often regarded as boilerplate until a problem arises. At this kind of meeting, everything that isn't spelled out can be discussed. For example, parties can agree on the mechanics of handling claims, how many copies will be needed, and to whom they should be delivered. Types of costs that will be allowable can also be discussed. Parties might also agree on how these costs will be calculated so that there won't be arguments about methodology later on.

What profit will be acceptable when a change order occurs? What kind of mark-up is going to be acceptable if the change involves subs? There are various manuals that can be helpful in answering these questions, such as those published by the Associated General Contractors of America, and the *Rental Rate Blue Book* (published by Dataquest, Inc.) often used to compute costs for heavy equipment. The contract may indicate the use of a certain manual; if it doesn't, parties might discuss this point.

Is the specified date for job completion sacred? If so, parties should talk about it right off. If an amicable relationship is set up, some necessary claims and changes can be handled expeditiously to insure that sacred completion date. Given the myriad construction-related problems that can occur during an ordinary project—weather, strikes, materials, delays, and so forth—the worst thing that can happen is to argue over every change when the job is in danger of being delayed. If parties cooperate, the job can come in on time and possibly within budget, too. That sacred date can, on the other hand, also create an emotionally charged atmosphere in which potential conflicts are escalated into real ones. Preconstruction meetings can prevent such situations.

PRECONSTRUCTION DISAGREEMENTS

It is possible that parties can't agree at a preconstruction meeting on the details of how changes will be handled or how costs of changes will be calculated. If this happens, the best the parties can do is to agree to deal with problems on a case-by-case basis. If and when these situations arise later on during the work, things may change. The parties involved may have a better relationship; one side or the other may concede a point in the interest of getting the job done, or other factors may enter the equation. Or, of course, they may have a knock-down-drag-out fight that delays the project further. It's in everyone's best interest to avoid that situation. Under these circumstances, it may be a good idea to choose a mutually agreeable neutral to evaluate changes.

TYPES OF CHANGES

There are two basic kinds of changes: (1) those that are directed and acknowledged as changes by owners, and (2) those that are not. First, the "easy" kind, those that are acknowledged. There are several possibilities for negotiating these:

1. *The Ideal Situation.* The owner and the contractor both agree that there is a change, and they then agree as to how much money or time need be adjusted because of that change. This is an ideal that can be achieved in a good working climate. If both sides agree on the cost and the time, they need only work out the arithmetic.

2. *Unit-Priced Work.* If the change involves a portion of the work that has been unit priced, then simple multiplication of units and costs is all that might be necessary. However, there is an element of a gamble in unit pricing. The contractor may have set a high or a low unit price on one element of the work for other reasons. He may have put a low price on one portion of the job to keep his bid low, hoping that the item would never involve an overrun. In the case of an overrun, of course, he would lose money. He may have put a high unit price on another portion of the work to balance out his risks. In that case, the owner could be forced to spend an inordinate amount of money if a change arises that involves more work in this category. The contractor may also have unbalanced his unit prices to cover the front-end costs of mobilization. By doing so, early items of work would have a higher

unit price than strictly necessary to perform that work. Correspondingly, unit prices for work done late might be lower than what the work actually costs the contractor.

Parties should be able to adjust these unit prices in the case of change where one party or the other might suffer. One way for the owner to do this is to go back to the original bids. He can compare unit prices of other bidders for the same portion of the work. Of course, the other bidders may have done the same thing. However, if some prices are substantially lower, this can be a good guide. The owner can also check his own estimate for this portion of the work.

3. *Variation in Quantity.* Some contracts include a variation in quantity clause that sets forth a percentage of change after which unit prices will be adjusted. In those cases, the potential for conflict is considerably diminished. For example, a contract could provide for adjustment if unit-priced elements of the work change more than 25 percent. Some enlightened public owners have incorporated such contract provisions; San Francisco's BART and the Baltimore subway are examples. Chapter 2 discusses this approach. At least one federal agency allows for split quantities, whereby a contractor can place one unit price on a certain percentage of the work in the original bid and another price on the remainder. This would usually be a higher price for the first portion and a lower price for the rest, so the contractor recovers his fixed overhead after performing the first part. All these approaches minimize risk and the potential for disputes.

4. *Forward Pricing.* If unit prices are not applicable to the changed work in question, the best approach is to forward price. This means both parties sit down and agree to a firm price for performing the change. This has great advantages to both. The owner knows how much he is going to pay. The contractor can logically integrate the changed work into his schedule. He can still make a profit—both sides agree on how much. And the arrangement retains some element of risk that is congenial to the construction industry. Once the price is set, either side may gain or lose a bit; that is, the work may come in for a little more or a little less than the forward price. But at least the parties have set the parameters of that risk.

How do parties go about setting that forward price? The best way is to have both parties estimate the cost of the change and then negotiate their differences. At least the owner should thoroughly analyze the contractor's estimate. The amount agreed on may either be a lump sum, a unit-price setup, or a cost-of-the-work setup with a guaranteed maximum (upset price).

Impact, or ripple, costs must be considered in forward pricing. These are added costs incurred in performing the other items of the unchanged contract that have been affected by the change. Either the forward price should include impact costs or the agreement should specifically state that the impact costs of the change are reserved and will be provided for at a later date. The more complex the project or the change, the more difficult it is to assess impact costs. This is a tricky area. Sometimes owners may refuse to allow impact costs on the grounds that the costs are virtually indeterminable. This is one of the areas that should be addressed in a preconstruction meeting when the relationship is cooperative and congenial. At the very least, items of allowable impact cost may be agreed on, such as labor and material escalation, extended job overhead costs, out-of-sequence work, and so on.

Sometimes the emergency nature of a change precludes immediate calculations of impact costs. When the walls of a tunnel are caving in, no one is going to sit around estimating impact costs of the repair work. Work done under stress conditions, however, is more likely to have ripple costs; no one has planned for the emergency, schedules are changed, and equipment moved to accommodate such work. If forward pricing is attempted, it is possible to estimate impact costs and include them in the price. It is probably preferable, though, to state that these costs will be recoverable when the job is finished and a more accurate tally is possible.

5. *Time and Materials.* The remaining way to negotiate price changes is on a time and materials basis. When unit prices don't apply and the parties don't wish to forward price, the costs of time and materials are kept, and a profit and overhead figure added in. This is a likely approach in an emergency change. Impact costs can be considered when negotiating an acceptable profit allowance, but it is preferable to keep profit separate and to reserve the right to recover impact costs later when it is easier to judge what they will be.

The time and materials method of assessing the costs of a change is often prescribed by contract and by the regulations of state and federal agencies. First, both parties must agree on what is to be included in the change, and then must agree on what kind of records are to be kept to document the costs. Finally, they have to follow up and see that these records are kept.

The ideal situation would be one in which the contractor keeps daily statement-of-work sheets, which are verified by the architect or engineer for the owner. Supplementary records can include time cards, equipment cards, invoices, diaries, and field reports. Any differences encountered between the different sets of records should be settled at once. The resolution should be put in writing and all parties involved should get copies. It is difficult and risky to reconstruct a verbal agreement later on.

RECOMMENDATIONS

It might be useful at this point to refer to the recommendations of the National Academy of Science Committee on Tunneling Technology. That group, as we have mentioned elsewhere, was primarily concerned with contracting practices for underground construction, but their suggestions have merit for other types of construction.

The committee report states that "negotiations for contract price adjustments for changes and extra work are unnecessarily protracted under firm-fixed price contracts and needlessly occupy the time of owners, engineers and contractors management personnel, time that might better be spent in completing the work." Some of their recommendations are:

- More contract items should be priced, whether on a lump sum or a unit-price basis.
- Provide for lump-sum pricing to cover the costs of mobilization and demobilization and for items not apt to vary greatly in quantity.
- Provide for unit pricing where item quantity may vary to any substantial extent.
- Provide for adjustment of unit prices where unit costs are affected by substantial increases or decreases (e.g., 15 percent from the estimated quantities set forth in the bid schedule).
- Provide items, to be unit- or lump-sum priced, to cover work that may become necessary; for example, to overcome problems related to excessive water conditions. (In this case, it might be advisable to provide for unit pricing for successive increments of water that might have to be handled.)
- Provide a percentage amount, or formula, for determining overhead and profit to be paid on changes, or provide a bid item for such percentages with a specified basis for bid comparison purposes.

- Provide that interest will be paid by the owner, at a specified percentage, on the money expended by the contractor in the performance of changes and extras ordered or required under the contract, commencing with the date that the owner has verified, or by which he has had a reasonable opportunity to verify, the contractor's proof of costs expended for performance, and then ending with payment.

- Establish a formula or basis for payment to the contractor, covering the use of his own equipment on changes and extras, and which will provide an automatic basis for such payment without room or argument.

IMPACT COSTS

Impact costs merit serious attention. Most owners would like to have the change order release read thus: "This change order is in full compensation for all time and costs arising out of this change, and any additional claims are specifically released..." This language absolves the owner of responsibility for unforeseen ripple costs owing to the change.

In contrast, most contractors would prefer this wording: "This change order includes only time and direct costs of the changed work and does not include any allowance for resultant delay or increased cost in performing the unchanged portions of the work, claim for which is specifically reserved." This wording protects the contractor in the event of unforeseen impact on other areas of the work.

Some owners are wary of such clauses, thinking that these give the contractor carte blanche to charge anything to impact. However, forward pricing, in which ripple costs are provided for, is the best approach for both parties when the cost of the change and its impact can be reasonably estimated ahead of time. Both parties are spared the annoyance and the expense of keeping detailed records of the cost of the changed work and can prevent the potential disputes over items of allowable costs. The owner has a guaranteed price for the work; the contractor can be assured of payment for work as it is performed and will not have to finance the work. He should, of course, be compensated when both parties can agree on the price of the changes, but that might be years later after completion of the job.

NEGOTIATION BREAKDOWNS

Sometimes both parties recognize that there is a change but cannot consummate agreement as to what an equitable adjustment of time or money might be. This can happen when there is a breakdown between the parties after negotiations are completed. Parties can walk away from the negotiating table and change their minds. They can get home and decide that they gave away the farm, and back out of that agreement. Sometimes there is an honest misunderstanding; parties have different perceptions of what went on at the bargaining table, and it is only when the written version is presented that they realize their perceptions were not shared. Notes and minutes of all meetings should help prevent the latter kind of breakdown. The change of heart or mind is less easy to deal with.

What happens next, in either case? The contractor can accept such a change order with reservations. Those reservations should be down on paper. A good solution is to agree on a draft-level change order, if the parties appear to be pretty close to agreement. Both sides can assemble notes and list those items on which there is accord. They can then agree to work out the rest as the work progresses. A fair amount of good faith is assumed here. In some cases, negotiations break down totally. What often happens is that the contractor stalls to see what his costs are really going to be and how much time the change entails. This means he won't be paid until he is finished and the owner assesses the cost. On the other side, the owner, or his architect or engineer, might find it in his own best interest to wait in order to get a better fix on the contractor's cost. Neither approach is conducive to a least-cost completion because the contractor has no incentive to be efficient without a predetermined price.

An owner can issue a unilateral change order when parties cannot agree on costs. The contracting officer can say: "Here is my final decision. I will give you X dollars, and X period of time to perform the work connected with this change. Your recourse is through the disputes clause." This will not enhance the morale of the contractor's team. Nevertheless, it does provide for some payment and allows for the next step in an orderly claims process.

If agreement cannot be reached, that draft-level change order is preferred to the unilateral order, of course. However, good faith is not always present.

CONSTRUCTIVE CHANGE

Constructive changes are those caused by an owner that are not acknowledged as such. Up to now, we have discussed ways to deal with changes when both, or all, parties agree that there is a change. In a constructive change, the contractor asserts that an action of the owner or his representative amounts to a change, one that involves an adjustment of the time and money accorded under the contract terms. The owner argues that there is no change.

The term "constructive change" evolved in federal contract usage in a somewhat tortured semantic effort to deal with such disputes under the contract's changes clause. In the private sector, or at other levels of government, the term is less likely to be used. There, ordinarily, a breach of contract would be claimed if an owner refused to concede that a change had been ordered.

Typically, a constructive change situation might arise from differing interpretations of the contract language. Obviously, if an owner orders that the diameter of a tunnel be doubled, we would have an acknowledged change of the nature discussed earlier in this chapter. A constructive change is likely to be a more subtle kind of change; examples would be unreasonable inspection practices or unduly rigid interpretations of specifications. One party might take the position that these practices effect a change; the other would undoubtedly disagree. "Reasonable customs and standards of the industry" are usually applied when such a dispute arises. The owner might insist, for example, that the concrete work meet a higher standard than the contractor believes he is obligated to meet. That standard may be more rigorous than that customarily accepted, say by the American Concrete Institute or another commonly used concrete standard. Unless the specifications clearly spell out what those special standards are to be, an owner's insistence on this standard would impose on that owner the obligation to pay extra costs arising from that higher standard.

However, a warning: Contract language often imposes on the contractor the duty to request clarification when there are patent ambiguities in the specifications. If there is doubt about the concrete work specifications, and the contractor didn't inquire about them, it could be held against him.

Another situation that could lead to a constructive change might be improper rejection of work. Suppose a functioning pipe has an inconsequential hairline crack. An order to replace such a pipe would be deemed a constructive change if replacement were unreasonable in terms of normal trade practices or if such cracks were not proscribed in the specifications. Multiple or differing inspections by an owner's team might also create a constructive change. This would be so if those inspections resulted in inconsistent directives and actual physical hindrance with performance of the work.

Delays in the delivery of owner-furnished equipment could cause a constructive change. The same could be said for owner-furnished equipment that arrives on site in an inoperable condition. If the owner refuses to acknowledge the impact on the work, we have a constructive change situation. Other kinds of interference, such as restricted access to the site, also come under this category. In one such instance, an electrical contractor was working on a job where he was to install conduits and cables in a New York City tunnel, working at night so that traffic interference would be minimal. However, on nights when there was a ball game in the nearby stadium, he was not permitted to work. Thus, he lost several hours of work on such nights. The contractor was entitled to be reimbursed for the costs of the delay caused by the ball-game traffic delays, including the nonproductive hours of his employees.

Defective or ambiguous specifications can also lead to what the contractor would contend are constructive change situations if the owner refuses to acknowledge any problem with the specs. While the contractor has the duty to request clarification when specifications are unclear, generally courts have held against the owner when ambiguities can be proved. The reasoning is that the owner made up the specification and had the opportunity to make them clear. The contractor need only demonstrate that his interpretation is a reasonable one. This rule applies to the contract documents in general, not just to the specifications.

Defective or ambiguous specifications can lead to other kinds of disputes. To take a classic example, suppose the roof of a building leaks. The owner would undoubtedly say that this is due to poor workmanship on the part of the contractor. The contractor would contend, typically, that the specs were defective, that a watertight roof could not have been built following those specs.

A combination of detailed and performance specifications in one set of documents can lead to problems. For example, suppose a contractor is obliged by the specs to meet local building codes and at the same time the specifications detail work that would not meet code requirements. Unless it were an obvious discrepancy, the contractor could not ordinarily be expected to know local code requirements and would be entitled to rely on the designer's expertise.

An example of this would be a case where the plans show the number of fire sprinkler heads, but the building code requires more sprinkler heads per square foot than provided for in the design. Under such circumstances, it would ordinarily be grossly unfair to compel the contractor to furnish the extra heads without additional compensation. Consequently, the extra work needed to satisfy the local code requirements would constitute a constructive change.

More often than not, the two kinds of specifications are mixed by the designer in an attempt to exculpate himself from responsibility wherever he has doubts about the adequacy of the design. Sometimes, though, detailed and performance specifications are mixed for a valid reason. One federal agency, for example, gives a performance specification for the moisture content of an embankment. The agency then specifies how many passes with a certain type of roller should be made over the embankment. The agency has discovered that they need fewer quality-control checks as a result of this mix. If performance and detailed specifications are mixed, the reason should be stated and the contractor clearly warned that the detailed spec is only a minimum and that, in any event, the performance must be fulfilled.

Assertions of impossibility or impracticality of performance can also lead to constructive change. An example: An architect specifies a particular decorative finish on a concrete structure exterior. The specification is basically a performance spec; the architect notes that the finish can be accomplished by distressing the formwork. The contractor tries that, and many other methods as well, to achieve the desired result, all in vain. Hand finishing might achieve the architect's vision, but the cost would be inordinate. It would be, in effect, practically impossible to achieve the desired result by the use of formwork, contrary to what the specifications stated. If the owner forces the contractor to proceed with specified performance without any consideration of time and expense, we have a constructive change situation.

Constructive change can also arise from nondisclosure of technical information on the part of the owner. The owner's action could constitute fraud or misrepresentation. It could also result from an innocent mistake; for example, certain information may be mislaid or incorrectly judged to have been inconsequential. Therefore, an allegation of withheld technical information is a sensitive area. Allegations of fraud are serious matters. However, if the owner's withholding of technical information has been unintentional, but he still contends that the information in question will not change the work, we might have a constructive change situation. This sort of problem might arise in subsurface construction, where that withheld information, whether or not the withholding was intentional, may be considered to be very important by the contractor's engineers.

ACCELERATION

Acceleration, the speeding up of the job schedule, is obviously a change. A contractor may realize that he is lagging, and be forced to accelerate the work to finish the job on time. This can be quite costly; extra men may have to work overtime and so forth. A contractor may do this and accept those costs because he will have to pay liquidated damages (or the owner's actual damages) on work not completed on schedule. He may have contractual obligations elsewhere that require the same men and equipment and thus find it in his best interest to accelerate. If the contractor is accepting the extra costs, there should be no conflict.

An owner may also direct an acceleration. If he does this, then he is shouldering the extra cost. Again, there should be no dispute, except about the actual costs of the acceleration. Good documentation of those costs, as always, is essential.

The next category of acceleration is laden with potential conflict. Suppose an owner makes threats to the contractor about the lateness of the work; that is, he makes reference in thinly veiled language to terminating the contract because of the failure of the contractor to perform according to agreement. Suppose the contractor believes that any delay on the job is due to the owner's actions and that the owner should pay for any necessary acceleration. Or suppose the owner has refused to grant legitimate time extensions, or has granted them in such untimely fashion that the contractor is forced into acceleration.

All of these situations might drive a contractor into what can be called "constructive acceleration." The acceleration is due to an action of the owner's that he does not acknowledge. In most cases, the contractor is entitled to recover these costs. (Chapter 6 includes a sample calculation of a contractor's claim for acceleration costs.)

In order to avoid such situations, an owner should do several things. One is to accept the fact that if acceleration is necessary, the owner may have to pay. Threatening letters can be held against an owner if the case goes to litigation. The best action for the owner to take if he believes the contractor is lagging is to write as neutral a letter as possible, alluding to the contract agreement and to job schedules without threats; for example, he might say, "I am concerned..." The contractor cannot claim that he didn't know anything was wrong, but he also cannot say that he was threatened. On the positive side, an owner can be reasonable about granting time extensions when they are justified.

These should be processed promptly so that the contractor is not driven into an acceleration situation when time is actually due him. When an owner has a sacred completion date, he should be working with the contractor to meet that date. If acceleration is in order and the owner caused some delay, then he should be willing to pay for that acceleration. This is true whether there is a sacred date or not. Threats to terminate the contract are out of order if the contractor is entitled to time extensions. Deduction of liquidated damages from the contractor's payments when there have been excusable delays on the project will further exacerbate the situation. It should be apparent to all concerned that a lawsuit will cause even more delay.

Chapter 5 deals with this subject in more detail. It should be clear that documentation of the cause and cost of delays are extremely important.

CHANGED SUBSURFACE CONDITIONS

Changed conditions, "differing site conditions," or "differing subsurface or physical conditions" are the changes during work that can most easily lead to litigation. We stated in Chapters 2 and 3 that it is best to include such clauses and Geotechnical Baseline Reports in a contract. Inclusion enables parties to deal with changed conditions on a rational basis and under the terms of the contract.

There are two kinds of changes that come under this heading: those not indicated by prebid data but that were "known or should have been known by the owner," and those that could not have reasonably been anticipated. It's easy to see how the first category can lead to disputes and litigation. If the owner knew about the differing conditions, it is possible that he is guilty of fraud or misrepresentation. On the other hand, if he says he didn't, it is difficult to prove otherwise. We are out of the constructive change category here in that we are discussing fairly dramatic change. For example, New York City's Third Water Tunnel is the classic example of a complicated suit that grew out of a dispute over changed conditions. There was no differing site conditions clause in the contract for that tunnel. Rock conditions proved different from what the contractor believed they would be, based on prebid data furnished by the city. The result was that vastly more support-steel was needed, and the contractor sought compensation for that steel. The subsequent legal struggle went on for six years.

A change that could not have been reasonably anticipated by either party is a more benign category. Typically, changed conditions occur in subsurface construction. It is well known that owner-furnished borings can never be absolutely correct, and contractors under ordinary circumstances do not have time within the bidding framework, nor the necessary resources, to perform their own site investigation. So a changed conditions clause sets up a reasonable way to stop and renegotiate time and costs on the basis of such a change.

RELEASE

In all releases, the language should be as tight and as clear as possible, and parties should be absolutely sure of what they are when they sign off. Recovery of additional costs can be difficult once a release is signed. In particular, this goes to the point-of-impact costs. Parties should make certain that a change order either expressly includes consideration of impact costs or expressly reserves the right to claim these later on. Neither the contractor nor the owner should sign a release unless each is certain that he accepts the terms of that release.

On the other hand, in at least one instance, the contract appeals board of a federal agency upheld the contractor's right of recovery for a change after the work was accepted and a release signed. The ruling was based on the fact that the agency officials continued to correspond with the contractor about further costs after the release was signed. Those officials had acted as though the release were not in effect, and the board ruled along those lines.

That ruling may be highly unusual, but it makes a point: An owner should act as though that release is in force once he has the contractor's signature. The contractor, for his part, should not sign a release unless he is certain that he accepts the terms of that release. Further recovery is almost certainly foregone by that signature.

COSTS OF CHANGES

Changes often result in extra costs, though sometimes they can save money. However, when extra costs are incurred, either direct or impact, those costs should be recoverable. (Costs are discussed in Chapters 4, 7, and 11.)

Examples of Change Orders

A. A change order that includes allowance of impact costs (see items 2 and 3, page 67).

B. A change order with the same parties, in which the contractor reserves the right to claim impact costs at a later date (see wording on page 68).

C. A change order in which no provision is made for impact costs, thereby opening up the over-a-time extension and the impact cost resulting from the change.

A.
UNITED STATES
DEPARTMENT OF THE INTERIOR
BUREAU OF RECLAMATION

Central Utah Project

Smithtown, Utah, June 22, 1995

ORDER FOR CHANGE NO. 9

Jones, Jones, Inc.
PO Box
Fairview, Utah 84078

Gentlemen:

Pursuant to clause No. 3 of the General Provisions of Contract No. XXXXX, dated April 14, 1993, for construction and completion of No-name Dam in accordance with Specifications No. XXXXX, the following changes in the specifications and/or drawings as related to the general damsite are hereby ordered:

1. The mass haul diagram available to bidders was not corrected to reflect changes in the construction of the dam access roadway as specified in Supplemental Notice No. 2 dated January 7, 1993. Accordingly, you are directed to distribute materials as required during construction of the dam access roadway in lieu of using the distribution shown on the original mass haul diagram.

The adjustment in contract price provided for above has been determined to be the equitable adjustment to which you are entitled for increased costs incurred through May 26, 1993, as a result of changes to the general dam site. The adjustment includes allowances for the following:

1. All increased direct costs incurred through May 26, 1995, for performance of the work directed.

2. All impact costs through May 26, 1995, related to the performance of the work directed.

3. All ripple costs to construction of the general dam site through May 26, 1995, resulting from previously directed changes.

4. All indirect costs and profit through May 26, 1995.

No adjustment in the time required for the performance of the contract will be made by reason of the changes and additions ordered herein.

Very truly yours,

Signed
_____
Contracting Officer

- -

Sir:

In accordance with section 1-3.807-3 of the Federal Procurement Regulations, I hereby certify that, to the best of my knowledge and belief, cost or pricing data submitted to the contracting officer or his representatives in support of the cost of the work provided for herein are accurate, complete, and current as of June 22, 1995.

The foregoing Order for Change No. 9 is satisfactory and is hereby accepted. In accepting Order for Change No. 9, the contractor acknowledges that he has no unsatisfied claim against the Government arising out of this change through May 26, 1995, and the contractor hereby releases and discharges the Government from any and all claims or demands whatsoever arising out of this change through May 26, 1995.

Jones, Jones Inc.

By _____ Signed _____

Title _____

Contract No. XXXXX
Specifications No. XXXX

B.

UNITED STATES
DEPARTMENT OF THE INTERIOR
BUREAU OF RECLAMATION

Central Utah Project
Smithtown, Utah, June 22, 1989

ORDER FOR CHANGE NO. 9

Jones, Jones Inc.
PO Box 1228
Farview, Utah 84078

Gentlemen:

Pursuant to clause No. 3 of the General Provisions of Contract No. XXXXXX, dated April 14, 1987, for construction and completion of No-name Dam in accordance with Specifications No. XXXX, the following changes in the specifications and/or drawings as related to the general damsite are hereby ordered:

1. The mass haul diagram available to bidders was not corrected to reflect changes in the construction of the dam access roadway as specified in Supplemental Notice No. 2 dated January 7, 1987. Accordingly, you are directed to distribute materials as required during construction of the dam access roadway in lieu of using the distribution shown on the original mass haul diagram.

2. Perform additional items of extra work, as directed by the contracting officer, incidental to construction of the dam and in support of Government operations in the general damsite. Such items of extra work may include but are not limited to digging test pits for geologic investigations, moving Government-owned drilling rigs, and performing additional excavation to expose parts of the foundation for investigation by Government personnel.

All necessary labor, equipment, and material required for completion of the work covered by this order shall be furnished by the contractor.

Except as modified above, all work shall be performed in accordance with the provisions of Specifications No. XXXX, where these are applicable as determined by the contracting officer or otherwise as directed by the contracting officer.

Adjustment of the amount due under the contract by reason of the changes ordered will be as follows:

a. Distributing materials as required during construction
   of the dam access roadway in accordance with item 1
   above for the lump sum of $125,229

b. Assisting the Government in performing support work
   in accordance with item 2 above for the lump sum of $14,484

The total net adjustment in the amount due under the contract by reason of
the changes ordered herein is an increase of $139,713.

~~No adjustment in the time required for the performance of the contract will
be made by reason of the changes and additions ordered herein.~~

Very truly yours,

_____
          Signed
       Contracting Officer

- -

Sir:

In accordance with section 1-3.807-3 of the Federal Procurement Regulations, I
hereby certify that, to the best of my knowledge and belief, cost or pricing data
submitted to the contracting officer or his representatives in support of the cost
of the work provided for herein are accurate, complete, and current as of June
22, 1989.

~~*The foregoing Order for Change No. 9 is satisfactory and is hereby accepted.
In accepting Order for Change No. 9, the contractor acknowledges that he has
no unsatisfied claim against the Government arising out of this change through
May 26, 1979, and the contractor hereby releases and discharges the Government from any and all claims or demands whatsoever arising out of this change
through May 26, 1979.~~

Jones, Jones Inc.

By _____ Signed _____

Title _____

*The total net adjustment of $139,713 contained in this Change Order No. 9 is
acceptable for the direct cost resulting from the changes described herein. However, the contractor hereby reserves our right to claim impact costs and time
resulting from these changes at a later date.

_____

Note: When the contractor received the form from the owner, he crossed out the
two paragraphs indicated and added the bottom paragraph to reserve his rights
to claim impact costs and time.

Contract No. XXXXXX
Specifications No. XXXX

C.
UNITED STATES
DEPARTMENT OF THE INTERIOR
BUREAU OF RECLAMATION

Central Utah Project

Smithtown, Utah, June 22, 1989

ORDER FOR CHANGE NO. 9

Jones, Jones Inc.
PO Box 1228
Farview, Utah 84078

Gentlemen:

Pursuant to clause No. 3 of the General Provisions of Contract No. XXXXX, dated April 14, 1987, for construction and completion of No-name Dam in accordance with Specifications No. XXXXX, the following changes in the specifications and/or drawings as related to the general damsite are hereby ordered:

1. The mass haul diagram available to bidders was not corrected to reflect changes in the construction of the dam access roadway as specified in Supplemental Notice No. 2 dated January 7, 1987. Accordingly, you are directed to distribute materials as required during construction of the dam access roadway in lieu of using the distribution shown on the original mass haul diagram.

2. Perform additional items of extra work, as directed by the contracting officer, incidental to construction of the dam and in support of Government operations in the general damsite. Such items of extra work may include but are not limited to digging test pits for geologic investigations, moving Government-owned drilling rigs, and performing additional excavation to expose parts of the foundation for investigation by Government personnel.

All necessary labor, equipment, and material required for completion of the work covered by this order shall be furnished by the contractor.

Except as modified above, all work shall be performed in accordance with the provisions of Specifications No. XXXX, where these are applicable as determined by the contracting officer or otherwise as directed by the contracting officer.

Adjustment of the amount due under the contract by reason of the changes ordered will be as follows:

    a. Distributing materials as required during construction
       of the dam access roadway in accordance with item 1
       above for the lump sum of $125,229

    b. Assisting the Government in performing support work
       in accordance with item 2 above for the lump sum of $14,484

The total net adjustment in the amount due under the contract by reason of the changes ordered herein is an increase of $139,713.

No adjustment in the time required for the performance of the contract will be made by reason of the changes and additions ordered herein.

Very truly yours,

_____
      Signed
    Contracting Officer

5. Environmental Regulation and Contract Claims

Environmental regulation can dramatically affect a construction project. Not only must projects be designed in accord with regulations, but unknown environmental conditions can cause major delays, work stoppage, remediation, and changes in the scope of the project.

A BRIEF HISTORY OF U.S. ENVIRONMENTAL REGULATION

Most people are familiar with the Environmental Impact Statements (EIS) mandated by the National Environmental Policy Act of 1969 (NEPA). As the name suggests, these statements set out the environmental impact of a proposed project. They also require exploration of alternatives to the project itself and proposals that would mitigate damage to or loss of resources. All "major Federal actions significantly affecting the human environment" require an EIS. Because many private projects are funded with some federal dollars, the scope of that mandate has been tremendous.

For example, if you plan to build a road, the impact of the road on the environment must be calculated; then alternative designs must be developed to the extent that their impact can be calculated and compared to the original plan. Mitigation efforts also can affect design and construction efforts. The EIS must be made available for public review and can be challenged at any time. At best, a challenge may require further study or additional alternative plans; at worst, it can result in a very long legal battle. Meanwhile, the project is at a standstill. Environmental science is extremely complex, so it is worthwhile to assure that a reputable firm with a good track record has prepared the project's EIS. It may prevent delays later on.

Beyond NEPA

Since NEPA, a host of regulations have been enacted that set forth complex federal environmental guidelines and liabilities. In some cases, such laws enable (or require) states to promulgate their own environmental regulations and permitting procedures. Some states simply rubber-stamp the federal examples; but because regulations differ, it is important to familiarize yourself with applicable law. It is not be enough to attempt to contractually allocate environmental liability away; the laws themselves designate responsibility.

This myriad legislation includes the Endangered Species Act, Clean Water Act, Clean Air Act, Solid Waste Disposal Act, the Comprehensive Environmental Response, Compensation, and Liability Act, to name just a few. In addition, statutory amendments such as the Hazardous and Solid Waste Amendments (HSWA) to the Resources Conservation and Recovery Act (RCRA) change and add requirements and procedural criteria. The relatively new RCRA lists detailed procedural criteria relating to underground storage tanks, including new tank standards, reporting and record keeping for existing tanks, corrective action, financial responsibility, compliance monitoring and enforcement, and approval of state programs. The law also required EPA to develop a comprehensive program for the regulation of underground storage tanks systems "as may be necessary to protect human health and the environment." Such provisions preclude cost-benefit analysis and variances in safety standards.

Liability and CERCLA

Parties must be aware of liabilities associated with regulatory compliance and be prepared to address changes, delays, and environmental obstacles. The Comprehensive Environmental Response, Compensation, and Liability Act (CERCLA), or Superfund, was passed to actuate the clean-up of serious toxic sites throughout the United States. The act includes funding mechanisms for Superfund clean-ups and sets forth rules under which responsible parties are held to pay for environmental remediation. Clean-up costs can be tremendous.

The act has come under considerable criticism for unfairly or inaccurately assigning potentially massive financial liability. Under the act, potentially responsible parties (PRPs) can be liable for the entire cost of remediation, regardless of their initial contribution to it. Current owners, past owners, and those who have arranged for the transportation and disposal of waste are all potential PRPs. A stock horror story involves a firm that moved hazardous materials from one area of a site to another part of that same site. As liability is assigned to anyone who transports hazards, the firm was liable for the entire toxic mess.

Not surprisingly, CERCLA and other federal regulations can completely transform the scope of construction project and be the starting point of contractual disputes. While the 1986 amendments to the act allow for contribution claims, (cross claims between PRPs allowing for more equitable assignment of monetary liability) and a defense of due diligence (which allows a PRP to prove that it was not aware nor responsible for hazards despite technically sound best efforts). This second addition has fostered a cottage industry of environmental firms that conduct detailed "phase 1" site studies to ensure the cleanliness of sites. Many projects, however, have simply gone undone due to understandable fear of unbridled risks.

Efforts are being made both contractually and legislatively to overcome the chilling effect environmental regulation has had on the construction industry. New initiatives will be addressed later in this chapter.

DEALING WITH ENVIRONMENTAL CHANGES UNDER TRADITIONAL DIFFERING SITE CONDITIONS CLAUSES

Unanticipated regulatory problems can lead to extraordinary changes in costs and scheduling. The disposal of hazards may require studies, environmental consultants, extra licensing, extensive health and safety measures, and other distinctive measures, which translate into major delays, dollars, and disputes. Significant changes in the scope of remediation and abatement projects are not uncommon. Likewise, the discovery of unknown hazardous materials can completely alter a planned project. In either case, it is important that the contract include a Differing Site Condition clause to adequately protect against such changes.

Standard versus Specific DSC Clauses

Changes required by hazardous materials can be handled under the generic differing site conditions clauses. As explained in Chapter 3, the two types of differing site condition claims apply.

An example of a type I claim might be a construction project entailing the removal of asbestos that becomes more involved after construction commences. Though the presence of asbestos may have been disclosed in the bid documents and factored into the bid price, if the scope of the removal differs materially from that specified in the contract documents, a type I claim should prevail. Again, the contractor is not required to conduct extensive inspection of the site during the bidding process and may rely on the owner's representation of project conditions. This is the easier claim to settle because it merely involves comparing conditions specified in the contract to the actual conditions encountered and quantifying the resultant change in construction costs.

A type II claim might involve encountering unexpected hazardous materials at the project site. In these cases, the contract includes no indication of the presence of hazardous conditions. The trick is the added burden of proving that the condition is "unusual" and differs "materially from those ordinarily encountered and generally inhering in the work of the character provided for in the contract." It is arguable that conditions such as the presence of asbestos or other toxic contaminants are ordinarily encountered in certain projects. For example, if a building is of a certain type and was built during the time when asbestos was commonly used in construction, the presence of asbestos may not be considered unusual.

Both types of claims may give rise to allegations of nondisclosure. In such cases where the owner does have superior knowledge of contaminants or the possibility of hazards, he does have the duty to disclose those conditions.

The discovery of unknown or increased hazards at a job site can be so costly and protracted as to obliterate the price and design set forth the contract documents. It is prudent, therefore, to include a differing site conditions clause in the contract that will adequately protect against the incidence of environmental issues.

Hazardous Environmental Conditions at the Site

The latest EJCDC and AIA documents provide detailed provisions regarding hazardous materials. It is worth noting, however, that in addition to CERCLA, other regulatory acts such as the Clean Water Act and Endangered Species Act, may also present delays related to differing site conditions.

The 1996 revisions of the EJCDC Standard Conditions of the Construction Contract contains significant changes in its hazardous environmental conditions clause. The new standard forms indicate that the Differing Subsurface or Physical Conditions clause shall not apply to Hazardous Environmental Conditions and sets forth very detailed procedures and liabilities with regard to such conditions. In addition, the EJCDC guide to preparation of supplementary conditions suggests that parties define the term hazardous waste themselves or by reference to applicable federal and state environmental laws.

Parts A and B of the provision limit contractors' reliance on technical data provided by the owner. While contractors are entitled to rely on such data, they may not make any claims related to the completeness of such reports or opinions or interpretations of such reports.

Part C places responsibility for hazardous materials uncovered or revealed at the site with the owner. However, it explicitly holds the contractor responsibility for environmental hazards created by the contractor or any of the contractors subs or suppliers.

If a contractor encounters or creates hazardous environmental conditions, he is to immediately isolate the conditions, stop work, and notify the owner and the engineer. The owner must promptly consult with the engineer as to the need for a qualified environmental consultant. Contractors are not required to continue work until the owner has secured and delivered any necessary permits and that the area has been made safe. The provision allows the contractor to refuse to resume work where he has a reasonable belief that it would be unsafe or under special conditions.

If the parties cannot agree to changes in contract times or prices, or deletions of work related to such hazardous conditions, either party may make a claim.

Additionally, provisions G and H require the party responsible for the hazardous conditions to indemnify and hold harmless all other parties to the contract.

No doubt these detailed provisions have been forged through costly and protracted litigation. The relevant portions of the new provision follow:

EJCDC No. 1910-8 (1996 Edition)

4.06 Hazardous Environmental Conditions Site

C. CONTRACTOR shall not be responsible for any hazardous Environmental Condition uncovered or revealed at the Site which was not shown or indicated in Drawings or Specifications or identified in the Contract Documents to be within the scope of the Work. CONTRACTOR shall be responsible for a Hazardous Environmental Condition created with any materials brought to the site by CONTRACTOR, Subcontractors, Suppliers, or anyone else for whom CONTRACTOR is responsible.

D. If CONTRACTOR encounters Hazardous Environmental Conditions or if CONTRACTOR or anyone for whom CONTRACTOR is responsible creates Hazardous Environmental Condition, CONTRACTOR shall immediately: (i) secure or otherwise isolate such condition; (ii) stop all Work in connection with such condition and in any area affected thereby (except in an emergency as required by paragraph 6.16); and (iii) notify OWNER and ENGINEER (and promptly thereafter confirm such notice in writing). OWNER shall promptly consult with ENGINEER concerning the necessity for OWNER to retain a qualified expert to evaluate such condition or take corrective action, if any.

E. CONTRACTOR shall not be required to resume Work in connection with such condition or in any affected area until after

OWNER has obtained any required permits related thereto and delivered to CONTRACTOR written notice: (i) specifying that such condition and any affected area is or has been rendered safe for the resumption of Work; (ii) specifying any special conditions under which such Work may be resumed safely. If OWNER and CONTRACTOR cannot agree as to entitlement to or on the amount or extant, if any, of any adjustment in Contract Price or Contract Times, or both, as a result of such Work stoppage or special conditions under which Work is agreed to be resumed by CONTRACTOR, either party may make a Claim therefor as provided in paragraph 10.05.

F. If after receipt of such written notice CONTRACTOR does not agree to resume such work based on a reasonable belief it is unsafe, or does not agree to resume such work under special conditions, the OWNER may order the portion of the work that it is the area affected by such condition to be deleted from the work.

The 1997 revisions of AIA document A201 likewise contain a hazardous conditions clause. The clause provides for work stoppage in the event that reasonable precautions would not prevent foreseeable injuries or deaths related to hazardous materials. The materials include but are not limited to asbestos and polychlorinated biphenyl (PCB). The site is to be examined by a licensed expert agreed upon by the owner and the contractor. When the site is rendered harmless work may continue after mutual written agreement, and any reasonable adjustments in contract times and costs will be provided. If the site is not rendered harmless the owner must indemnify and hold harmless the contractors, subcontractors, and others against any claims.

The AGC of California has gone further to give the contractor an unconditional option to refuse to continue work, and sets forth the Owners responsibilities in the event the Contractor does agree to continue work. The relevant language follows:

Section 4.3.2   The Contractor shall have no obligation to perform any corrective or remedial work that would require the handling or exposure to hazardous material. However, if the Contractor agrees to perform such work:

(a) The Owner agrees to indemnify, hold harmless, and defend the Contractor from and against any claim, action or legal proceeding brought against the Contractor seeking to make the Contractor strictly liable for the performance of such work.

(b) The Owner shall provide specific instruction to the Contractor with respect to the handling, protection, removal and disposal of such material.

(c) An equitable adjustment in the Contract Price and the Contract Time shall be made for such work.

4.4 The Owner shall have the sole responsibility for furnishing all written warnings, notices or postings required by state or federal law regarding the use or existence of hazardous or potential hazardous substances.

The AGC provision allocates all of the risk of discovery, work stoppage, and compliance to the owner. The final provision is a good example of state-specific contractual protection. It provides the contractor with protection against California's Proposition 65, which mandates strict public notice requirements related to hazardous materials. The notification requirements of law include posting signs in areas where workers of members of the public may be exposed to hazards. In some cases, more extensive and expensive measures must be taken, such as mass mailings and media announcements.

It is preferable, where possible, to shift the burden of local compliance to owners who are in a better position to arrange compliance with local rules and regulations. In such cases, it may be prudent to contact local chapters of professional organizations for guidance.

Brownfields

Brownfields are abandoned, idled, or underused industrial and commercial facilities where expansion or redevelopment is complicated by real or perceived environmental contamination. The EPA estimates that there are from 100,000 to 450,000 brownfields sites in the United States. The complicated regulatory, liability, and safety issues associated with their clean-up have kept developers away from these sites, which may not be contaminated at all. The resultant physical, economic, and social deterioration of urban areas has led to proactive initiatives to ease the way for construction and development.

The EPA recently formed Brownfields National Partnership Action Agenda, which hopes to more effectively link environmental protection with economic development and community revitalization programs. Its Interagency Working Group includes:

- Department of Agriculture (USDA)
- Department of Commerce (DOC)
- Department of Defense (DOD)
- Department of Education (ED)
- Department of Energy (DOE)

- Department of Health and Human Services (HHS)
- Department of Housing and Urban Development (HUD)
- Department of the Interior (DOI)
- Department of Justice (DOJ)
- Department of Labor (DOL)
- Department of Transportation (DOT)
- Department of the Treasury (Treasury)
- Department of Veterans Affairs (VA)
- Environmental Protection Agency (EPA)
- Federal Deposit Insurance Corporation (FDIC)
- General Services Administration (GSA)
- Small Business Administration (SBA)

Working together, they hope to remove some of the barriers associated with such remediation projects and to provide incentives for owners and contractors to embark on redevelopment projects. The project includes grants for pilot projects; clarification of liability issues; tax incentive plans; partnerships between federal, state, and local municipalities; and local job training. Such partnerships will, under specified circumstances, relieve prospective purchasers, lenders, contractors, and property owners of Superfund liability.

Regulatory bodies and local governments have developed a flurry of Brownfield initiatives in recent years. Programs include alternative cleanup standards based on planned reuse, covenants not to sue, limits on lender liability to encourage investment, and expedited review and oversight. This is an area that could provide for financially and socially beneficial working relationships that effectively control costs and litigation.

SUMMARY

Criticism, legislative reform, and cooperative initiatives promise to bring about continued change in environmental regulation and its practical application. While new initiatives such as Brownfields offer hope of mutually beneficial progress for owners, contractors, government agencies, and communities, compliance and its accompanying risks will continue to have significant impact on construction disputes.

It is safe to say that the occurrence of hazardous materials will continue to wreak havoc on the even the most mindful planning. While efforts to limit liability and community partnership can help to control risks and liability, changes in site conditions and resultant delays should be anticipated and protected against.

6. Delays

Delays are a way of life in the construction industry. Construction claims dealing with delays are among the most complicated and difficult to analyze. Delays do not occur in a vacuum. Sometimes there are overlapping or concurrent delays. On one project in the course of a few months, there may be a strike, late delivery of critical materials, and a change in design. It often takes considerable skill to analyze these delays and to separate the numerous factors that contribute to the overall delay in completion of the project. Determining the origin of the delay and the impact on the job, and even more important, the responsibility for the delay, can easily lead to conflict.

That potential for conflict is heightened by the fact that some delays are compensable only by time extensions. For others, costs may be recoverable by the contractor or the owner. In still other cases, a contractor may have no choice but to accelerate and to pay for acceleration costs, or later to put in a claim for recovery of those costs.

How do we begin to analyze delays on the job in order to avoid conflict? Throughout this book, we emphasize two precepts: the first is to understand the contract thoroughly; the second is to keep thorough, accurate records. Here, as elsewhere, these two principles apply.

CONTRACT PROVISIONS RELATING TO DELAY

A number of contract provisions speak to delay, and it's a good idea to separate these at the beginning of a project. Two concepts are useful in this analysis: time and money.

All contract clauses that deal with time or money relating to delays should be consulted when a delay does arise on the job. We'll begin by discussing the standard clauses. However, there are other portions of the contract that can be relevant to delays. AIA document A201-1997, Article 8, at 8.3.1, reads as follows:

> If the Contractor is delayed at any time in the commencement or progress of the Work by an act or neglect of the Owner or Architect, or of an employee of either, or by any separate contractor employed by the Owner, or by changes ordered in the Work, or by labor dis-

putes, fire, unusual delay in deliveries, unavoidable casualties or other causes beyond the Contractor's control, or by delay authorized by the Owner pending mediation and arbitration, or by other causes which the Architect determines may justify the delay, then the Contract Time shall be extended by Change Order for such reasonable time as the architect may determine.

The federal government has a standard contract form for construction. Formerly known as 23-A, the form is now part of the Federal Acquisition Regulations (FAR), where Section 52 is the relevant portion. Many in the industry refer to Section 52 of FAR and Form 23A interchangeably. Section 52.249-10 contains this clause:

The Contractor's right to proceed shall not be terminated nor the Contractor charged with damages under this clause if—(1) The delay in completing the work arises from unforeseeable causes beyond the control and without the fault or negligence of the Contractor. Examples of such clauses include (i) acts of God or of the public enemy, (ii) acts of the Government in either its sovereign or contractual capacity, (iii) acts of another Contractor in the performance of a contract with the Government, (iv) fires, (v) floods, (vi) epidemics, (vii) quarantine restrictions, (viii) strikes, (ix) freight embargoes, (x) unusually severe weather, or (xi) delays of subcontractors or suppliers at any tier arising from unforeseeable causes beyond the control and without the fault or negligence of both the Contractor and the subcontractors or suppliers.

The clause goes on to define the proper notice limits (ten days) for informing the contracting officer of such delays.

The Engineers Joint Contract Document Committee's standard form, EJCDC No. 1910-8 (1996 edition), Article 12, at Clause 12.3, is somewhat briefer:

Where Contractor is prevented from completing any part of the Work within the Contract Times (or Milestones) due to delay beyond the control of the Contractor, the Contract Times (or Milestones) will be extended in an amount equal to the time lost due to such delay if a claim is made therefor as provided in Paragraph 12.02.A. Delays beyond the control of Contractor shall include, but not be limited to, acts or neglect by Owner, acts or neglect of utility owners or other contractors performing other work as contemplated by Article 7, fires, floods, epidemics, abnormal weather conditions or acts of God.

We have presented three different versions of standard contract provisions dealing with delay. All three provide that the contractor must give written notice in a certain number of days beginning with the commencement of the delay to the proper party: the architect in the AIA version, the contracting officer in the federal government's FAR, and the engineer in the EJCDC's form. This is the modus operandi in all cases. The designated individual examines the facts and determines whether the delay warrants an extension of time. In all cases, however, his determination is subject to appeal.

Even in these standard versions, wording differs slightly and parties should be knowledgeable about what their contract states. Words such as "unforeseeable," and phrases such as "beyond the control of" and "shall include but not be limited to" have some latitude of interpretation. Language concerning subcontractors varies notably. The federal government's FAR Section 52 specifically limits time extensions for delay caused by subcontractors to the same causes that would be allowable for the general contractor—that is, those that are "unforeseeable." In other words, if a subcontractor uses poor judgment or undermines the job, in spite of the care and diligence exercised by the contractor, and even if the contractor warns the sub about this situation, the delay caused by the sub would not entitle the prime contractor to a time extension.

Other Delay-Related Clauses

Parties to the contract must look further for other clauses relating to time, delay, and to compensation in order to get the total contractual picture. There will be several clauses or phrases in addition to the standard delay clause (or variation thereon). Some examples are as follows: the fixed completion date (or the number of consecutive calendar days or working days assigned for completion of a portion or specified portions of the whole of the work), clauses that authorize the owner to suspend work, and clauses that permit the owner to order that work be accelerated. "Time is of the essence" is a phrase inserted in contracts that gives time clauses the force of law, that is, that the contractor will assume liability to the owner for delayed completion unless it is excusable.

Recently, clauses relating to delays caused by subcontractors or trades have become quite common. Such clauses apply in cases where there is an owner (usually with a CM) with multiple primes or a general contractor with a group of subs. They state that if one sub or trade delays another, under no circumstances can the affected party go after the GC or the CM with the owner for relief. His sole remedy is against the offending trade contractor or subcontractor.

A reciprocal clause states that if one trade or subcontractor delays another, the delaying contractor will be directly liable to the party affected for such delay, and he will indemnify the owner, the CM, or the GC from such claims.

As to money, all three standard contract documents allow, or at least do not preclude, recovery by either party of damages for inexcusable delay caused by the other party. In other words, under these standard contract forms, compensation to the contractor is not limited to time extensions only; there is the possibility of recovery of monetary damages. Of course, as we have stated, any owner may write a contract as he wishes; the standard clauses are guides only.

NO DAMAGES FOR DELAY

In the public sector, however, many owners do include language pertaining to equitable compensation for delays out of a sense that it reflects a consensus as to good contracting practices. Yet there are many exceptions. New York City, for example, includes a "no damages for delay" clause in its contracts. This inclusion, however, has been rejected by courts in instances of unreasonable, uncontemplated owner-caused delay when disputes have reached that level. The clauses have also been overruled where fraud, malicious intent, and intentional wrongdoing have been proven.

Courts in some states, though, have strictly enforced such clauses despite their perceived inequity. Parties to contracts should look for exculpatory language such as "no damages for delay" clauses. The owner risks that such language will not be honored by the court; the contractor risks that such wording will be strictly observed. Nonetheless, there are compelling arguments for and against inclusion of no damages for delay clauses in public sector construction contracts. At this point, the inclusion arguments still prevail. Proponents argue that public owners want to know the total cost of a project from the beginning. All owners, public or private, would like to know that cost from the outset, but where public funds are used, the general feeling is that the taxpayers have a right as well as a desire to have the final cost known at the outset of the project.

Public works expenditures are closely scrutinized for improprieties, and a fixed-cost arrangement, with no possibility for additional outlays, can eliminate unfortunate interpretations, or so the arguments allege. The question of fairness to all bidders is also raised by those who defend no damages for delay clauses. Public work almost always goes to the lowest bidder. If damages for delay were possible, the argument goes, contractors would submit artificially low bids. Instead, contractors, all of whom know there will be unexpected costs, include sufficient contingency funds to allow for these. All bidders know that they have to do this and, it is alleged, are at least as likely as owners to know that delays are likely in any major project.

The analogy of a warranty for construction equipment has been offered in defense of these clauses. The purchaser is not entitled to negotiate the terms of the warranty. He accepts the conditions when he buys the equipment. In a construction contract, inclusion of such conditions serves to warn the contractor in advance that he will not get any more money when delays occur. So the reasoning is that the contractor accepts the terms of the no damages for delay clause and prepares for the likelihood of delay with his contingency inclusion. In some cases, the public owner wants to complete the project even more than the contractor does. So there is a disincentive to delay work. For example, a sewage treatment plant may be constructed under court order, and heavy fines may be incurred by the municipality each day a deadline is violated. The cost to the public owner may be far more than damages to the contractor for delay.

Other arguments are that differing site clauses and provisions for change orders take care of unforeseen problems and include reimbursement provisions. Even those arguing on behalf of the clauses agree that this low bid/contingency approach means that bids are higher than need be. Protecting the public is worth the extra cost, in the eyes of defenders of these clauses. On the other side of the argument, opponents cite the language in many of these clauses about "causes of delay that were not contemplated at the time the contract was made." This is a criterion subject to too many interpretations. Another argument against these clauses, the owner's interest in immunizing himself from liability for delay, does not outweigh contractor hardship. Disallowing damages for delay can destabilize construction costs, drive up bid prices, and foster an adversarial relationship, say those who oppose inclusion of such clauses.

A strong opposition argument is that the clauses reduce public employee incentive to resolve job problems expeditiously. The public employee often feels that his wrong decision will be criticized and his correct action ignored in any case. The no damages for delay provisions reinforce this perceived tendency.

We believe, on balance, that in the majority of cases, the opponent arguments outweigh those of the proponents of no damages for delay clauses, for cases within the owner's reasonable control. It is fundamentally unfair to give one party to a contract carte blanche to violate the terms of the contract, to not live up to its contract obligations, to put the other party at its mercy, or to impose added costs on the other party by its acts or omissions. In general, the guiding principle is that the party best able to control risk should bear the cost of that risk. Certainly, reductions in cost of public construction can better be achieved by other means, such as taking steps to improve performance of public sector employees. No damages for delay clauses are tantamount to an admission that public sector delay cannot be controlled, that delay is inevitable. We are not so fatalistic. Other clauses serve to allocate risks in public sector work, including bonuses for early completion, liquidated damages, and differing site conditions.

No damages for delay is not an uncommon clause in public construction contracts, but other provisions relating to monetary recovery for delay may be in the contract. Payment of liquidated damages by the contractor to the owner if the contractor fails to complete the work by the contract date is one. (See the section on liquidated damages in Chapter 7.) There may be a clause stating that there will be no payment to the contractor for suspension of work (by the owner) or for acceleration. Or there may be a clause about no payment to the contractor for delays caused by other prime contractors or by subs. The latter would be an onerous situation for a contractor, but if there are such clauses, he should know about them from the start and plan accordingly. Likewise, he should consult with legal counsel to determine how the courts in his jurisdiction customarily deal with enforcing such clauses.

To sum up, parties to the contract must be aware of the total delay package. They should have examined all contract clauses relating to time and to money. When a delay does occur, these clauses should be referred to first, in order to deal rationally with remedies.

There are, as we have said, applicable court and contract appeals board precedents as to compensation and time extensions for delay. Once the contract has been consulted, it is advisable to check out these precedents. If in-house staff counsel is not available, it may be wise to seek outside aid before proceeding, depending on the extent of the delay or the money involved.

EXCUSABLE/NONEXCUSABLE DELAYS

Within the framework of the contract wording and the precedents of interpretations, delays may be analyzed as excusable and nonexcusable, that is, analyzed on the basis of whether or not the contractor would be entitled to a time extension. If delays are excusable, they can be further broken down into excusable/compensable, and excusable/noncompensable categories.

"Compensable" is understood to mean compensable to the contractor. In effect, nonexcusable to the contractor means the delay should be compensable to the owner, either as liquidated or as actual damages. Further discussion and examples are given later in this chapter.

It is essential that these distinctions be understood when analyzing delays. It follows that parties cannot negotiate rationally without understanding these differences.

Excusable Delays

Put as simply as possible, these are all delays not caused by the contractor. In the technical sense, these are all delays for which the contractor is entitled to an extension of time under the contract.

Examples of excusable delays could be failure of the owner to provide site access, a change in design by the owner, or delays stemming from that list designated as beyond the control of the contractor in the delay clause: unusual weather, strikes, acts of God, and so on.

In order to be excusable in the technical sense (that is, in order to warrant an extension of contract time or other recovery), the delay must be on the critical path for completion of the project. In other words, the delay must directly affect the ultimate completion of the job. Correspondingly, if the delay, whatever the cause, is not on the critical path, thereby not affecting the ultimate completion date of the work, there will be no compensation.

There are exceptions, in which a project component not on the critical path can result in total project delay and where compensation would be possible. For example, consider a situation where a small item of work, say a pump house for a large hydroelectric project, has been put on hold because the owner wishes to redesign it. While the pump house is not on the critical path, the contractor had planned to construct in good weather. If he has to do it later and in adverse weather, he would be entitled to winter protection, loss of productivity in concrete construction in winter, and so on. So there can be compensation for noncritical path delays. It must be noted here that the critical path may shift. A delay that is not immediately seen as being on the critical path may ultimately affect the completion date. For example, some concrete work may not be on the original critical path; changes in the work and a delay in that concrete work may push other work off schedule. This is a good reason to track the critical path of a job on a regular basis.

In order to determine whether the completion date will be affected, parties should have progress schedules, bar charts, graphs, or some other visual presentation of the work components so as to document delays. The more complex the job, the more advisable it is to use a CPM chart to track work progress.

Excusable/Compensable

These delays are due to some act or omission of the owner, for example, lack of site access or late arrival of owner-furnished material or equipment. In such cases, the contractor would be entitled to damages for extra costs incurred unless there is an enforceable contract clause barring such recovery.

In general, a no damage for delay clause will not be applied to an owner-caused delay that could not reasonably have been contemplated by a contractor at the time the contract was entered into. For example, assume that totally unbeknownst to the contractor, the owner did not have title to the property, and there was no way the contractor could reasonably have known that. It seems doubtful that any U.S. court would bar recovery of delay damages with a no damage for delay clause under such circumstances.

On the other hand, a contract document might specifically state that the owner does not hold title to part of a site at the time of contract award. The owner may provide for this situation by stating that he is attempting to secure title as quickly as possible and expects that title will be obtained by a date set forth in the contract. If title is not obtained by that date, a no damage for delay clause would be enforceable. The contractor would have to show willful conduct or gross negligence on the part of the owner in his efforts to obtain title to recover damages.

Excusable/Noncompensable

These are delays for which neither party is at fault: acts of God, epidemics, and so on, as set forth in the delay clause. Time extension is the only remedy for such delays.

Nonexcusable Delays

These are delays caused by the contractor. These could include failure to coordinate the work, too few men on the job, equipment furnished by the contractor that is delivered late, low productivity, defective work that must be removed and replaced, and so on.

Such delays could be compensable to the owner in the form of liquidated or actual damages paid by the contractor for late completion, or could be the basis for contract termination by the owner or for an order to accelerate the work.

CONCURRENT DELAYS

The term "concurrent delays" is used to describe two or more delays that occur at the same time, each of which if it had occurred alone would have affected the ultimate completion date. These can be difficult to sort out. A critical path method (CPM) chart or other visual representation of job progress is almost essential in order to analyze overlapping delays.

There are three points to bear in mind when looking at concurrent delays. The first is to see if other work could have been accomplished during the delay period. The second is to determine whether both delays impact the critical path. The third is to analyze all delays in the framework just described. Is each delay excusable, nonexcusable, excusable/compensable, or excusable/noncompensable?

There are several possible permutations here. For example, suppose the owner failed to supply certain materials on time. At the same time, the workers who would have installed the material were on strike. The contractor cannot claim damages for work that would not have been accomplished had the owner-caused delay not occurred. He would, however, be entitled to a time extension for the period of the strike unless it can be proved that the strike was due to his own actions. This is an example of concurrent compensable and noncompensable delays. The result is a time extension, but no damages for the contractor.

Another example of a concurrent delay would be if the contractor failed to submit shop drawings on time while the owner failed to provide access to the site; in this case, the owner ought not to penalize the contractor for that late submission. If the workers had been able to gain access to the site, they might have performed other critical path work unrelated to those shop drawings. In other words, site access was on the critical path, but the shop drawings were not. Therefore, in this case, the contractor was entitled to both a time extension and delay damages.

Suppose instead that the owner failed to turn the site over in time *and* instituted a redesign, both affecting the critical path. Here the contractor has been foiled on two counts by the owner, and would be entitled to a time extension and to damages. Both delays were compensable.

Another example of concurrent delay might be a situation in which there is an owner-caused delay in gaining access to the site. This prevents excavation work from taking place. At the same time, the contractor should have been getting his structural steel shop drawings in, but there was plenty of float time on that activity. Clearly, the steel work was not going to be critical until much later due to the delay in site access. It would be hard to penalize the contractor for a delay to an activity no longer on the critical path. So the contractor should be entitled to both a time extension and delay damages. If the contractor, on the other hand, caused two simultaneous delays, both affecting the critical path, he would be liable to assume the costs of acceleration, or for payment of liquidated damages to the owner for late completion.

These are simple examples. Concurrent delays often are more complex. If we look at an overlapping delay situation, we can see the difficulties involved. For example, if an owner blocks access to the job site between August 15 and October 1, and the contractor doesn't get his equipment to that site until September 15, is the delay compensable? Ordinarily, the belated site access would be an excusable and a compensable delay. However, the contractor could not have worked anyway for the first four weeks of that delay. Therefore, he is entitled to a time extension, but not to damages for the period of August 15 to September 15, unless he can show that his equipment was available on August 15, and that he held up delivery only due to the fact that the site was not accessible.

The contractor would, however, be entitled to damages and a time extension for the last two weeks of that delay period. The possibility of the work being forced into winter months would also arise, and impact costs for delay might be requested.

Generally speaking, when an excusable and a nonexcusable delay are concurrent, the contractor ought to be entitled to an extension of contract time.

In the case of concurrent compensable and noncompensable delays, the contractor should be entitled to a time extension, but not to damages. In order for the contractor to claim damages, the owner would have had to cause both (or all) compensable delays. As always, the contract may stipulate otherwise. Some contracts state that there will be no damages for delay and that time extensions will be granted for concurrent delays only if both delays are excusable and otherwise compensable.

COMMON CAUSES FOR DELAY

Delays are often attributed to subcontractors or vendors and to unusual weather. As usual, the first step in establishing the nature of the delay is to look at the contract. In the case of subcontractor- or vendor-caused delay, the contract clause may have stipulated general contractor responsibility for the subs that he retains. As we pointed out, the federal standard clause indicates that the only delays excusable for a subcontractor would be the same ones that excuse the prime—those "beyond his control, and including acts of God," and so on. Some contracts require the contractor to use a single-source supplier. It might be expected that an owner would be (or should be) more lenient when a delay is due to the actions of a subcontractor dictated by the contract. The better rule is that only if the contractor chose his own subs, should he be held responsible for delays caused by those subcontractors. In a complex job, with many subcontractors, the owner should make certain the contract speaks to such delays. The contractor, as always, should know what is in that contract.

Weather is often cited as an excusable delay that justifies a time extension under the contract. Contract clauses often speak of "abnormal" or "unusually severe" weather, or in some cases weather that is "unforeseeable." The contractor must prove that these descriptive terms apply if he wishes a time extension to be granted. One way to do this is to check meteorological records for past years. These might have to go back five or ten years in order to establish that the weather condition was unusual.

Documenting Weather-Caused Delay

Three basic rules can be used in determining if weather justifies a time extension under clauses stipulating abnormal or unusually severe conditions.

1. There must be identification of the work impacted by the weather as controlling the overall completion of the project; that is, the work affected must be on the critical path.
2. It must be established that the controlling work was delayed by the weather.
3. It must be established that the weather was unforeseeable, that is, abnormally severe.

The key to time extensions for unusually severe weather is not the cause per se, the weather, but the effect of the unforeseen weather on the work being performed. For example, an exceptionally heavy one-day rain could have a serious adverse effect on a construction site highly subject to erosion. However, the same exceptionally heavy rain would affect exterior painting less than would a lighter rain falling over a longer duration of time.

Another example of an excusable weather-caused delay is that of a light, but dust-laden, wind. Such a wind would preclude painting or the installation of sensitive electronic equipment. Yet another example might be the effect of unusually low temperatures on paving or masonry construction.

OTHER CAUSES OF DELAY

Similarly, not every fire or quarantine or strike would excuse delay. The contract might be one to excavate for a building in an area where a coal mine has been on fire for years, well known to everybody connected with the project, including the contractor, and where a large element of the contract price was attributable to this known difficulty. A quarantine or freight embargo may have been in effect for many years as a permanent policy of the controlling government. A strike may be an old and chronic one whose settlement is not expected. In any such situation where the contractor could have been expected to anticipate these difficulties and provide for them in his bid, no extension is warranted. The same rule could be applied to floods: if normally expected high water in a stream over the course of the year is foreseeable, flooding would not justify a time extension.

ACCELERATION

We have referred, in Chapter 4, to acceleration as a means of putting lagging work back on schedule. This can mean that the contractor puts extra men on the job, puts his men to work overtime and on weekends, and acquires more and better equipment, or whatever else the contractor thinks he has to do in order to catch up.

The cost of acceleration can be considerable. Contractors are not likely to take this step unless it is absolutely necessary. One reason that it might become necessary is that the contract calls for payment of liquidated damages if the completion date is not met. It may be less costly to pay the acceleration expenses than the damages, and the contractor would maintain some control over his work. He is, after all, arranging the accelerated work. He can estimate the accelerated costs; he can then accelerate if he thinks this will be cheaper for him. He knows what the liquidated damages costs will be; that figure is in the contract.

Another reason that a contractor might elect to accelerate is that he stands to lose money on another job if he does not. Or he might need men and equipment from this job to move to another. He might also choose to accelerate because he risks moving critical portions of the work into winter months.

The owner ordinarily has the right to terminate the contract for unreasonable, nonexcusable delay, and if the contractor fears that this might occur, this would be a most compelling reason for his assuming the costs of acceleration.

In some cases, the owner may direct acceleration even though all delay is excusable. When an owner does this, then he must bear the extra costs. He might do this because he himself has caused the project delays, and he would rather pay for acceleration to achieve an earlier completion than grant an extension of time. He may have many reasons for preferring that the project stay on schedule, even if it does cost more.

Other owners might order an acceleration no matter which party has been responsible for delays because they truly have a "sacred" completion date. The owner may be building an urgently needed military facility; or if the facility is in the private sector, profits lost may be so great that acceleration costs diminish by comparison.

Naturally, not many owners would prefer to take this step. Sometimes, owners and contractors agree to share the costs of acceleration rather than argue over who is responsible for the delay. The trade-off might be a waiver of all claims one against the other. (See Chapter 7 for a sample calculations on an acceleration claim.)

Constructive Acceleration

A more problematic situation arises when an owner "requests" rather forcefully that his contractor put the job back on schedule, but without actually using the word "accelerate." For example, he may remind the contractor that liquidated damages will be assessed for later completion without actually threatening him or ordering acceleration. The contractor may nonetheless seek to recover acceleration costs. This situation is known as "constructive acceleration." The term "constructive" is more or less equivalent to "de facto." In other words, the owner is insisting that the job be completed by the date specified (or a different date) in the contract, but does not produce a written order to accelerate. He may write letters that pointedly refer to the liquidated damages clause and termination clauses and to the job schedule, yet not mention the term "acceleration." In most cases, the contractor has to pay the extra costs himself. He can then invoke the disputes procedures to recover those costs or go to court.

The contractor is most likely to seek restitution if he does accelerate and if he believes delay was caused by the owner. It is human nature, though, for parties to consider that fault is on the other side of the relationship. Here are some points that can help avoid a constructive acceleration situation:

1. The contractor's and the owner's representatives on the job should be in communication throughout the duration of the project, and ample records should be both kept and discussed so the situation doesn't get to the point of constructive acceleration. If the contractor receives a letter that seems threatening, he should ask the contracting officer for clarification. The contracting officer may then have to come to terms with the situation and recommend that a written order to accelerate be given.

2. An owner can, when writing a letter about a lagging schedule, offer specific instances in which delay was caused by the contractor. That way, the contractor has to stop and think before totally disavowing all responsibility. If indeed the contractor delayed the project, he will recognize that he has to bear the costs of acceleration. If he feels that the information in such a letter is in error, he can again request clarification and discuss the matter with the contracting officer.

3. Sometimes owners let legitimate claims for time extensions get bogged down in paperwork. The result might be that the owner is prodding the contractor to move quickly, even though a time extension is actually due him. If the owner persists and the contractor accelerates, the contractor is most certainly going to seek relief. The resulting ill will and possible litigation could have been easily avoided by prompt processing of time-extension requests. Accurate CPMs or other visual representation of the work progress can help avoid such misunderstandings. If these are on display and the causes are documented, constructive acceleration situations should not arise.

4. The contractor can help avoid adversarial situations by promptly submitting requests for time extensions. He should not wait until he feels the owner is suggesting that he accelerate to decide that he has some time coming. The owner would be justified in refusing to consider the request; it would be "untimely" under the contract language.

A contractor can also respond to a constructive acceleration situation by complying, with reservations. He can write to the owner and tell him that acceleration is underway and that the contractor will bill the owner for the extra costs. The owner can think things over one more time and accede, or he can refuse and invite a claim by the contractor.

SUSPENSION OF WORK

Suspension of work clauses state that the owner has the right to suspend or delay the work. Such clauses usually provide that the owner allow for adjustment in cost or time required to complete the work. Thus, under such clauses the owner can, in a sense, acknowledge a delay caused by him and provide compensation by a change order under the terms of the contract, thereby eliminating the need to invoke contract dispute procedures. The length of allowable suspension or delay may be limited by a contract provision.

The Engineers Joint Contract Documents Committee clause, EJCDC No. 1910-8 (1996 Edition) Article 15, at 15.01, states:

> At any time and without cause, OWNER may suspend the Work or any portion thereof for a period of not more than 90 consecutive days by notice in writing to CONTRACTOR and ENGINEER which will fix the date on which Work will be resumed. CONTRACTOR shall resume the Work on the date so fixed. CONTRACTOR shall be allowed an adjustment in the Contract Price or an extension of the Contract times, or both, directly attributable to any such suspension if CONTRACTOR makes a Claim therefor as provided in paragraph 10.05.

The federal government FAR form contains a suspension clause that states:

> The contracting officer may order the contractor in writing to suspend, delay, or interrupt all or any part of the work for such period of time as he may determine to be appropriate for the convenience of the government.

The clause continues:

> If the performance of all or any part of the work is, for an unreasonable period of time, delayed, suspended or interrupted,…an adjustment will be made for any increase in the cost of performance of this contract (excluding profit) and the contract will be modified accordingly.

The government form does not define "reasonable," but states that if the owner does not specify a time duration for the suspension, then the adjustment may be considered after that period of time deemed "reasonable." The clause also disallows profit as part of any adjustment made for suspensions ordered by the owner. The contractor must submit a claim for the costs owing to the suspension "as soon as practicable after the termination of such suspension, delay or interruption, but not later than the date of final payment under the contract." The EJCDC suspension of work phrasing refers the parties back to paragraph 10.5, which outlines the procedures under which adjustments of cost and time may be made.

If we refer back to categories of delay noted at the beginning of this chapter, a suspension would be an excusable, compensable delay. Under the federal government FAR wording, adjustment of price or time cannot be made if performance during that period of time (the owner's suspension) would have been "so suspended, delayed, or interrupted" through other causes, including fault or negligence of the contractor. In other words, this reasoning is similar to that governing concurrent delays. A contractor cannot ask for adjustment of time or money during a suspension if he causes another delay at the same time that would also prevent work being performed.

All contracts do not include suspension of work clauses. As we have said, the inclusion of such clauses provide an administrative remedy for owner-caused delays and the means by which the contractor can recover by change order rather than by claim. In the absence of such clauses, the contractor would discuss the delay with the contracting officer, and if an agreement is not forthcoming, he would have to pursue recovery through the disputes clause.

RESERVATIONS OF CLAIMS

Many owners in granting time extensions or costs for owner-caused delay or suspension require the contractor to waive any further damages attributable to the items covered in the change order. Unless impact has been specifically considered and included in the negotiation, the contractor may wish to put a reservation clause in change orders, stating that the orders cover the direct time and costs of the changed work, but not include any impact or ripple costs, or further extension of time. These are to be "reserved until such time as they can be finally ascertained." It is recommended, though, that parties price out (i.e., forward price) the impact costs as part of the change order wherever possible. (See examples of change orders in Chapter 4.)

FLOAT TIME

Float time often becomes an item of contention when any kind of delay occurs on a project. For example, let us say that two work tasks, A and B, must be completed before task C can begin. Suppose that task A takes two months and that task B takes one month to complete. Generally speaking, the contractor considers that the extra month, defined as "float time," is his to dispose of. He may begin tasks A and B together and then move workers from task B to another portion of the project during that month. Or he may find it convenient to continue work elsewhere and then bring in the necessary workers for task B one month after task A has begun, assuming they will finish simultaneously and that task C can then commence.

But suppose the contractor takes the second choice, and during that second month when workers are assigned to task B, there is an unavoidable strike. Task C will not begin on time. If the contractor had begun tasks A and B at the same time, and pursued them both continuously to completion, he would have avoided delay caused by the strike during the second month. (The strike affects only the task B workers; for example, they may be electricians.)

Is the contractor the cause of the delay because he did not prosecute the work diligently? Or is the strike considered an excusable delay warranting a time extension? The latter decision would assume that the contractor is free to schedule float time as he sees fit. Generally speaking, float time should be deemed to "belong" to the contractor, so in the instance described, he would be entitled to a time extension. The reasoning is that the method of running the project is usually up to the contractor, who also defines the critical path. However, some contract clauses specifically state the contrary, that float time belongs to the owner or that it can be shared.

CHARTING JOB PROGRESS

To this point, we have discussed various types of delays. A most important aspect of this subject is record keeping. The means by which delays are noted and documented depend on the job schedule chart. Such charts may be of several types, and are sometimes specified in the contract documents. They range from simple bar charts to elaborate schedules incorporating computer printouts of job progress.

Following is an example of a contract provision, quoted in part, used for the Lacey V. Murrow bridge replacement in Seattle, Washington. We are not suggesting that this is unique, but it is an example of provisions governing progress schedules. In practice, such provisions are not always followed or enforced. This provision includes language requiring the contractor to use a computerized scheduling system. Item 10 is quite specific about the acceptable procedures. The document was developed by the Washington State Department of Transportation, which has been requiring computer use since the mid-1980s. At that time, the requirement was relatively rare in a state agency, but since then public agencies increasingly ask for computer documentation of project progress.

PROGRESS OF WORK, CONSTRUCTION TIME ANALYSIS

(February 3, 1994)

Because the schedule of the SR 90. Jct. SR 5 to Jct. SR 405 project of which this contract is a part, is critical and delays encountered due to late completion of each contract will create delays to subsequent contracts and ultimately to the project to a whole time is of the essence in this contract. The State, therefore shall require the Contractor to submit a progress schedule in order for the Engineer to assess the Contractor's pursuit of the work and to ensure successful timely completion of the contract. The approved progress schedule shall become the Engineer's primary tool for evaluating progress, determining the effect of contract changes and delays, and documenting the actual timing of the construction activities.

Section 1-08.3 is supplemented by the following:

1. The Contractor shall submit the progress schedule to the Engineer for approval no later than 10 calendar days after execution of the contract. The progress schedule shall be an arrow or precedence time scaled logic diagram (TSLD) which clearly shows the critical path. The TSLD shall show for each activity: the node or activity code(s), description, duration, early and late start and finish dates, and float or slack. Imposed dates shall be identified on the activity listing and a written explanation of their use provided. If precedence scheduling is used, a precedence logic listing showing predecessors and successors shall be provided in addition to the activity listing. A written explanation shall be provided for each lead and lag time used. Activity durations shall be in working days as defined under the Special Provisions titled TIME FOR COMPLETION. The progress schedule shall be in sufficient detail that progress of the work can be evaluated accurately at any time during the performance of the contract. The Contractor may be required to further subdivide activities with durations longer than 30 working days to facilitate evaluation of progress. The schedule shall portray the Contractor's proposed sequence of construction and completion of work as required to meet intermediate and final contract completion times.

2. In addition to key construction activities, separate activities shall be shown for the following:

a. Procurement of material or equipment that could affect progress.

b. Submittals requiring the Engineer's approval that could delay progress. Submittal activities shall be broken down to show the Contractor submittal preparation and the State review separately. Durations for State review shall be in accordance with the Special Provision titled ENGINEER'S REVIEW.

c. Proposed traffic detours and closure periods.

d. Any work performed by a subcontractor, agent or any third party.

e. Delivery of State furnished material.

f. Any testing, training, system check-out, and system start-up called for in the contract.

3. The estimated contract value of each work activity shall be shown. However, it may be shown on a separate listing rather than on the schedule at the discretion of the Contractor.

4. The progress schedule shall reflect and be consistent with the time of any interim completion dates, availability dates, be consistent with points of interface with work by other contractors or other restrictions for portions of work as established elsewhere in the contract.

5. The Engineer will review with the Contractor the progress of actual work as compared to the progress schedule. The Contractor shall submit a monthly Progress Report to the Engineer within 5 days after the scheduled end date for each monthly progress payment period. The Progress Report shall show status as of the end date of the pay period. The report shall include the entire time scaled logic diagram form of the approved progress schedule, clearly marked to indicate progress to date. Actual start and completion dates for all completed activities shall be shown and progress on partially completed activities shall be graphically represented. If the status time scaled logic shows the Contractor to be behind schedule, the Progress Report shall include a narrative identifying causes of delay or slow progress and the Contractor's plan to get back on schedule.

6. If the Contractor desires to make changes in the approved progress schedule, the Contractor shall notify the Engineer in writing stating the reasons for the change. If the Engineer considers these changes to be of a major nature, the Contractor shall revise and submit for approval, without additional costs to the State, supplemental progress schedule(s), in the form prescribed herein, to show the effect on the entire project.

7. If, in the opinion of the Engineer, the Contractor's actual progress falls behind the progress schedule, the Contractor may be required by the Engineer to submit for approval a supplemental progress schedule(s) at no additional costs to the State. The submittal will be in the form prescribed herein, and shall demonstrate the manner in which the necessary progress will be regained. The Contractor shall implement such steps as necessary to improve progress. Should the Contractor fail to improve progress as needed the Engineer may require the Contractor at no additional cost to the State, to increase the applied resources, including labor, equipment, materials, facilities, number of shifts, overtime operations, days of work, or all of these to improve progress.

8. Extensions of time will be granted pursuant to Section 1-08.8. The analysis will be performed as follows:

When Change Orders or delays are experienced, the Contractor shall submit to the State a written time impact analysis illustrating the influence of each change or delay on the contract completion time. Each time impact analysis shall include a fragment smaller, more detailed section of the network demonstrating how the contractor proposes to incorporate the change order or delay into the logic diagram. The time impact analysis shall be based on the date the change is given to the Contractor and the status of construction at that point in time. The impact to the scheduled dates of all affected activities shall be shown. Upon approval by the State, the time impact analysis shall be incorporated in a supplemental progress schedule to be submitted with the next monthly progress report.

9. The form and methods employed by the Contractor to make monthly progress reports and schedule supplements shall be the same as the approved progress schedule and all in accordance with this Special Provision. Schedule revisions shall be accompanied by a narrative which describes all revisions.

10. The Contractor shall utilize a computerized schedule system which shall operate on an IBM AT with PC-DOS version 3.x, 512K RAM, and a hard disk.

The Contractor shall furnish to the state a copy of the complete scheduling software package and plotting software package utilized by the Contractor. The software package provided to the State shall be new, complete, unopened and licensable by the State.

All schedule submittal required by this Special Provision shall be accompanied by a 5 1/4 inch floppy disk containing a copy of all of the Contractor's progress schedule data files. The data files shall be complete so that independent analysis of the schedule may be performed using the scheduling software package provided to the State.

11. As provided in Sections 1-08.3 and 1-09.9 the State may withhold progress payments if the Contractor does not diligently pursue timely submittal of the progress schedule, supplemental progress schedules or monthly progress reports.

From an owner's standpoint, the most important thing is to require scheduling information that the owner will find practical for use in monitoring the contractor's progress and scheduling of the work. A detailed specification coupled with computer printout are worthless if no one understands them and tucks them away in a drawer when received. It is better to require a simple bar chart.

DOCUMENTING DELAYS

To sum up, many contracts will be specific about the type of progress schedule required on the job and the procedures for submittal, updating, and revision. Others, of course, may not. There are several types of charts commonly used for indicating the progress of the work.

Bar Charts

These are the most simple visual presentation of work and time. A bar chart separates the work into categories, with the intended starting date, the duration, and the intended completion date for each activity. However, bar charts have this weakness: they do not show the interrelationships between the individual work activities.

CPMs and Network Diagrams

The critical path method (CPM) as the name indicates, shows the interrelationship of the portions of the work. In other words, portion B of the work cannot begin until portion A is completed. Portion Q can't begin until portions M, N, O, and P are completed. A delay on any one of these lines impacts the others.

S-Curves

An S-curve represents project progress: slow at the start, a subsequent pickup of momentum, a tailing off at the end. The actual S-curve can be plotted from the information on the monthly payment requisitions. The horizontal distance between the intended S-curve and the actual S-curve represents how far the job is off schedule at any point in time. The S-curve represents dollars spent to time.

Sometimes a job is so complex that a CPM is a virtual spiderweb. The intended S-curve can be drawn from the information on the bar chart and the trade payment breakdown. The intended S-curve can give some rough idea of progress; it is not as accurate as the CPM, but it is more accessible. The case history at the end of this chapter includes illustrations of S-curves and other progress schedules. Increasingly, the S-curve is prepared by a computer from the CPM and the payment input.

Scheduling Software

Computer software for scheduling, project management, cost estimating, accounting, and other aspects of construction have proliferated over the last decade. There may be hundreds of such packages currently available, some in relatively narrow areas. For example, in response to events in the mideast in 1990, several software packages were introduced that forecast the impact of changing costs and availability of fuel during a construction project. There are probably a dozen software developers concentrating on scheduling packages alone at this writing, and since the software field is a volatile industry, the proliferation will continue.

It is difficult to evaluate the wide array of software tailored to the construction industry. There are many publications and organizations that review software, but none that deal exclusively with engineering and construction-related software.

An important development regarding software was the 1991 issuance by the U.S. Army Corps of Engineers of performance specifications for scheduling software. The Corps requires the use of computerized scheduling on their projects, and the specs preclude the need to require a brand-name software product. Other public entities may follow suit and develop their own specs, or adapt or adopt the Corps specs. This chapter includes the Washington State Department of Transportation contract document requirements for computerized scheduling, and many other public agencies nationwide do the same. This trend will proliferate.

SCHEDULE APPROVAL

When delays and subsequent claims occur, parties will look to the schedule for a reference point, so it is absolutely imperative that these schedules be understood and used properly. CPMs and other progress charts are like motherhood: more people respect the idea than pay actual attention to the individuals. Here are some points worth remembering about CPMs and other visual representations of project work.

1. Most contracts require that some sort of progress schedule be kept, but they are not necessarily clear on the checking-off or approval process. What legal significance does the owner's check-off of a progress schedule carry? Typically, when a lawsuit occurs, the owner looks at the schedule and says that the contractor could never have built the job in the manner indicated. The contractor's rejoinder is likely to be that the owner "accepted" the schedule.

   Both parties should look carefully at the progress charts, and they should understand what the contract says. Some owners "acknowledge" receipt of the schedule without approving it in order to attempt to avoid responsibility for acceptance. There is no hard and fast rule about the legal significance of a schedule. If the owner has the contractual duty to accept the chart, then he has the duty to state his objections if he has them. If the contractor submits his schedule for approval and the owner does express his objections, then the schedule should be revised and resubmitted before the issue becomes moot over the passage of time. If the schedule is submitted for approval, parties should not regard that approval as a casual matter.

2. The critical path itself may shift during the course of the work due to change orders, delays, or other variations in the progress of the work. Unless the schedule is updated to reflect the variations, the original schedule will be of little value in the completion of the work.

3. When the critical path changes, the rationale should be recorded. Sometimes, when cases reach the courts years after the fact, the CPM shows that the path shifted, but parties can no longer recall why.

4. The owner may sometimes keep his own CPM or other progress schedule. This might not be shown until a dispute arises, so the contractor should be aware that there may be a progress schedule being checked by the owner.

5. Progress schedules can be mounted defensively. Either side can construct a chart after the fact, purporting to show that its position is correct. Thus, it is advisable to require periodic check-offs. A "new" schedule can't materialize overnight when both parties are regularly checking off and revising the schedule as a mutual effort.

6. Scheduling consultants can be called in to analyze these charts when disputes arise, or to give advice during the course of the work. In very complex projects, it is a good idea to utilize the services of these consultants.

7. Any CPM or network diagram is only as good as the logic used to set up the critical paths. These charts should not be regarded as scripture. They can be changed and revised as the job progresses to give the clearest possible picture of what has happened on the job, and why. Some agencies, as a matter of policy, include the in-house logic diagram used for project design along with contract documents to assist the contractor. Agencies typically point out that this is not necessarily the best or right way, but simply the way designers thought the project might be built. Agencies should include exculpatory language indicating that this is not necessarily the right way to construct the project, but

it is simply the way designers envisioned the project would be built.

8. When disputes reach the courtroom, a more succinct summary of job progress would probably be in order. It would be unlikely that parties walk through every intricacy of a vast CPM network. This is another good reason to use a scheduling consultant if a courtroom appearance is necessary. That consultant should be able to condense the CPM into an effective summary.

CASE HISTORY

The following hypothetical situation is an extended case history showing how a tangle of overlapping delays can be analyzed. It was originally a chapter that appeared in *Construction Contracts,* published by the Practising Law Institute. It is reprinted here with their permission. While this section is addressed to attorneys, it is not technical, and the analysis and examples are useful to all parties in the construction process.

In this chapter we will deal with how to analyze construction delays. We will consider only the interval of time from the moment the client enters the attorney's office until the claim is prepared for submission to the owner or the complaint is drawn.

The client tells the following story: he had a $700,000, thirteen-month contract to construct a five-story, steel-frame office building. Completion was delayed five months; he lost $250,000 and, to add insult to injury, the owner assessed him $9,000 liquidated damages ($100 per day for three months) plus actual damages of $10,000 representing loss of rentals in the building.

The client lost two months as a result of a change in design from closed-end pipe piles to large diameter step-taper piles. One month was lost due to the architect's delay in approving shop drawings; two months due to a roofing workers' strike and one month as a result of working during the wintertime. The owner ordered the client to accelerate the interior finishing work by working nights and weekends, enabling him to make up one month's delay.

The client wants to know whether he has a case. How much can he collect from the owner and does he have to pay any liquidated damages?

What does the attorney do to answer the client's questions? Several steps should be followed. The first step is to assemble the mass of papers that have survived the construction of the project. (An enumeration of records is omitted; these include, for example, daily reports, payment requisitions, progress schedules, and so forth.)

The second step in analyzing the client's problem is to understand the job thoroughly. The attorney should barrage the client with questions about how the job was actually constructed. He should ask to see the plans, have them explained to him and examine all the available photographs of the job. The attorney should continue to ask the client questions about the job until satisfied that he understands it thoroughly.

The third step in analyzing the client's problems is to read the contract documents, particularly the agreement and general conditions. Usually, a completion date is specified, or a period of completion in consecutive calendar days from the date of a notice-to-proceed from the owner. The contract usually states that time is of the essence. There is often a contract clause that sets forth liquidated damages for each day's delay in completion. There is often a clause seeing forth justifiable causes for which the owner will give an extension of contract time. This clause is often coupled with a clause which states that an extension of contract time is the contractor's sole remedy for delay and that in no event will the owner be liable for damages. The legal effect of such a provision will be discussed later.

TYPES OF DELAYS

There are two main categories of delay—nonexcusable delays and excusable delays. Nonexcusable delays are generally specified as those that are the contractor's fault, such as his failure to coordinate his subcontractors, his failure to submit shop drawings on time, the necessity of remedying defective work, and his failure to furnish a sufficient number of workmen.

Excusable delays are generally those that are not the contractor's fault. They fall into two major subcategories: compensable and noncompensable delays.

Compensable delays are those for which the contractor is entitled to claim damages from the owner. These are generally delays caused by fault of the owner, such as his failure to turn over the site on time, suspensions ordered in the performance of the work, defective plans, and the failure to approve shop drawings on time.

Noncompensable delays are generally those which are deemed to be no one's fault, such as strikes, fire, and acts of God.

The two major issues that arise in the excusable delay context are delays caused by weather and delays caused by subcontractors and vendors.

Weather Delays

Some time extension clauses are silent as to weather delays. They merely contain a general category of delays caused through no fault of the contractor. Other time extension clauses limit excusable weather delays to those caused by unusually severe or extraordinary weather delays conditions. These would usually have to be proved by reference to meteorological records that show the weather encountered during a particular interval was more severe than that experienced during the preceding five or ten years.

In the absence of such an express limitation, it is an open question whether the contractor should be entitled to an extension of time for ordinary weather delays.

Subcontractor Delays

The second major excusable delay issue is delay caused by subcontractor or vendor default. A contractor will contend that if he exercised good judgment in selecting a reputable supplier or vendor, he should not be charged for delays caused by their default or failure to deliver on time because he had no effective control over them. Many time extension clauses restrict a general contractor's excusable delay on account of subcontractors' delays to those causes which themselves would constitute excusable delay for the general contractor. The contractor's argument for excusability has been advanced more strongly in sole-source-of-supply cases where the owner specifically designated the vendor of a particular item. There, the contractor argues that he had no choice in selecting a vendor, and that by restricting his choice, the owner in effect, warranted and represented that the supplier would deliver on time. The authorities are divided on this point.

Concurrent Delays

Another problem in dealing with time extension clauses involves concurrent delays. Concurrent delays are two delays occurring at the same time, such as, both a strike and a failure to submit shop drawings on time. Where an excusable and a nonexcusable delay are concurrent, the contractor ought to be entitled to an extension of contract time. However, there are contract clauses to the contrary, that state that a time extension will be granted only where both concurrent delays are excusable. In the case of concurrent compensable and noncompensable delays, the contractor ought to be entitled to an extension of contract time but no delay damages.

PREPARING CHARTS AND GRAPHS

The next step in evaluating a client's delay problems is to organize and assemble the facts presented. A most expeditious way is to prepare a number of charts and graphs.

Chart #1—Bar Chart—Intended Schedule

Bar Chart #I shows the contractor's intended schedule for performing the work. This is known as a bar chart, which divides the work into its major elements or activities.

Chart #1 is greatly simplified. It divides the work into only five major activities: the foundation, structural steel, masonry, roof, and interior work. Adjacent to each activity is a bar representing the intended starting date, the duration, and the intended completion date. Ordinarily, in practice, a bar chart for a project of the scope set forth in the hypothetical example (i.e., a $700,000 five-story, steel-frame office building) would contain anywhere between 20 and 40 separate activities. The problem or weakness with a simple bar chart is that it does not show the interrelationship of the various activities nor the logic behind the contractor's intended performance nor the percentage of completion at any given time.

Chart #2—CPM Progress Schedule—Intended Schedule

From such a simple bar chart one is not able to ascertain which activities are independent of others and which activities are dependent for their start or completion upon the start or completion of other items. This information can best be represented on another type of progress chart, known as a CPM (Critical Path Method) schedule. Chart #2 shows the same construction activities as Chart #1, but laid out in somewhat different manner. Each activity has the same duration; however, the interdependence of the activities is shown. For example, the structural steel is shown to start only upon the completion of the foundation work. Then the chart splits into two segments. The masonry work can be performed at the same time as the roofing work. However, the masonry work takes three months to perform whereas the roofing work takes only two months. It is then shown that the interior finishing work cannot proceed until the completion of both the roof and the masonry work. The one month excess time on the path of the roofing work is known as "float time." This means that although the roofing work is shown to start at the same time as the masonry work, in fact, the roofing work could start one month later without and delay in the completion of the entire project.

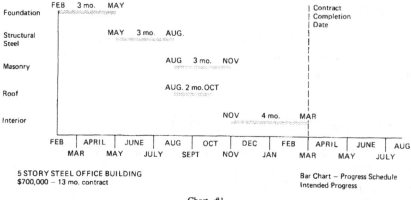

5 STORY STEEL OFFICE BUILDING
$700,000 – 13 mo. contract

Bar Chart – Progress Schedule
Intended Progress

Chart #1.

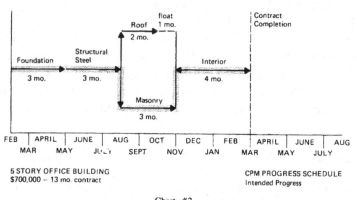

5 STORY OFFICE BUILDING
$700,000 – 13 mo. contract

CPM PROGRESS SCHEDULE
Intended Progress

Chart #2.

The ▬▬➤ line represents the critical path through the project which, by definition, means that the prolongation of any single activity along the critical path will prolong the completion of the entire project. The roofing work is not on the critical path because of the one month of "float time."

Chart #3—Bar Chart—Actual Progress

The following chart requires a meeting with the client and use of his records. Chart #3 is prepared and superimposed upon Chart #1 indicating the periods when the various construction activities were actually performed. This information is often obtainable from the contractor's progress payment requisitions, which indicate the percentage of work completed for each such activity during each month of the job. Information regarding the dates of actual performance can also be obtained from the foreman's time card, the superintendent's daily reports, the correspondence, the job meeting minutes, and the photographs.

An explanation is then requested from the client for each delay, prolongation of work, or change in sequence of the work. For example, we see on Chart #3, that although the foundation work was scheduled to begin February 1, it did not commence until March 1. The client then recalls that there was a strike at the outset of the job which prevented the start of foundation work for one month. Then we see that the foundation work proceeded for one month during March and then was suspended for one month in April. The client says that this suspension was ordered by the owner after it became apparent that the specified closed-end pipe piles were being driven much deeper than anticipated. The architect took one month to conduct these tests to redesign the work. We then see that the work was resumed in May and proceeded three months until completion in August. This indicates that the aggregate time for performance of the foundation work was four months rather than the three months intended. Your client explains that the driving of the large diameter step-taper piles proceeded more slowly than the closed-end pipe piles as a result of the different pile driving equipment required to be used.

We then see that the structural steel work did not begin until September, one month after the completion of the foundation work, although it was originally scheduled to have begun immediately upon completion of the foundation work. The client explains that while he submitted structural steel shop drawings at the time required, the owner's architect unreasonably delayed approving them, thereby preventing the start of the structural steel work until September. We see that the structural steel work was performed in a duration of three months, as it was originally scheduled to be performed.

We next see that upon completion of the structural steel work, the masonry work commenced immediately, but had a duration of four months rather than the three months originally scheduled. Your client explains that the additional month of performance time was caused by the work being pushed into the winter months. Originally, the masonry work was to be performed from August to November, during good weather. However, it actually had to be performed in the middle of the winter which resulted in one month's lost time. We next see that the roofing work, while originally scheduled to begin at the same time as the masonry work, did not begin until one month later. Your client explains that, as originally scheduled, there was one month of float time, which he decided to take advantage of because he knew that the masonry work would be prolonged during the winter. However, after proceeding for a month, the roofing work was suspended for two months during February and March.

Your client explains that there was an industry-wide roofers' strike, which prevented the performance of any work during this period. Thereafter, we see that the roofing work took two months to complete from April to June, a total performance time of three months, whereas only two months was originally scheduled. Your client explains that defects were found in the roofing work by the architect which took your client approximately one month to repair.

You then see that the interior work started in May and was completed in a duration of three months rather than the four months scheduled. Your client explains that in May the owner demanded that your client accelerate performance by working evenings and weekends in order to complete the project as quickly as possible and that one month's time was made up by performing the work in this manner.

Chart #4 CPM Progress Schedule—Actual Progress

Your next step in analyzing your client's claim requires you to insist that he prove each and every delay to you from the available records rather than rely upon his memory. Some clients have a tendency to accentuate the positive and eliminate the negative. Time dulls memories, especially unpleasant memories. As noted previously, it is apparent that your client has forgotten or neglected to tell you several relevant facts regarding the case, such as the initial one-month delay caused by the strike. Now you must use the records in order to test the validity of each of his contentions.

This is best done with the aid of another form of CPM progress schedule. Actual progress is laid out in a ▨▨▨▨ line, in much the same format as was the original CPM schedule. The original critical path and critical path activity duration, is indicated by the ▭▭ line. The ▬▬ line represents the actual critical path through the job. It is significant to note that whereas, as originally scheduled, the critical path went through the masonry work, now the critical path shifts and goes through the path of the roofing work. It is also noted that the critical path is not through the entire roofing work, but only a portion of it. The ▨▨▨ line and the ▨▣▨▣▨ line represents delays. A ▭▭ line represents a total suspension of work. A ▨▣▨▣ line represents a delay that occurs over an interval of time. For example, the one-month prolongation of the masonry work occurred during the four-month interval from December to April.

Using Records to Verify Delays

In reviewing your client's records you find that the one-month initial strike is substantiated. You next find that, in fact, the architect did order a one-month suspension in the foundation work and that the foundation work did, in fact, take one month longer to perform using the different equipment required for the step-tapered piles. This is ascertained by comparing the production records of the closed-end pipe piles for the month of March, prior to the suspension, with the pile driving records from May through August.

You then find that the one-month steel shop drawings delay is not properly attributable to the owner, but, rather, was your client's fault, albeit unbeknownst to him. You find that your client's steel erection subcontractor did, in fact, submit shop drawings on time. These were promptly reviewed by the architect and returned to the subcontractor for correction because they contained substantial errors. The subcontractor then delayed in correcting the errors and the architect finally approved the revised shop drawings promptly after resubmission to him. This information is ascertained both from an examination of the shop drawing log maintained by your client and a close examination of the shop drawings themselves.

Frequently, architects use the vehicle of shop drawings to effect substantial design changes. It is therefore necessary to examine carefully each comment and correction made by the architect on the shop drawings to determine whether he is merely correcting errors made by the contractor or whether he is, in effect, redesigning the work.

You then see that the one-month delay in starting the roofing work was simply a matter of your client's exercise of choice not to commence the roofing work until that time. However, he did not know that there would be a roofers' strike which, in fact, delayed the roofing work for two months. Thereafter it took two months to complete the roofing work, of which, one month was due to repairs required by the architect because of defects in the workmanship—the fault of the client. However, it is noted that the repair of the defects did not prevent the interior finishing work from commencing, i.e., the repairs did not fall on the critical path.

Your client's records indicate that in fact he did work overtime and weekends and that the one month shortening of the intended four-month interval for interior work was directly attributable to that overtime work.

Deciding on Excusability and Compensability

Consider whether each of the delays was excusable or nonexcusable and, if excusable, whether it was compensable or noncompensable. The one-month initial strike is excusable but noncompensable. The one-month suspension due to a redesign of the foundation is both excusable and compensable. The one-month prolongation of the foundation work directly attributable to the change in design would also be both excusable and compensable.

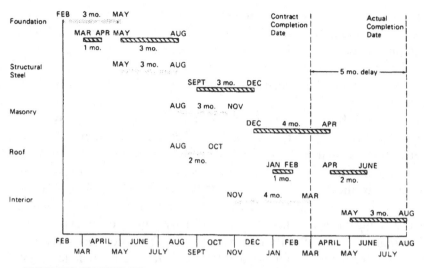

5 STORY STEEL OFFICE BUILDING
$700,000 – 13 mo. contract

Bar Chart — Progress Schedule
Intended vs. Actual Progress

Chart #3.

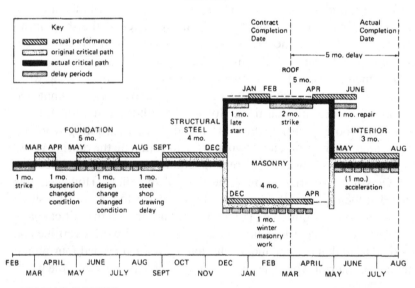

5 STORY OFFICE BUILDING
$700,000 – 13 mo. contract

CPM Progress Schedule
Intended vs. Actual Progress

Chart #4.

Change Orders

Here, however, it is important to note that your client's records must be examined carefully to find out whether he waived any further rights to compensation on account of the change in design. Frequently, in granting change orders, an owner requires the contractor to waive any further costs attributable to the work encompassed by the change order and specifies a total extension of time that will be allowed for the change. Since the ultimate effect on final completion of the entire job often cannot be determined at the time the changed work is being performed, it is advisable for contractors to execute change orders with a reservation clause. This clause should state that the change order is for the direct cost of the changed work, and, and does not include any impact costs or extension of contract time, which are specifically reserved until such time as they can be finally ascertained.

The one-month steel shop drawing delay is nonexcusable.

Analyzing Concurrent Delay Effects

Now you come to the split paths of masonry/roofing work. These include concurrent delays and the analysis is a rather complicated one.

Start with the masonry path. The one-month prolongation of the masonry work as a result of the necessity of performance in the winter would ordinarily be inexcusable, since, in our case, the winter was an ordinary one, not an unusually severe one. However, in our case, there were three preceding months of excusable delay. If your client had started masonry work three months earlier, in September, it would have been completed prior to the onset of winter. Therefore, the masonry delay, considered alone, ought to be excusable. There were, however, only two months of precedent compensable delay. (See the earlier discussion concerning when delay is excusable, compensable, nonexcusable or noncompensable.) If he had started the masonry work two months earlier, in October, the work would have been performed in two months of good weather and one month, December, of winter weather. In such event the masonry would undoubtedly have been prolonged for a period less than the one month it was actually prolonged. Therefore, considered alone, before analysis of the concurrent roof delays, there would be somewhat less than one month of compensable delay for the masonry work, even though the entire one month delay is excusable.

Next, we consider the delays in the roofing work. The initial one-month late start of the roofing work by your client's choice raises the issue: Who owns the float time? While this has not been raised in many cases, the prevailing view is that the contractor owns the float time. The two-month delay in the roofing work, therefore, ought to be excusable. However, if there was precedent inexcusable delay attributable to your client, pushing him into the strike, the strike ought then to be inexcusable. In the same way, if there was precedent compensable delay which pushed the contractor into the roofing strike period, the time of duration of the roofing strike ought to be both excusable and compensable. Here we find that there were three months of precedent excusable delay. Had the roofing work been started three months earlier, the strike would have been avoided. Here, also, we find that there were two months of precedent compensable delay. Had the roofing work been started two months earlier, it would have been completed before the roofing strike. Therefore the roofing strike delay should be both excusable and compensable, when considered in the absence of the concurrent masonry delay.

In considering the masonry path and the roofing path, we find that there was an aggregate delay of two months along these paths. There were two months of excusable delay in the roofing path and one month of excusable delay in the masonry path, with the remaining month in the masonry path being float time. The net result should thus be two months of excusable delay.

In the masonry path, there was somewhat less than one month of compensable delay and there were two months of compensable delay in the roofing path. Therefore, the aggregate of these concurrent delays would be something less than one month of compensable delay. (See above discussion of concurrent delays.)

DETERMINING RESPONSIBILITY FOR DELAY DAMAGES

Now you are in a position to answer the questions posed by your client. Was the order to accelerate a valid one? Yes. As of May, when the order was given, there was a six-month delay, only five of which were excusable. Therefore, the owner was entitled to demand that your client make up the one-month of nonexcusable delay by accelerating performance.

If the order were not a proper one, the owner would be liable in damages for the overtime wages paid your client's employees plus the loss of productivity attributable to the increased number of hours per day and the increased number of days per week that the employees worked. Studies have shown that men working under these conditions are less productive than when working a normal week.

May Contractor Recover Damages From Owner?

Is your client entitled to recover any damages from the owner? In spite of the contract clause stating that the contractor's sole remedy is an extension of contract time, it has been held in New York and elsewhere that such a clause is not enforceable and that an owner will be liable in damages for the unreasonable delays he causes. The delay resulting from the change in foundation design aggregated something less than four months—two months while the foundation work was being performed and then some period less than two months during the concurrent performance of the masonry and roofing work. Your client is entitled to recover mobilization and demobilization costs and equipment standby rental values attributable to the one-month suspension while the architect redesigned the work. He is entitled to labor and material escalation costs attributable to the prolongation of the work.

He is entitled to recover the loss of productivity, aggregating something less than one month's delay, during the masonry work. This is best computed by comparing actual records of masonry work performed in early December, before the winter conditions set in, with the work performed during the winter conditions. If this is not possible, other methods which may be used include a comparison of the actual cost of the masonry work with the reasonably estimated cost of the work, or the application of productivity reduction factors from studies that have been conducted in the industry. The client is also entitled to recover any additional costs of winter protection of the masonry work.

The client is entitled to be compensated for the extended period costs of something less than four months. This would include extended superintendence and supervision, trucking, vehicles and equipment, insurance, miscellaneous field expenses such as telephone, light, shanty, and an allocable proportion of central office overhead for the extended period.

Assessing Liquidated Damages

The next question is whether the owner was entitled to assess liquidated damages for three months. No liquidated damages were properly assessable because the aggregated delay was five months and there were five months of excusable delay.

In the absence of excusable delay, liquidated damages will generally be enforced, as the owner's sole remedy, if they are found to be "reasonable and not a penalty." The sum specified as liquidated damages must be found to be reasonably related to the damages the owner might expect to incur as of the date that he entered into the contract. In the event that the liquidated damages are deemed to be a penalty, the owner may only collect the actual damages he can prove.

The owner is not entitled, in most jurisdictions, to collect both liquidated damages and actual damages. Therefore, if the nature of the project is such that the owner's damage could be easily computed, it would prob-

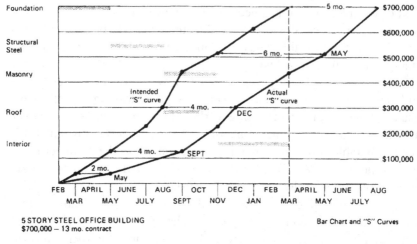

5 STORY STEEL OFFICE BUILDING
$700,000 — 13 mo. contract

Bar Chart and "S" Curves

Chart #5.

ably be better for an owner not to include a liquidated damages clause. Excluding the clause would permit the owner to collect all of his actual damage in the event of inexcusable delay in completion. If, however, delay damage is not easily computable, such as damage that would occur to a school district from delayed completion of a school, or damage that would accrue to a congregation from delay in completion of a church, it would be advisable to insert reasonable liquidated damages into a construction contract.

Chart #5—Bar Chart With "S" Curve

As you can see, even with the use of an oversimplified example, the CPM method of analysis can get quite complicated. In the example chosen, the analysis of the concurrent roof and masonry delay, in light of precedent delays, was exceptionally complicated. One can imagine that if some 50 different activities had to be charted, rather than five, each with different periods and causes of delay, a CPM analysis could become impracticable. Therefore, another method of analysis is often used. This is represented by Chart #5 and includes the super-imposition of an "S" curve over the contractor's intended bar-chart schedule, Chart #1. It should be noted that on the right-hand side of Chart #5 the contract price is plotted in intervals of $100,000. It is possible to compute the amount of money that would be earned per month if the job had proceeded on schedule (from the intended bar chart and the trade payment breakdown). This is represented by the line at the left hand side of Chart #5 labeled Intended "S" Curve. One might think that this ought to be a straight line.

However, experience has proven that construction projects often begin somewhat slowly, pick up their rate of progress during the middle of the job, and then tail off towards the end. Thus, the resulting curve is in the shape of an "S." The actual payments made to the contractor each month can be plotted against this curve. Care should be taken however, to exclude change order work from the payments, so that the comparison is valid. Then, at any point along the actual "S" curve, the horizontal distance between the actual curve and the intended curve will indicate the duration that the job is behind schedule at that date.

Chart #5 indicates that; as of May, the job was behind schedule two months: in May, at the time of the owner's order to accelerate the work, the job was six months behind schedule, confirming our prior CPM analysis. "S" curves are helpful in ascertaining whether the job is behind schedule for purposes of termination of contract and orders to accelerate the work. They are also helpful to explain delays in a simplified manner when a CPM analysis is not practicable.

CONCLUSION

There are several conclusions that can be drawn from our discussion. First, the importance of using actual records to analyze delays cannot be overemphasized.

Second, the validity of a CPM analysis is entirely dependent upon the validity of the assumptions made. It is like a deck of cards. If one assumption is proved incorrect, then the entire analysis may collapse. For example, in our hypothetical case, if, in fact, the interior work could be started independent of the completion of the roofing work, the entire analysis and result would be changed. It is important, therefore, in most substantial cases, to retain an outside expert consultant to evaluate the assumptions of the critical path logic to ascertain whether or not they are correct.

7. Contractor's Costs of Delays

Once the contractor decides that he has a legitimate claim for costs due to owner-caused delays, he must next quantify these. The owner, for his part, should be knowledgeable about costs for which he might be liable and about the amounts that are justifiable. If parties disagree about these amounts, they may have to defend their positions in a formal disputes resolution proceeding. Therefore, it is important for both to have accurate records and accurate calculations.

Costs can be approached in two ways: (1) actual costs and (2) estimated. With actual costs, the contractor keeps records as the work progresses. Depending on the complexity of the situation and the level of communication between the parties, an agreement may be reached as these costs accrue on the job. There is, however, a basic flaw in this approach. The contractor frequently has no incentive to be efficient during such an operation, and thus the costs may be greater than they might otherwise have been. Estimated costs are subject to question, too. The logic on which the estimates are based may be in error. There are many ways to estimate the costs; none may be actually incorrect, but disputes can arise over which is the most accurate or acceptable.

In some areas, it is common practice to accept a formula-based cost calculation. The Eichleay formula, which will be discussed in more detail later in this chapter, is widely used to allocate overhead costs. The AGC manuals and other similar books are used to calculate the cost of contractor-owned equipment on the site. Sometimes contracts contain clauses that specify how costs attributable to delay are to be computed.

TYPICAL DELAY-CAUSED COSTS

Costs that may increase due to a delay on the project may be many. They can include direct labor costs and costs related to labor, such as social security, unemployment taxes, union benefits, compensation, and other insurances. Other costs could include materials, tools and supplies, equipment, fuel and repair on equipment, subcontracting costs, bonds, the cost of capital, and the main office, district and job-site office overhead.

The following are examples of how some of these costs can increase: standby costs, extended job supervision, extended field expenses, extended equipment costs, increased labor costs due to moving into a period of higher labor rates, and increased costs of labor and materials owing to moving the work into winter months. This list includes items that are usually recoverable. Among the items that are almost never recoverable are attorney's fees. Only a very few states allow recovery for those fees; otherwise, they are rarely recoverable. The federal government allows attorney's fees in certain circumstances. And, as always, there may be a contract clause that specifically allows recovery.

Reduced ability to take on other profitable work is another item that is rarely ground for recoverable damages. The contractor would have to prove that he was offered other work that was precluded because of the delay on the project in question, and what is more, he would have to prove that the other work would have been profitable. Such claims are generally considered too speculative to be valid.

Contractors sometimes request compensation for the cost of the claim preparation. This, too, is frequently not granted. Those preparing the claim do not typically keep time records, for one thing. In cases in which recovery was granted for the cost of claim preparation, the argument has been that the cost was due to the contract requirement regarding the submission of claims. In other words, the cost was a result of the contract wording. Finally, there have been claims based on the devaluation of money during an inflationary period. Any recovery in current dollars is worth less than the dollars the contractor originally laid out, the argument goes, and therefore should be increased to reflect inflation. As far as we know, there has not been a decision favoring such reasoning, although in periods of high inflation rates, it certainly is a real damage. Generally, interest on the amount recovered is all that is allowed; this is usually adjusted to current interest rates.

ALLOWABLE COSTS

The following lists costs that are commonly allowed. The amount and the method of estimating that amount may be questioned, however. Here are some points to bear in mind when calculating these costs:

Main Office Overhead

This includes every cost involved in operating an office: building rent; depreciation (if owned); depreciation on equipment such as a computer; lights; heat; telephone; stationery; and payroll costs with their add-ons such as taxes, insurance, and benefits. Most contractors keep such office overhead records for tax purposes, and these can be used to allocate costs due to the delay. If we are talking about a major project, particularly if it is a joint venture, there will be a full staff at the job site that handles the project with little input from the main office for daily oversight of the project. In such cases, the main office overhead of each partner may come out of the job profit, or the joint venturers will have previously agreed on a lump-sum payment to each main office on a monthly or project basis.

In other cases, where the project receives sizable input and supervision from the main office, including payment of invoices, cost accounting, and payroll, a main office cost will be assigned to the project as either a fixed sum or as a percentage of the work performed.

For claims purposes, the Eichleay Formula is one method used to apportion overhead costs to a particular project. This formula was used in a 1960 Armed Services Board of Contract Appeals decision, Eichleay Corp. 60-2 BCA; pg 2688, and was widely accepted. Subsequently, the method fell out of favor and became somewhat controversial. Currently, however, the Eichleay Formula has regained acceptance and is used by the federal courts and some state courts. Overall central office overhead is allocated on the basis of the ratio of the dollar amount of the contract in dispute to all other contracts billed during the same period of time. Therefore, an "extra," any additional work, should carry the same percentage mark-up for home office overhead as the original contract carries. The rationale is that central office overhead is a necessary expense of doing business. To put it another way, the contractor's office is a "benefit conferred" on the owner. For instance, estimators, purchasing departments, data processing equipment, and the like are all part of job expenses.

Like many formulas, the Eichleay Formula is an oversimplification and cannot be applied indiscriminately. Another method for allocating home office overhead states that an increase in home office overhead directly traceable to the delay should be allowed. That is, if a contractor can demonstrate that his home office overhead escalated as a result of the delay, then the increase is an allowable expense of the delay. The following variables illustrate how difficult a task it is to determine what home office overhead would or would not have been without the delay: variations in dollar volume or work; variations in complexity of work, in type of work, and in form of contracts and risks related thereto; variation in location of the work, in the management capability, and in escalation, and state-of-the-art.

When the contractor has the burden of proof, which is usually the case, he uses historical records adjusted for the factors just listed, and any other factors applicable, to attempt to measure the amount of the increased office overhead. The Eichleay Formula and similar formulas have been accepted because of the problems of quantifying the actual increased office overhead costs and of proving that they were attributable to the delay.

A determination must be made as to the involvement of the main office in the project and whether the main office is used strictly for bidding new work and/or maintaining corporate headquarters. In some cases, one job that is actually a small part of the contractor's total billed work may be the cause of disproportionate overhead costs. These costs might be due to more secretaries typing letters, more accounting, and so on. because of delays and disputes on that one job. In such a case, the Eichleay Formula would not reflect true overhead costs for delays on that one job.

It should be noted that formulas akin to the Eichleay Formula have been rejected in some courts and accepted in others. For example, a 1978 New York decision held that the Eichleay Formula could not be applied, and that in order to use any formula, proof would have to be given that the chosen formula was a reasonable measure of the actual damage sustained by the contractor. Calculating "actual" damage is not easy. One approach would be to carefully analyze the effect of the underlying impact problems caused by the delays. The contractor may have had to write more letters, hold more meetings, prepare more shop drawings, and so on. It should be clear that allocating the time spent by the contractor's employees and proving that some of that time was directly attributable to the delay can be a formidable and costly effort.

So while many jurisdictions will accept the Eichleay or other similar formulas as fair assessments of central office overhead, a contractor should check to make sure that this is the case. Any such formula should be applied with some common sense. Excessive claims for office expenses are sure to be questioned. On the owner's side, he should recognize that central office overhead is itself a legitimate expense. The amount claimed and the method by which a figure is derived may be questioned, however.

Intermediate Office Overhead

Some contractors might claim intermediate or field office overhead expenses owing to the delay. These can be valid if the contractor is large enough to maintain regional offices, for example, those that oversee several projects. Similarly, overhead expenses may accrue at the job site or field office. Supervisory and other overhead staff including the project manager, office manager, accountants, time-keepers, the master mechanics, and drivers of pick-up trucks and other vehicles not directly connected with a work item may be considered as overhead. Also, the cost of site buildings, light, heat, telephone, water, sewage disposal, office supplies, watchmen, maintenance of haul roads, and other such facilities and items can be charged to job overhead.

Obviously, the Eichleay or other such formulas would not apply unless the intermediate office is used for more than one project. If the office is overseeing only one project, then the contractor has to show that his office overhead was increased as a result of the delay. As we have shown, it is difficult to show how the delay did or did not increase overhead.

Labor Costs

The actual labor costs can generally be computed by using payroll records to show what was paid during the period of time in question. However, when a delay is involved, the problem consists of showing the increased cost of performing the work due to the delay. This is more complex. The contractor must prove the actual cost and what the work would have cost if there had been no delay. Escalation of labor costs may be relatively easy to prove; for example, a new union contract may have been signed as the result of a strike settlement.

Loss of Productivity

Loss of efficiency or productivity (the two terms are used interchangeably) can be more difficult to document. Either term can be used to apply to numerous situations where the work has been interfered with in some manner, must be performed in a different manner, or must be performed under different conditions from originally planned. These can be situations where work is forced to proceed in adverse weather, where there is a change in the sequence of work, or where the work is repeatedly stopped and started.

Additional costs from such situations can arise as a result of the loss of actual worker productivity, such as due to fatigue from working overtime. Losses due to shift changes, time lost in getting one crew in and another one out, and time lost in explaining the work and training new personnel are all possible items under the heading of loss of productivity. Time lost in explaining and training is often referred to as impact on the learning curve.

Excessive numbers of workers in an area can contribute to lowered productivity, as can the performance of more than one operation in the same area. These situations can arise as a result of delay, from acceleration, from a change of sequence in the work owing to delay, or from numerous changes to the work. A contractor can prepare charts to demonstrate what the job would have cost in man hours had there not been a delay. It is imperative that he have a baseline of work accomplished during some "neutral" period; otherwise, there is no reference point with which to compare the work done under the different conditions caused by the delay. The differences can document the actual costs of the delay. Such a comparison can be convincing, and is sometimes accepted as proof of loss of efficiency. Recovery for inefficiency as a percentage of actual costs has been obtained in the wide range of between 5 percent and 50 percent of the actual labor costs paid to the contractor during the delay period and interference to the work.

Figures based on industry statistics are available through construction industry associations. These are often used to document loss of productivity, but should be used with caution. The data may be questionable, and the conditions under which they were assembled may be different from those in one's claim.

Expert witnesses can be used to prove the contractor's argument as to increased costs arising from the loss of efficiency. There are also cost consultants who specialize in analyzing cost data. Sometimes a large owner has the in-house staff to make judgments of losses due to inefficiency on a percentage basis. The actual increased costs, however, rarely can be tracked with any real accuracy.

Another method sometimes used is referred to as the total cost method. This involves taking the total cost versus estimated cost of labor and the total versus estimated man hours of labor. But there are plausible arguments against using percentage calculation. For example, the contractor's estimate might have been wrong, or the conditions on the job in question might have differed from the assumed conditions on which the statistics were based.

So, to sum up, in most jurisdictions, recovery is allowable for loss of productivity or efficiency due to owner-caused delays, interferences, and acceleration. Determining the proper amount of recovery, however, can be difficult.

Insurance and Bonding

Fringe benefits and payroll-associated insurance costs are relatively simple once the labor costs have been agreed on. These costs are based on the billings to the contractor from the insurance carrier and are easily proven by audit.

Property insurance and insurance for theft, fire, flood, and so on, are time-linked. If the job is delayed through compensable delay, the contractor can ask the owner to pick up the premium after the contract time has expired. Liability insurance is also part of the increased costs; such insurance is based on the labor force on the job.

If the claim results in an increase to the amount payable to the contractor, there will be an additional bond payable. Again, proof is readily available in the bond agreement between the contractor and the bonding company. While the percentage will undoubtedly be small, in the range of 1 percent to 1 1/2 percent, it is a legitimate part of the claim for extra costs to the contractor. There may also be specialized insurance necessary, and the additional costs for these may be part of the claim. For example, water-borne equipment may be involved for the portion of the work affected by the delay, and thus marine insurance may have to be extended. (See Chapter 9 on Bonding.)

Materials

A delay can affect the cost of materials. The most obvious effect would be escalation of prices due to the delay. The price may also be increased due to shortages that would not have been encountered were it not for the delay or because additional materials were needed owing to the delay.

In the case of acceleration, costs of materials could be higher because of a delivery earlier than that originally required or from a different supplier. This idea can also apply to a delay that stems from a change in the sequence of the work. Some materials that might have been used later in one sequence might need to be used earlier in the changed order of work. There can be a premium for buying and/or delivering materials in smaller quantities than those originally planned. For example, a contractor may purchase water pumps to dewater a site. Because of a delay, he must buy new pumps because the original pumps wear out before the original work can be completed, and the smaller number of new pumps cost more than the larger number of the original purchase (per pump).

Unusual transportation costs owing to the changed sequence of work can be included in the claim, too. Tight scheduling of deliveries may add to transportation delivery costs.

Equipment

If the equipment used on the job has been rented from an independent source, records of bills paid will ordinarily suffice as proof of the cost of such equipment. Difficulties could arise if the renter had committed the equipment elsewhere at the completion of the agreed-on rental period. Other equipment might have to be located for the delayed portion of the work at an increased rental cost. However, in either case, paid invoices should suffice as proof, provided the arrangement is with an outside rental company as opposed to a wholly contractor-owned equipment subsidiary.

It is rare in rental equipment disputes for expert witnesses to be retained. Paid invoices or the industry standard rental publications rates are usually accepted; and federal agencies sometimes have their own rates stated in the contract. Occasionally, though, when a contractor and owner cannot agree on the value of equipment in order to calculate a daily expense rate, an expert might be called in.

If the contractor owns the equipment, the situation is quite a bit more complicated. Some jurisdictions will accept a rental value as set forth in standard trade books such as the so-called green book or blue book rates or a percentage of such rates. Sometimes, the contract includes a specific rate for equipment rental and mentions one of those manuals by name. Rates should be specified if the project requires the use of specialized or unusual equipment.

Other jurisdictions will allow recovery on a computed ownership expense basis, such as the Associated General Contractors method, which takes into account factors such as age, initial cost, useful life, operating hours, and maintenance and repair costs for each piece of equipment. Still other jurisdictions limit recovery to added operating expense, unless the contractor can prove that the equipment could have been used elsewhere if the delay had not occurred.

In some cases, the contractor owns a separate equipment owning/leasing company and uses this company to bill the job for the leasing of equipment. Generally, this equipment will be considered as being owned directly by the contractor. Another complication that can arise from a dispute over contractor-owned equipment involves the theory of "benefit conferred." The contractor confers certain benefits on the owner by making the equipment available to the owner for an extended period of time. The equipment represents a sizable capital investment by the contractor.

Suppose, for example, a fully depreciated, specialized, and costly piece of equipment owned by a contractor sat on a job six months longer than anticipated and that this was due to an owner-caused delay and that the owner wished to reimburse the contractor for out-of-pocket expenses only. The contractor would claim that he was entitled to a fair and reasonable rental for the period. The equipment was there for the owner's benefit, whether he used it or not.

Suppose the contractor had no other use for the specialized equipment during those six months, and it could not have been rented to others. The contractor could still seek to assert a claim on the ground that he was providing something of great value to the owner. The fact that the owner did not actually use the equipment and that the contractor could not have gained other income from the equipment ought not to make a difference. The owner had the benefit of the equipment; the fact that he could not take advantage of it was due to his own delay.

Establishing the actual value of equipment related to delay is difficult, and reimbursement of the out-of-pocket direct expenses may not be fair compensation for the contractor. Thus, the benefit-conferred theory of compensating for these costs may be more equitable. Another example is apt. An owner, operating under the damages or actual expenses theory might seek to diminish the amount of recovery because the equipment had been paid off. It had been completely depreciated, and therefore, in the eyes of the owner, the contractor should be entitled to less compensation. What the owner ignores in such situations is that despite the fact that the equipment was completely depreciated, it has value and is usable by the contractor for the benefit of the owner.

A final note on equipment costs as related to delays: this category need not be limited to major and costly equipment; lesser items such as tools would also be recoverable if their use is related to the delay.

Interest Expenses

Interest on claims is recoverable depending on local statutes and on the contract language. In jurisdictions where interest expenses are allowable, the rate can be below the market rate. The date after which such interest accrues is often well after the date the added expense was first incurred. For example, in the state of Pennsylvania, no interest on a delay claim is recoverable prior to judgment, and thereafter only at a rate of 6 percent. In claims against the City of New York, interest starts to run when the increased costs are incurred, but only at the rate of 3 percent.

The federal government under the Contract Disputes Act of 1978, allows recovery based on the interest rate that is set every six months by the Secretary of the Treasury, starting from the date when the claim was formally presented to the contracting officer and certified by the contractor as to its validity. That rate varies, but has typically been below market rates. In some cases, as we have noted, such low interest rates act as a disincentive for an owner to settle expeditiously. If an owner had to pay 12 percent interest on any money due a contractor, he would undoubtedly be more anxious than when the rate is 9 percent. However, the point here is that interest can be allowable. Parties should check the contract and the local applicable statutes or other relevant law to see what the rate and starting date (for the interest to begin running) should be.

Sometimes a contractor has to borrow money to finance added costs imposed on him by the owner's delay. If he can prove that he had to borrow this money, some courts have allowed recovery of the added actual interest expense. But proving that the contractor was actually forced to borrow that money for a special claim item can take considerable effort, and may be futile in the end.

Profit on the Amount Claimed

Profit on delays may be difficult to recover depending on whether the delay is considered to be pursuant to the changes clause or a suspension of work clause. Under the Federal Acquisition Regulations, profit is specifically excluded on a suspension of work. However, if the contractor is not clearly suspended by the owner, it can be argued persuasively by the contractor that profit is allowable because the delay is really a change pursuant to the changes clause, wherein profit is rightfully a part of the equitable adjustment. Common practice varies widely depending on the form of contract, the owner, and the forum of litigation.

Finally, if a project is large and complex, it would be wise to retain a cost consultant to help prepare and document a claim for delays. The contractor should be familiar with the applicable law before setting his staff or computer to work to calculate his damages and costs. There may be several ways to calculate different items, and different jurisdictions may allow some items and not others.

The law as to recoverable damages changes constantly. It makes sense for a contractor and his staff to include all reasonable, legitimate, and provable items of damage in a claim. The law or a decision in a similar case may change during the litigation period. Unreasonable or extraordinary claims are never recommended, as they tend to discredit the validity of the entire claim, and may result in an onerous decision.

What follows are two sample cost calculations for a claim, which should help illustrate many of the points made in this chapter. The first involves changed conditions; the second is a claim for costs due to acceleration.

SAMPLE CLAIM AND CALCULATIONS/CHANGED CONDITIONS

Rock Differing Site Conditions, Reach Two

During the early days of an excavation, it was found that the location and amount of rock within the limits of the trench cut were not as predicted based upon the geological data available. The contractor gave written notice that the unexpected rock was considered to be a differing site condition for which additional time and money was due.

During the bidding process, the geologic data were quantified into cost of excavation for Reach Two. To enable quantification, certain assumptive extrapolations from the exploration logs had to be made. These assumptions were based on the contractor's understanding of the geology of the area as described in the specifications.

The geologic descriptions stated that some of the bucket auger holes encountered caliche and/or limestone layers, which they could not penetrate.

The overall picture of the materials to be excavated was that they would be generally unconsolidated, soft, and granular, with occasional caliche layers that would be hard and difficult to penetrate with an auger. This meant that most of the material would be easy to excavate and that the only excavation difficulties would be with the layers of caliche and limestone, which would classify as rock. Estimating effort was therefore concentrated on predicting how much of such rock would be encountered. This meant that the contractor would have to use "site-specific" exploration data and extrapolate from there.

The main source of information for determining the expected amounts and locations of rock at bid time were the logs of augered holes taken in the vicinity of the pipe line. In total, there were 70 explorations made along Reach Two, 28 of which were made with a 24"dp bucket auger, 40 with a 6"dp flight auger and 2 which were core-drilled. The bucket-augered holes, in most cases, would not penetrate the rock layers; therefore, their most significant contribution in regard to information pertaining to the rock was to indicate the depth at which it could be expected. The 6"dp flight-augered holes usually penetrated the rock layers, therefore the thickness of the rock layers was known. The combination of the two sources provided what the contractor believed to be a reasonable basis from which to estimate the thickness at points of auger refusal.

The horizontal extent of the layers of rock was extrapolated from the boring data provided in the vicinity. In some instances, the layers could be predicted to extend from borehole to borehole. In other instances, it was obvious that the rock layer terminated between auger holes. In these cases, the horizontal extent was estimated, based upon the thickness of the layer, the distance to other auger holes, and the breadth of other similar layers whose extent was more accurately determinable. Based upon these extrapolations, it was estimated that 25,000 cubic yards of rock would be found in the trench.

The contractor's extrapolation of the horizontal extent of the rock layers at bid time was, we believe, reasonable and logical, considering the origin of the sediments. As indicated by the specifications, the sediments were very lenticular and discontinuous, with localized lenses of caliche and limestone. This fits with the fact that the sediments were laid down by braided streams flowing from the higher areas down into the valley basin. Their routes of flow were wandering, so rates of flow were highly variable. This produced channels of more pervious sediments heterogeneously interlayered and irregular in trend, and each channel became buried by some later deposition. The pervious lenses produced in this manner tended to hold mineral saturated ground water for long periods of time and more frequently than the surrounding, less pervious sediments. This produced the carbonates and sulfates that subsequently precipitated and formed the concentrated cementing agent that created caliche lenses. The limestone layers were the product of localized ponding and precipitation of carbonates in quiet water ponds and lakes. All of this resulted in rock lenses of variable length and direction.

The location of the material in the trench had a significant bearing upon the excavation production. When there was one layer of rock in the trench above invert, it could usually be excavated with the backhoe without severely slowing the production. However, when the layer was a thick one, greater than 2 feet, the productivity was greatly reduced because it usually required the use of the pumpkin ball to break up the rock so that the backhoe could remove it from the trench. When the rock layer was in the invert of the trench, it was extremely hard to excavate because it could no longer be undercut to allow easy breakage with the pumpkin ball. Therefore, the rock had to be worn down to grade by backhoe, pumpkin ball, tractor and dozer, hoe-ram, drilling and blasting, pavement breakers, or a combination of the above. Other combinations of these conditions had combined effects on the productivity.

In the actual case, there were two major differences encountered relative to the rock. The first case was in the quantity, where a total of 85,000 cubic yards was actually encountered. This represents about three and a half times the quantity that could have been reasonably anticipated. This excessive overrun in quantity demonstrated that the conditions encountered differed materially from that expected. The second major difference was the location of the material within the limits of the trench excavation. There were two instances where the rock layers were indicated to be above the invert and were found to be in the invert. Similarly, an area where one layer of rock was indicated sometimes contained two or more layers. In some cases, the condition actually improved from that expected.

The production rates for the actual conditions encountered contained the usual everyday occurrences such as utility interferences, setting trench jacks, and so on. However, for those occasions when unusual or prolonged delays occurred, such as broken water line, floods, relocating past tunnels, crossing intersection, and major repairs, the data had been excluded from the production rate development so as not to distort or overlap costs charged elsewhere, and to be able to exclude noncompensable costs, as well!

An excavation production analysis showing the historical production rates achieved for the various conditions encountered is included for review and used in evaluating this proposal. Application of the data included in this analysis yields a performance time of 3,565 hours for the expected conditions and 4,507 hours for the actual conditions for a net extra of 942 crew hours. For the two-crew setup employed, this represents 471 job hours, 59 workdays, and 83 calendar days.

A summary of the increased costs resulting from the extended performance period and some typical examples of calculations of these costs are included in the following sections.

COST SUMMARY

	LABOR	EQUIPMENT	SUPPLIES	TOTAL
Extended Operations	$534,071	$350,340	$118,744	$1,003,155
Cost Escalation	57,005		42,770	99,775
Consultants' Expense			11,101	11,101
Subcontractor Costs			79,455	79,455
Backfill Costs			132,408	132,408
SUBTOTAL	$591,076	$350,340	$384,478	$1,325,894
Overhead @ 61.3% of Labor				362,330
SUBTOTAL				$1,688,224
Profit @ 10%				168,822
SUBTOTAL				$1,857,046
Home Office Expense @ 5.16%				95,824
SUBTOTAL				$1,952,870
Additional Bond @ $5.09/1000				9,940
SUBTOTAL				$1,962,810
Less Reimbursed Labor Escalation*				(102,523)
TOTAL				$1,860,287

*Contract provides for escalation on labor including changes. Therefore, previous payments are a credit to the change order.

EXTENDED OPERATIONS

	LABOR	EQUIPMENT	SUPPLIES	TOTAL
Steel Pipe Crew				
Nov. 9, 1989 to June 30, 1990				
Labor $596.40 × 380 Hrs.	$226,632			
Equipment $386.92 × 380 Hrs.		$147,030		
Supplies $132.86 × 380 Hrs.			$ 50,487	
July 1, 1990 to Sept. 4, 1990				
Labor $645.31 × 76 Hrs.	$ 49,044			
Equipment $386.92 × 76 Hrs.		$ 29,406		
Supplies $146.07 × 76 Hrs.			$ 11,101	
Second Shift Crew				
May 14, 1990 to June 30, 1990				
Labor $252.88 × 46 Hrs.	$ 11,632			
Equipment $264.25 × 46 Hrs.		$ 12,156		
Supplies $50.21 × 46 Hrs.			$ 2,310	
July 1, 1990 to August 13, 1990				
Labor $272.13 × 9 Hrs.	$ 2,449			
Equipment $264.25 × 9 Hrs.		$ 2,378		
Supplies $55.41 × 9 Hrs.			$ 499	
RCP Crew				
January, 1989 to June 30, 1989				
Labor $530.44 × 69 Hrs.	$ 36,600			
Equipment $363.86 × 69 Hrs.		$ 25,106		
Supplies $116.70 × 69 Hrs.			$ 8,052	
July 1, 1989 to Sept. 25, 1989				
Labor $562.91 × 72 Hrs.	$ 40,530			
Equipment $363.86 × 72 Hrs.		$ 26,198		
Supplies $125.46 × 72 Hrs.			$ 9,033	
Sept. 26, 1989 to Mar. 27, 1990				
Labor $562.91 × 297 Hrs.	$167,184			
Equipment $363.86 × 297 Hrs.		$108,066		
Supplies $125.46 × 297 Hrs.			$ 37,262	
TOTAL	$534,071	$350,340	$118,744	$1,003,155

CREW HOURS
SUMMARY

	CREW HOURS RQ'D W/O EXCESSIVE ROCK	CREW HOURS RQ'D WITH CHANGED CONDITION (ROCK)	ADDITIONAL CREW HOURS
Steel Pipe			
Nov. 9, 1989 to June 30, 1990	1,420	1,840	420
July 1, 1990 to Sept. 4, 1990	939	1,023	84
SUBTOTAL STEEL	2,359	2,863	504
January, 1989 to June 30, 1989	179	248	69
July 1, 1989 to Sept. 25, 1989	328	400	72
Sept. 26, 1989 to Mar. 27, 1990	699	996	297
SUBTOTAL RCP	1,206	1,644	438
PROJECT TOTAL	3,565	4,507	942

CALCULATION OF SECOND SHIFT EXCAVATION CREW HOURS

A second shift excavation crew was worked between May 14 and August 13, 1990. A total of 63 second shifts were worked during this period with a production of approximately 20,600 cy of excavation for the steel pipe. This excavation quantity represents 9.6% of the total steel pipe excavation.

The additional steel pipe crew hours shown above are therefore adjusted as follows:

	CREW HOURS	FACTOR	2ND SHIFT WORK HOURS	PAY HOURS	FULL CREW HOURS
11-9-89 to 6-30-90	420	9.6%	$40 \times 8/7 = 46$		380
6-1-90 to 9-04-90	84	9.6%	$8 \times 8/7 = 9$		76
TOTAL	504	9.6%	$48 \times 8/7 = 55$		456

SUMMARY
EXCAVATION PRODUCTION

CLASS CODE	DESCRIPTION	WORK DAYS	HOURS	QUANTITY CY	PRODUCTION CYH
1	Wet Ground/Sheet Pile Area			9,680	N/A
2	Misc. Disruptions			48,569	N/A
3	Common Exc.	52	416	47,589	114
4	Caliche in Invert	92	736	39,623	54
5	Caliche in Trench < 2'	69	552	54,256	98
6	Caliche in Trench > 2'	131	1,048	90,564	86
7	Caliche Inv. & 1 Layer < 2'	40	320	15,072	47
8	Caliche Inv. & 1 Layer > 2'	59	472	21,686	46
9	Caliche 2 or more Layers in Trench	55	440	38,598	88
	TOTAL	N/A	N/A	365,637	N/A

N/A → Not available

PRODUCTION SUMMARY
EAST HEADING

SHEET NO.	CASE 1 WD	CASE 1 CY	CASE 2 WD	CASE 2 CY	CASE 3 WD	CASE 3 CY	CASE 4 WD	CASE 4 CY	CASE 5 WD	CASE 5 CY	CASE 6 WD	CASE 6 CY	CASE 7 WD	CASE 7 CY	CASE 8 WD	CASE 8 CY	CASE 9 WD	CASE 9 CY
1	7	1,258																
2	22	4,433																
3	21	278																
4	19	1,607																
5	23	782																
6	20	1,322																
7	3	0	17	8,360	3	3,238												
8	5	0	17	8,112														
9			9	5,317	11	9,586	3	1,489										
10			2	506	5	4,334			7	6,113	5	3,975						
11			5	2,211														
12			6	4,165			3	978			9	2,159	1	546				
13					3	2,413			3	2,194	10	6,236					3	1,552
14					2	1,553	1	330	3	2,312	9	5,913					8	5,344
15			3	797			1	518			8	6,445			7	3,266	3	2,120
16			1	394	1	1,077			3	2,325	6	3,912			4	2,157	7	4,044
17			2	946														
18					1	1,443	7	3,621			7	3,879	3	1,463	5	1,920		
19							8+2	3,652			5+2	3,944	7+7	5,094			2	1,209
20							1	552	6	5,536	11	9,066					1	1,009
21			1	-0-	5	4,744	6	4,441	1	1,135	3	3,436	3	912			2	1,290
22									12	7,465	18	12,019	1	742	7	352		
23							2	962			5	3,648						
EAST TOTAL	120	9,680	63	30,808	31	28,388	34	16,543	35	27,080	98	64,632	22	8,757	23	7,695	33	21,065

PRODUCTION SUMMARY
WEST HEADING

SHEET NO.	CASE 1		CASE 2		CASE 3		CASE 4		CASE 5		CASE 6		CASE 7		CASE 8		CASE 9	
	WD	CY	WD	CY	WD	CY	WD	CY	WD	CY	WD	CY	WD	CY	WD	CY	WD	CY
1			2	532									5	796	8	851		
2			4	0			3	1,102	1	1,050			12	5,463	4	1,117		
3			1	0			3	1,342	2	1,344								
4			3	0			13	6,272			4	1,987						
5			5	1,126	1	1,065	13	5,892			4	1,674						
6			16	2,200	3	2,725	2	1,163										
7			10	4,903	5	4,668	2	1,748	3	1,828								
8			5	4,156	6	5,474	1	762	7	6,647	2	1,814			2	1,231		
9			4	0			15	3,513										
10					1	935	6	1,286	2	1,637	2	1,757			5	578	8	6,657
11					1	1,067					1	599	1	56	13	7,408	3	2,423
12									1	857	14	12,743			3	2,806		
13			6	1,682					5	5,200	4	3,531					8	6,556
14			11	2,653	4	3,267			3	2,339							3	1,897
15			7	509					10	6,274	2	1,827						
WEST TOTAL	—	—	74	17,761	21	19,201	58	23,080	34	27,176	33	25,932	18	6,315	35	13,991	22	17,533
EAST TOTAL	120	9,680	63	30,808	31	28,388	34	16,543	35	27,080	98	64,632	22	8,757	23	7,695	33	21,065
JOB TOTAL	120	9,680	137	48,569	52	47,589	92	39,623	69	54,256	131	90,564	40	15,072	58	21,686	55	38,598

STEEL PIPE CREW HOURS WITHOUT EXCESSIVE ROCK
What Should Have Been (typical reach)

BEGINNING STA	END STA	LENGTH	AVG WIDTH	AVG CUT	CY	CLASS NO	CY/HR	3	4	5	6	7	8	9	TOTAL
9-25-89 to 6-30-90															
	1009+70	146	11.6	20.3	1,273	3	114	11							
	1010+70	100	11.8	22.2	970	5	98			9					
	1020+75	1,005	11.9	21.6	9,125	3	114	85							
	1022+75	200	11.9	21.5	1,895	4	54		35						
	1033+00	1,025	11.7	21.8	9,683	3	114	85							
	1036+50	350	11.5	18.1	2,698	6	86				31				
	1048+50	1,200	11.7	14.2	7,394	3	114	65							
	1051+50	300	11.6	17.5	2,256	6	86				26				
	1061+50	1,000	11.9	16.1	7,096	3	114	62							
	1064+50	300	12.0	16.9	3,004	6	86								
	1075+00	1,050	12.4	16.0	7,716	3	114	68							
	1078+00	300	12.3	14.8	2,023	6	86				24				
	1094+00	1,600	12.4	16.5	12,124	3	114	106							
	1095+10	110	14.4	21.3	1,250	6	86				15				

CREW HOURS BY CLASS

136

Station		Station											
1096+14	Tunnel	1118+00	2,186	17.7	18.9	19,433	9	88					221
1118+00		1122+88	488	11.8	19.6	4,180	6	86				49	
1164+10		1171+00	690	12.0	15.2	4,661	3	114	41				
1171+00		1173+00	200	12.1	16.3	1,464	9	88					17
1173+00		1175+50	250	11.9	16.8	1,851	3	114	16				
1175+50		1177+00	150	11.8	17.0	1,114	9	88					13
1177+00		1181+00	400	11.6	16.4	2,818	3	114	25				
1181+00		1183+00	200	11.5	15.7	1,337	5	98			14		
1183+00		1194+00	1,100	11.7	19.0	9,057	3	114	79				
1194+00		1197+00	300	11.4	15.9	2,014	5	98			21		
1197+00		1202+22	522	11.5	16.8	3,735	4	54		69			
1204+14	Tunnel	1211+00	686	12.0	15.0	4,573	4	54		85			
1211+00		1224+90	1,390	11.0	17.5	9,910	6	86				115	
1228+20	Tunnel	1228+58	38	11.5	21.8	353	3	114	3				
TOTAL									646	189	44	260	251
													1,420

STEEL PIPE CREW HOURS REQUIRED WITH CHANGED CONDITION (ROCK)
Actual (typical reach)

NOVEMBER, 1989 TO JUNE 30, 1990

BEGINNING STA	END STA	LENGTH	AVG WIDTH	AVG CUT	CY	NO	CLASS CY/HR	3	4	CREW HOURS BY CLASS 5	6	7	8	9	TOTAL
915+50	1008+24	CCN#17 Area													
1008+24	1008+85	61	11.5	19.7	512	4	54		9						
1008+85	1009+45	60	11.7	21.0	546	7	47					12			
1009+45	1009+95	50	11.7	21.5	466	4	54		9						
1009+95	1012+10	215	11.9	22.8	2,159	6	86				25				
1012+10	1012+65	55	12.0	22.8	557	9	88							6	
1012+65	1013+45	80	12.0	22.4	796	5	98			8					
1013+45	1014+50	105	12.0	21.3	995	9	88							11	
1014+50	1016+15	165	11.7	20.0	1,429	6	86				17				
1016+15	1016+94	79	11.8	20.3	701	3	114	6							
1016+94	1017+75	81	12.0	20.6	742	5	98			8					
1017+75	1019+15	140	11.8	20.9	1,281	6	86				15				
1019+15	1020+05	90	12.0	21.2	848	3	114	7							

1020+05	1020+90	85	12.0	21.6	816	6	86			9	
1020+90	1021+80	90	12.0	21.6	864	3	114			32	
1021+80	1024+60	280	11.7	22.5	2,710	6	86	8			
1024+60	1027+10	250	11.7	22.9	2,483	5	98		25	20	24
1027+10	1029+35	225	11.7	22.0	2,136	9	88				
1029+35	1031+30	195	11.5	20.6	1,713	6	86				
1031+30	1033+10	180	11.6	20.1	1,553	3	114	14			
1033+10	1034+00	90	11.7	18.9	737	9	88				8
1034+00	1035+45	145	11.6	17.4	1,083	5	98		11		
1035+45	1037+65	220	12.0	15.9	1,557	9	88				18
1037+65	1039+45	180	11.8	14.7	1,157	6	86			13	
1039+45	1040+00	55	11.5	14.1	330	4	54		6	7	
1040+00	1040+95	95	11.5	14.1	571	6	86				
1042+40	1043+36	96	11.6	13.7	565	9	88				6
1043+36	1055+20	1,184	11.6	15.9	8,105	6	86		10	94	
1055+20	1055+95	75	11.5	16.2	518	4	54				
1055+95	1058+80	285	11.7	15.3	1,895	8	46				41
1058+80	1059+95	115	11.7	16.3	812	6	86			9	

COST ESCALATION DUE TO EXTENDED OPERATIONS
Summary

	LABOR	EQUIPMENT	SUPPLIES	TOTAL
Line Crews	$40,898			
Miscellaneous Crews	16,107			
Supplies			$32,350	
Backfill Material			10,420	
TOTAL	$57,005		$42,770	$99,775

ESCALATION DUE TO EXTENDED OPERATIONS

Line Crews

As a result of the delays described herein, the work force was moved into periods of wage escalation to a greater degree than originally anticipated. The extra costs resulting from this shift are as detailed below:

Hourly Cost	7/1/88 RATE	7/1/89 RATE	INCREASE	7/1/90 RATE	INCREASE
Steel Pipe Crew	$459.57	$492.08	$32.51	$540.99	$48.91
Second Shift Crew		185.96		205.21	19.25
RCP Crew	432.21	464.68	32.47		

Escalation Calculation

	HOURS	RATE	AMOUNT
Steel Pipe Crew			
7/1/88 to 7/1/89	35	$32.51	$ 1,138
7/1/89 to 7/1/90	749	48.91	36,634
Second Shift Crew	46	19.25	886
RCP Crew			
7/1/88 to 7/1/89	69	32.47	2,240
TOTAL			$40,898

MISCELLANEOUS CREWS WAGE ESCALATION

	7/1/88 HOURLY RATE	7/1/89 HOURLY RATE	7/1/90 HOURLY RATE
Laborers 4 Each	$ 56.36	$ 61.40	$ 65.96
Case 580 B/H Operator @ 1/3	6.21	6.50	7.31
Motor Grader Operator	18.75	19.65	22.06
Mechanics 4 Each	75.04	78.60	88.48
Service Truck Operator	18.23	19.13	21.53
Carpenter	17.02	19.84	20.04
Teamster @ 1/2	6.77	7.49	8.07
CREW TOTAL	$198.38	$212.61	$233.45
Hourly Rate Increase	N/A	$ 14.23	$ 20.84

Wage Escalation Calculation

	HOURS	RATE	AMOUNT
7/1/88 to 7/1/89 Rate	35	$14.23	$ 498
7/1/89 to 7/1/90 Rate	749	20.84	15,609
TOTAL			$ 16,107

SUPPLY ESCALATION

Supply Cost—One Year Period Ending March 31, 1989	$352,949
Annual Escalation Rate — 22%	
Annual Escalation Amount	$ 77,649
Delay Period — 5 months	
Delay Escalation	
$77,649 ÷ 12 months = $6,470/month × 5 months =	$ 32,350

Subcontractor Costs

As a result of the extra rock encountered on this project during construction of the steel pipeline, Ajax Welding incurred extra cost for which we request reimbursement. Excavation was slowed by the extra rock quantity which in turn slowed Ajax Welding crew and caused an inefficient operation in the field.

Based on the extent of the rock changed condition defined by a comparison of actual conditions with anticipated conditions, we have calculated a crew delay of 11 weeks. The cost of this delay, slowed production, and inefficiency is as follows:

11 weeks × 40 hours per week = 440 hours	
440 crew hours × $142.75 per hour	$62,810
Overhead, 15%	9,422
Subtotal	72,232
Profit 10%	7,223
TOTAL	$79,455

The following is a tabulation of the hourly crew cost:

GENERAL WELDING CREW COST

Labor	
1 — Foreman	$ 24.72/hr.
1 — Welder (D)	23.83/hr.
1 — Welder (S)	23.83/hr.
TOTAL HOURLY LABOR COST	$ 72.38/hr.
Equipment	
1 — Pickup	$ 7.20/hr.
1 — Pickup, 1 T	8.73/hr.
1 — Welding Tr., 1 T	14.25/hr.
2 — Fans @ $5.48/hr.	10.96/hr.
3 — Welders 200 A G @ $5.36/hr.	16.08/hr.
1 — Welder 250 A D @ $4.58/hr. × ½	2.29/hr.
TOTAL HOURLY EQUIPMENT COST	$ 59.51/hr.
S T & S @ 15% Labor	10.86/hr.
TOTAL HOURLY LABOR COST	72.38/hr.
TOTAL HOURLY CREW COST	$142.75/hr.

EQUIPMENT RATES

AJAX WELDING
---

1. Pickup, 1/2 Ton
    (AGC p. 45) $7,500

    Depr.   = 30%
    Tax & Ins. = 3%
    Annual Ownership 36%

    Hourly Ownership = 33% ÷ 1800 =   .0183%
    Hourly Repair & Maintenance Expense = .0310%
    Subtotal   .0493%

    Acquisition Cost = $7,500

    Hourly Ownership, Repair and Maintenance = $3.70/Hr.

    Estimated Operating Expense

    Fuel   1.50
    Oil and Grease   0.10
    Parts and Supplies 0.90
    Service Labor   0.75
    Tires   0.25

    TOTAL HOURLY COST = $7.20/Hr.

2. Pickup, 1 Ton $9,500

    Depr.   = 22.5%
    Taxes & Ins. = 3 %

    Hourly Ownership + 25.5% ÷ 1250   = 0.0204%
    Hourly Repair & Maintenance Expense = 0.0310%
    Total Ownership, Repair and Maintenance = 0.0514%

    Hourly O R & M = $9,500 Acquisition × .0514% = $4.88/Hr.

    Estimated Operating Expense

    Fuel   1.50
    Oil and Grease   0.10
    Parts and Supplies 1.00
    Service Labor   0.75
    Tires   0.50

    TOTAL HOURLY COST = $8.73/Hr.

3. Welding Truck $18,000

    Hourly O, R, and M (same as Item 2) = 18,000 × .0514% = $9.25/Hr.

    Estimated Operating Expense

    Fuel   2.00
    Oil and Grease   0.20
    Parts and Supplies 1.20
    Service Labor   1.00
    Tires   0.60

    TOTAL HOURLY COST = $14.25/Hr.

4. Blower $6,000
    (AGC p. 51)

                Depr.   ≈ 20%
                Taxes & Ins. = 3%
                Total   23%

    Hourly Ownership = 23% ÷ 2000 = 0.0115%
    Hourly Repair & Maintenance   = 0.0081%
    Total   0.0196%

    Hourly O, R & M Cost = 6,000 × .0196% = $1.18/Hr.

        Estimated Operating Expense

            Fuel   2.50
            Oil and Grease   0.10
            Parts and Supplies 0.90
            Service Labor   0.75
            Tires   0.05

    TOTAL HOURLY COST = $5.48/Hr.

5. Welders, 200 Amp Gas $3,000
    (AGC p. 52)

                Depr.   = 18%
                Taxes & Ins. = 3%
                Total   21%

    Hourly Ownership = 21% ÷ 1400 = 0.0150%
    Hourly Repair & Maintenance   = 0.0588%
    Total O, R & M Rate   0.0738%

    Hourly O, R & M = 3,000 × .0738% = $2.21/Hr.

        Estimated Operating Expense

            Fuel   1.80
            Oil and Grease   0.10
            Parts and Supplies 0.50
            Service Labor   0.75

    TOTAL HOURLY RATE = $5.36/Hr.

6. Welder 250 Amp Diesel $4,200
    (AGC p. 52)

        Hourly Ownership   = 0.0150%
        Hourly Repair & Maintenance = 0.0261%
        Total O, R & M Rate   0.0411%

    Hourly O, R & M Rate = $1.73/Hr.

        Estimated Operating Expense

            Fuel   1.50
            Oil and Grease   0.10
            Parts and Supplies 0.50
            Service Labor   0.75

    TOTAL HOURLY RATE = $4.58/Hr.

LABOR RATES

AJAX WELDING	FOREMAN	WELDER
Base	$15.96	$15.21
Vacation	1.00	1.00
Health & Welfare	1.18	1.18
Pension	1.25	1.25
Apprenticeship Fund	0.04	0.04
FICA 6.13%	0.98	0.93
SUI 4.8%	0.77	0.73
FUI 0.7%	0.11	0.11
Subsist $19/Day	2.38	2.38
W/C 6.58%	1.05	1.00
TOTAL RATE	$24.72/Hr.	$23.83/Hr.

REIMBURSED LABOR ESCALATION

CREW	HOURS	REIMBURSED HOURLY RATE	REIMBURSED AMOUNT
1) *Steel Pipe Crew*			
Nov. 9 to June 30, 1990	380	$60.76	$ 23,089
July 1 to Sept. 4, 1990	76	$92.58	$ 7,036
2) *Second Shift Crew*			
May 14 to June 30, 1990	46	$26.90	$ 1,237
July 1 to August 13, 1990	9	$39.28	$ 354
3) *RCP Crew*			
Sept. 26, 1989 to Mar. 27, 1990	297	$58.66	$ 17,422
4) *Escalation Section*			
a) Line Crew			
Steel Pipe Crew	749	$36.68	$ 27,473
Second Shift Crew	46	$14.44	$ 664
b) Miscellaneous Crew			
Sept. 26, 1989 to June 30, 1990	69	$17.58	$ 1,213
July 1, 1990 to Nov. 20, 1990	749	$32.09	$ 24,035
TOTAL			$102,523*

*This number is on summary page of cost.

STEEL PIPE CREW (typical)
Reimbursed Labor Escalation
July 1, 1990 through September 4, 1990

DESCRIPTION	#EACH	HOURLY RATE	HOURLY ESCALATION AMOUNT	HOURLY R & S LABOR AMOUNT
Laborers Group 3	13	$3.20	$ 41.60	
Operators	10.3	4.07	41.92	
Operators—HDR	2	4.12	8.24	
Teamsters	3.5	2.79	9.77	
R & S Labor				$104.32
TOTALS			$101.53	$104.32

Field Labor Escalation
$101.53 × 75% Reimbursement — $ 76.15
R & S Labor
$104.32 × 21% = $21.91 × 75% Reimbursement — $ 16.43
TOTAL Reimbursed Escalation Per Hour — $ 92.58

STEEL PIPE CREW COSTS (typical)
Rates Effective 7/1/90

	LABOR	EQUIPMENT	SUPPLIES	TOTAL
Labor:				
Labor Foreman	$ 14.55			
Laborers 12 @ $14.09	169.08			
Operator Foreman	19.54			
245 Backhoe Operator	18.62			
580 B/H Operator 1/3 @ $18.62	6.21			
Loader Operator 2 @ 18.62	37.24			
Crane Operator	18.76			
Travel-Lift Operator	18.62			
Roller Operator	18.62			
Dozer/Roller Operator	18.62			
Oiler 2 @ 17.40	34.80			
Mechanic 2 @ 18.76	37.52			
Teamsters 3½ @ 13.54	47.39			
SUBTOTAL	$459.57			
Supplies:				
Misc. Supplies @ 27% of Labor			$124.08	

STEEL PIPE CREW COSTS (continued)

	LABOR	EQUIPMENT	SUPPLIES	TOTAL
Equipment:				
Water Truck @ 1/2	$ 2.92	$ 5.51		
Mechanic Truck	1.21	6.43		
Compressor 175 cfm	1.70	6.48		
245 Backhoe	18.25	68.35		
580 Backhoe @ 1/3	2.84	5.18		
966 Loader—2 ea.	12.18	46.12		
LS-98 Crane	7.30	27.05		
Travel-Lift	4.39	31.78		
Roller, Single Drum	10.95	40.83		
Roller, Double Drum	7.30	29.28		
Dozer, D-3	6.09	18.75		
Dump Truck, 3 ea.	29.19	101.16		
TOTAL	$563.89	$386.92	$124.08	$1,074.89

This is a typical breakdown of a crew cost.

SAMPLE CLAIM AND CALCULATIONS/ACCELERATION

Example of Method of Computing Contractor's Claim

Description of Facility.   Commercial, light processing plant with some office and retail space. Site description and full set of plans and specifications were provided by the owner. The contract was based on FAR documents.

The contract provided for a lump-sum bid including grading, parking areas, landscaping, and a structure to be ready for manufacturing and occupancy on completion of the project. Completion time of construction was set in the contract at 18 months (547 calendar days) from receipt of the owner's notice to proceed.

Actual completion date was 740 days after notice to proceed. Within 30 days of start of work, the contractor submitted a changed condition request concerning a foundation problem. Within a short period, another such request was submitted for another changed condition. In both cases, the owner agreed to a "direct cost plus overhead" arrangement for the changes but gave no extension of time. As the project progressed, there were a total of 65 change orders issued for such items as:

- differing site conditions
- owner's changes
- plan errors and omissions
- delayed review of shop drawings

Of the 65 change orders (COs), the parties agreed that 50 involved extensions of time and that most of the situations giving rise to these COs occurred during the first six months of the project. Of the 50 COs involving time extensions, only four were issued before the original contract completion date. Another two were issued before the actual completion date. The remaining 44 were issued after the project was accepted and beneficially occupied by the owner. The actual days of extension granted by the owner were as follows: 35 days were granted during the original contract completion period; five days were granted during the 193 days of time overrun; and 330 days were granted after the job was complete and accepted. Total was 370 days.

The owner, at the end of the first six months of work on the project, issued a letter to the contractor informing him that the contract was behind schedule and directing him to take the necessary steps to put the project back on schedule. In this letter, the contractor was directed to take such steps including:

1. Put on additional shifts.
2. Assign overtime.
3. Establish longer workweeks.
4. Increase construction plant.

The contractor was directed to take these steps at no additional cost to the owner, and no mention was made of the pending change orders that involved extension of time. A similar letter was sent by the owner to the contractor at approximately the twelve-month mark. In both cases, the contractor formally protested, mentioning the unresolved change orders, but agreed to comply. At the completion of the project, the contractor filed a claim for:

inefficiency due to acceleration	$1,361,097
increased overhead	912,154
price increases	
labor	25,380
materials	62,100
subcontractors	211,000
finance costs	100,000
administrative costs 5%	133,612
profit 15%	400,835
Subtotal	$3,206,678
bond .5%	16,033
Total Claim	$3,222,711

This was in addition to the amount paid by the owner for the 65 change orders.

Method of Computing Acceleration Costs

Direct Labor. The contractor, based on the first letter from the owner directing him to accelerate, stated that during the first six months, he was in a preacceleration period and was working at the normal rate of efficiency for his organization. The contractor used a standard estimating book to compute labor man hours. The handbook was one accepted by contractors working in the same locale and performing work of a similar nature. The contractor computed the man hours required to perform the actual work completed during the first six months of the contract. This work included:

rough grading

foundation excavation

footing excavation

pile driving

forming, pouring, stripping pile caps, reinforcing

forming, pouring, stripping grade beams

excavating, placing and backfilling drains

grade, form, pour slab on grade; form, pour, strip columns

lay brick in walls

In the interest of simplicity, the contractor separated the man hours into two classifications: skilled labor and unskilled labor. He computed 15,200 man hours of skilled labor and 6,100 man hours of unskilled labor.

Then, using job time cards and other records, covering the same time period and the same work, the contractor found that he had spent 21,700 man hours of skilled labor and 10,200 man hours of unskilled labor. Dividing the handbook man-hours by the actual man hours expended by the contractor, he arrived at the following:

$$\text{Skilled Labor} \quad \frac{15,200}{21,700} = 70\%$$

$$\text{Unskilled Labor} \quad \frac{6,100}{10,200} = 59.8\%$$

The contractor stated that these percentage figures were his factors of efficiency as compared to the handbook figures for ideal productivity, for a period during which the contractor claimed that he was working at his normal rate of efficiency.

This process was repeated for the remaining work on the project during the time the contractor stated that he had been accelerating.

From the handbook, the contractor computed 30,100 man hours of skilled labor and 21,400 man hours of unskilled labor as the labor theoretically required to perform the work. Then, applying the efficiency factors previously computed, he obtained:

Skilled: 30,100 divided by 70% = 43,000 man hours
Unskilled: 21,400 divided by 59.8% = 35,790 man hours

The contractor stated that this is what it would have taken to complete the project if he were not forced into the inefficiencies of acceleration.

When he went to his time cards and other records for the same period, he found that the project had used 93,150 man hours of skilled labor and 54,400 man hours of unskilled labor. The difference between the time card figures and the adjusted handbook figures were as follows:

Skilled Labor	Unskilled Labor
93,150 man hours	54,400 man hours
43,000 man hours	35,790 man hours
50,150 man-hours	18,610 man hours

The contractor computed his average labor rates for skilled and unskilled labor including all add-ons applicable to labor for insurance, FOAB, union payments, unemployment, and vacation fund (but no labor rate increases). These figures were $21.50 and $55.20, respectively, which were applied as follows:

Skilled	50,150 MH × $21.50	=	$1,078,225
Unskilled	18,610 MH × $15.20	=	282,872
	Total		1,361,097

This, the contractor stated, was the increase in direct labor cost due to the inefficiency of acceleration. (This is the first item on the claim listed earlier in this section.)

Increased Overhead

By audit of his books, the contractor found that during the first six months (preacceleration period) his overhead was $2,500.00 per day based on a seven-day week. This was made up of 31 items, which included: travel expense, postage, utilities, job photographs, overhead salaries, watchmen, and office supplies. For the original eighteen-month (547 days) project, his overhead would have been:

$2,500.00 × 547 days = $1,367,500.

By audit of his actual costs for the original 18 months, plus 193 days overrun for a total of 547 days, the contractor found his actual overhead was $2,590,750. He subtracted:

$$\begin{array}{r} \$2,590,750 \\ -\ 1,367,500 \\ \hline \$1,223,250 \end{array}$$

with the difference being his increase in overhead due to acceleration and delay. However, the change orders issued by the owner included overhead costs totaling $310,596. He subtracted:

$$\begin{array}{r} \$1,223,250 \\ -\ 310,596 \\ \hline \$912,654 \end{array}$$

The result was the net claim for overhead due to acceleration. (Refer back to the claim on page 147 of this section.)

Price Increases

Labor. The average labor rates used in previous computations did not include any wage increases. Therefore, this portion of the claim was arrived at by calculating the number of man-hours expended in a later wage period than had been originally contemplated, by the labor escalation, including add-ons, in each such period. This included labor performed both during the contract period and afterward to the extent that it was performed in a higher wage period than it should have been, had it not been for the delay.

Material. This includes the cost of any increase in material costs after original completion date, using audited invoices and previous sales contracts or quotations. If the CPM spot-points the delivery of materials, and there was sufficient delay due to change orders or other causes, and material prices increased significantly during the delay, these also are included.

Subcontracts. Subcontractors computed increased labor and material cost as previously, and included overhead and profit.

Finance Charges. These included the cost of financing the additional work due to the change orders, together with delayed payments and cash flow accounts. An audit of the contractor's bank statements should support this.

Administrative Cost. This refers to main office costs, which are applicable to the project and which are based on the main office work load and estimate of involvement in the additional work claimed, or the use of a formula such as the Eichleay Formula.

Profit. This is the normal percentage applied by contractors.

Bond. This is based on terms of the bonding agreement between the contractor and the bonding company.

WORKSHOP PROBLEM: DELAYS AND ACCELERATION

The following problem is used as a classroom exercise during the Construction Claims course taught by the authors of this book. The example, involving highway construction, illustrates some principles about delay and acceleration, and is used in the course to elicit discussion. Therefore, there are no absolute "right" answers to the questions posed. The problems and the solutions offered are intended to provide some useful insights, and the exercise is a valuable one.

The Project

During the past ten years, the City of Warden had been constructing a six-lane urban highway across the southern portion of the city. Appropriately named Southern Parkway, the remaining segment needed to complete the route, which was across a narrow 50'-deep rock gully through which the Jones Falls stream flows.

The project consisted of the following:

1.   From Martin Avenue to Garden Drive (see pages 153 and 154).
2.   Project length: Sta. 50+25 to Sta. 59+10 = 885 lf.
3.   Six-lane divided urban highway.
4.   Triple-cell 15'-span × 9'-rise concrete box culvert.
5.   Relocation of a 36" high-pressure water line and construction of a pumping station.

The City of Warden retained the services of a consultant engineering firm to design the project, develop specifications, and provide construction estimates. The project was to be advertised by the Department of Public Works.

Although short in length, the project required a sequence of operations that must accommodate the temporary relocation of a major stream and existing 36;dp high-pressure water line (HPWL) through solid rock, the construction of a triple-cell concrete box culvert and new pumping station for the water line. Since this project was to be advertised in the Spring of 1991, the following completion dates were included in the contracts.

1.   Completion of the temporary relocation of a stream and HPWL: March 10, 1992.
2.   Completion of concrete box culvert and pumping station foundation: December 1, 1992.
3.   Completion of embankment and paving for the highway, and completion and testing of HPWL: December 15, 1992.
4.   Balance of work under contract: June 1, 1994.

The contract contained the following liquidated damages provisions:

(a) For failure to meet the March 10, 1992 date
(b) For failure to meet the December 1, 1992 date
(c) For failure to meet the December 15, 1993 date
(d) For failure to meet the June 1, 1995 date

The City of Warden budgeted the necessary funds and proceeded to advertise the contract for public bids on March 1, 1991, with bids to be received on April 14, 1991. Due to problems of coordination with the City of Warden's sewerage bond program, the bid opening was rescheduled for May 14, 1991.

The Gamma Construction Co. was the certified low bidder.

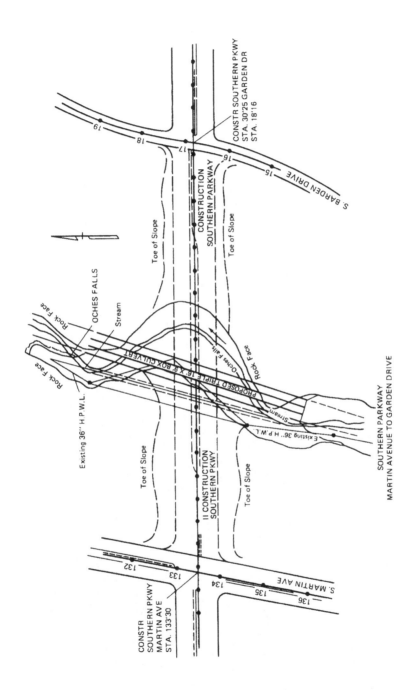

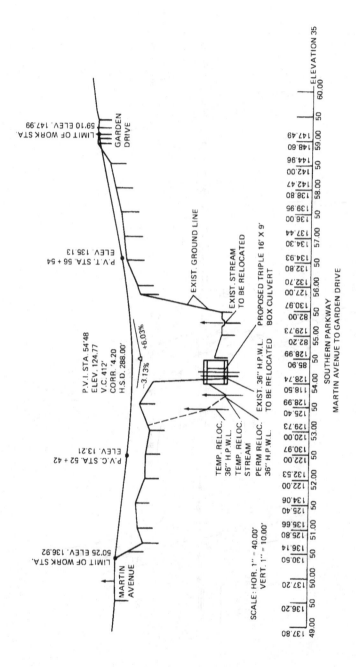

SCALE: HOR. 1″ = 40.00′
VERT. 1″ = 10.00′

P.V.I. STA. 54+48
ELEV. 124.77
V.C. 412′
CORR. 4.20
H.S.D. 288.00′

+6.03%
−3.13%

LIMIT OF WORK STA. 59+10 ELEV. 147.99
GARDEN DRIVE

P.V.T. STA. 56+54
ELEV. 135.13

EXIST. GROUND LINE

EXIST. STREAM
TO BE RELOCATED

PROPOSED TRIPLE 16′ × 9′
BOX CULVERT

EXIST. 36″ H.P.W.L.
TO BE RELOCATED

TEMP. RELOC.
36″ H.P.W.L.
TEMP. RELOC.
STREAM
PERM. RELOC.
36″ H.P.W.L.

P.V.C. STA. 52+42
ELEV. 13.21

LIMIT OF WORK STA. 50+25 ELEV. 136.92
MARTIN AVENUE

SOUTHERN PARKWAY
MARTIN AVENUE TO GARDEN DRIVE

ELEVATION 35

| 60.00 | 50 | 59.00 | 50 | 58.00 | 50 | 57.00 | 50 | 56.00 | 50 | 55.00 | 50 | 54.00 | 50 | 53.00 | 50 | 52.00 | 50 | 51.00 | 50 | 50.00 | 50 | 49.00 |

147.49
148.60
144.96
142.00
142.47
138.80
139.95
136.00
137.44
134.30
134.93
132.80
132.70
127.00
130.97
82.00
129.73
82.20
128.99
85.90
128.74
118.50
128.99
125.40
129.73
120.00
130.97
122.00
132.53
122.00
134.06
125.40
136.66
126.80
136.14
130.50
137.20
136.20
137.80

154

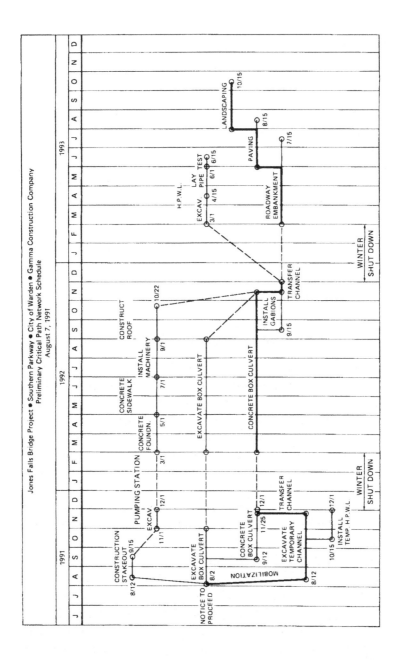

Jones Falls Bridge Project ● Southern Parkway ● City of Warden ● Gamma Construction Company
Preliminary Critical Path Network Schedule
August 7, 1991

155

Construction Progress

Notice to Proceed

Gamma was awarded the contract (using the EJDCC forms) and given Notice to Proceed on August 2, 1991.

Critical Path Method

In accordance with the contract proposal, Gamma submitted its preliminary CPM network schedule to the Department of Public Works on August 7, 1991, within 5 days of the Notice to Proceed.

This was a General Schedule, covering the major items of work, and was to be supplemented within 60 days by a detailed schedule containing at least 200 events.

Department of Public Works reviewed the schedule and returned it to Gamma on August 12, 1991, bearing the stamp "Reviewed and Accepted."

The engineer, in a transmittal letter, raised certain questions and asked for specifics to be included in the detailed schedule to follow.

Construction of Temporary Relocation of HPWL and Stream

On August 12, Gamma moved drilling and blasting equipment to the site that would be used to construct the temporary relocated HPWL and temporary relocated stream, which traversed a rock face. Work commenced on August 12, but was stopped on August 15, because of a lack of right-of-way clearance. Although the contract stated all right-of-way were available, a condemnation case became protracted over a title clearance of a property wherein lay most of the temporary relocation of HPWL and the stream.

The General Schedule submitted by Gamma and accepted by the Department of Public Works, had shown Gamma beginning excavation of both of the temporary relocations August 12, 1991. Although Gamma had proceeded with its mobilization and moved its drilling-blasting equipment on the site, its power shovel, loader, and back dump trucks did not arrive on the site until September 1, 1991.

The City obtained a clear title to the land on October 3, 1991. On October 4, 1991, Gamma was allowed to continue work. Gamma wrote the Department of Public Works informing them of its intentions to make a claim for additional time and damages because of failure of the City to turn over the property for the period of August 15, 1991, thru October 3, 1991.

Gamma's rock excavation proceeded at the rate indicated on the CPM. They were unable to improve on their progress as a result of the limited space within the gully. These conditions precluded adding additional equipment.

The temporary relocation of the HPWL was to be located on a bench cut into the west slope of the relocated stream channel. To maintain the horizontal and vertical alignment of the temporary relocated HPWL, the plans indicated vertical strap anchors every 50 feet and horizontal strap anchors every 100 feet. The straps were to be secured to anchors consisting of holes drilled 6 feet in the rock, filled with grout, in which a #6 bar was embedded. An item of "HPWL Anchors" was provided in the proposal bid on an "each" basis. No delineation was made between vertical and horizontal anchorages.

During the rock excavation, it became evident that the rock was more fractured and not as hard as anticipated. This condition resulted in a design change. Gamma was advised on November 15, 1991, that rock conditions required that "anchors" be placed vertically and horizontally every 25 feet.

Since the contract was a unit price contract, the Department of Public Works proposed to pay for the additional anchors at the existing price for "HPWL anchor." In the letter transmitting drawings to Gamma showing the changes, the Department of Public Works requested the contractor concur with these changes at the existing prices. Gamma did not concur, however, saying that its preliminary examination of the drawings indicated that there were changes which materially increased the unit price of the work. Gamma indicated it intended to submit a claim for added cost when it completed a thorough study.

Gamma's study indicated that the original conditions required twice as many "vertical" as "horizontal" anchors with the construction operations for a vertical anchor not as time-consuming as that of a horizontal anchor. The change conditions increased both items, however, and there were now as many horizontal as vertical anchors.

Gamma notified the Department of Public Works that its cost had increased materially, citing the conditions as stated and requesting a time extension to be determined. The engineer denied the request of added cost, again noting the unit price for HPWL anchors and stated that any request for time extension appeared to be without merit.

To expedite operations for the HPWL anchors, Gamma mobilized two Crane and Drill Rigs and one "Jury Rig," when compared with the initial rig, which was specifically designed for the operation.

The specifications for the contract had set a completion date of March 10, 1992, for the temporary relocation of the HPWL and stream. This completion date was predicted on the requirements of the State's Department of Wildlife Agency, which included a prohibition against working the stream during the trout fish spawning period from March 15 and June 15 of each year.

In December, 1991, Gamma realized that its present schedule of operations must be accelerated in order to meet the March 10, 1992 completion date. In light of the project's delayed start and the design changes for the HPWL anchors, Gamma instituted a six-day per week schedule, stretching work day by the use of flood lights. Gamma notified the Department of Public Works on January 2, 1992, of the delays and its added costs of premium time and special lighting equipment required.

This accelerated schedule resulted in the completion of the relocation of the HPWL on February 14, 1992. The HPWL was tested and accepted on February 18, 1992. The relocation of stream was completed on March 8, 1992.

Construction of the Triple-Bell Concrete Box Culvert and Pumping Station Foundation

Simultaneous to the construction of the temporary relocation of the HPWL and the stream, Gamma began drilling, blasting, and excavation of the alignment of the box culvert and pumping station foundation.

Gamma had submitted its shop drawings for the box culvert to the Department of Public Works on October 10, 1991, in anticipation of approval by October 20, 1991, which would allow a possible eight weeks of concrete placement, weather permitting.

The engineer advised Gamma of the approval of the shop drawings for the steel fabrication on October 19, 1991. Gamma advised its fabricator to proceed as rapidly as possible, in order that concrete might be placed as soon as possible.

On November 10, 1991, the engineering office of the Department of Public Works advised that a plan error had been discovered in the top slab transverse reinforcing. In an apparent drafting error, the plan indicated ;ns6 bars ;ca8;dp center to center, when it should have shown ;ns8 bar ;ca6;dp center to center. This fact was verified by the consultant's design computations.

Gamma had not as yet installed any of the incorrect reinforcing, and immediately advised its fabricator of the error. The fabricator advised that he had completed 10 percent of the top slab re-bars. Gamma advised the Department of Public Works of an anticipated time delay and the need to replace the incorrectly fabricated re-bars.

On November 12, 1991, the Department of Public Works advised Gamma that a design change would be made for the top slab where the lighter fabricated re-bars would be alternated with ;ns10 bars on 6;dp center to center.

On November 13, 1991, Gamma resubmitted the revised shop drawings and received approval of the modified re-bars design on November 15, 1991.

On December 1, 1991, the weather turned extremely cold and it snowed the entire day. Several days later, the weather warmed up and made working site extremely difficult. When it snowed again on December 8, Gamma petitioned for a partial shutdown, as it did not intend to concrete through the winter.

With the completion of the temporary relocation of the stream and the HPWL on March 8, 1992, Gamma now had the entire alignment of the box culvert ready for construction.

In order to meet the December 1, 1992 completion date for the concrete box culvert, it was necessary to get a good jump on the concrete operation. Weather conditions allowed work to begin on March 12, 1992. The December 1, 1992 completion date was critical for several reasons. Gamma had lost 4 weeks of concrete time in the fall of 1991, and needed a coordinated effort to complete this phase of the operation.

Although the winter temperatures in the Jones Falls Valley were such that excavation and some limited backfilling could be carried on in the daytime, temperatures at night dropped well below freezing between December 1 and March 1. Therefore, if concrete were to be placed during the period, winter protection would be necessary during mixing, transporting, placing, finishing, and curing. Gamma felt that its schedule allowed sufficient time to cover normal problems that might arise and that it could avoid the added costs connected with winter concrete.

Gamma, in the spring and summer of 1992, worked the box culvert operations six days per week, with two shifts. In general, form work was performed during the first shift and concrete placement during the second shift.

On July 8, 1992, the project site was swept by the worst flood that the Jones Falls Valley had experienced in over 40 years. Gamma's haul roads, stream crossings were washed out, and the work area inundated in spite of Gamma's diversion ditching and diking.

It was not until July 15, 1992, that Gamma was able to reestablish construction operations. Gamma notified the Department of Public Works of the delay by letter of July 16, 1992.

On October 16, 1992, the carpenters, who had been working without a labor contract, went on strike until October 23, 1992, when an agreement was reached. While they were on strike, all concrete operations came to a halt. Gamma notified the Department of Public Works of this delay in accordance with the contract.

As December 1, 1992 approached, it became quite evident that the contractor would still have some concrete to place.

On November 15, 1992, Gamma wrote the engineer informing him that it proposed to shut down concrete operations between December 1, 1992, and March 1, 1993. It further requested that the box culvert completion date and final completion date be extended each 64 days, citing as excusable reasons for the extension (1) failure of the owner to provide access, 50 days; (2) error in re-bar design and bad weather, 20 days; (3) storm in July 1992, 7 days; (4) carpenter strike, October 1992, 7 days.

The engineer denied the request for extension of time, saying that the specified completion dates were final and that Gamma must organize and schedule its remaining work so as to meet them. Gamma saw no alternative to working concrete through the winter and proceeded to mobilize for winter concrete. By December 1, 1992, Gamma had placed approximately 75 percent of the box culvert concrete; the pump station foundation was complete.

Gamma continued after December 1, 1992, but with noticeable reduction in production. The box culvert was completed on March 12, 1993. Gamma again requested an extension of time, this time requesting an additional 27 days for winter work. The engineer replied as before, that the specified completion dates must be met.

On March 15, 1993, the box culvert was inspected and accepted by the engineer. Diversion of the stream was not accomplished until June 16, 1993, in conformance with regulations of the State Wildlife Agency.

Relocation of HPWL and Pumping Station

On June 17, 1993, Gamma began excavation for the HPWL, utilizing the existing stream bed and the relocation channel for the temporary stream. This alignment required very little additional excavation and the first sections of pipe were laid from the upstream end. Gamma also commenced the construction of the pump station.

On July 2, 1993, the welders and pipe fitters walked off the job and remained on strike five days until July 7, 1993. During this period, no work was performed on the pipe line or the pump station.

Once the welders and pipe fitters returned to the site, the work progressed smoothly and the pipeline was completed on August 15, 1993, with the pump station completed on August 30, 1995. The HPWL was tested and accepted on September 9, 1993.

Embankment and Paving for the Highway

When it became apparent that it was not going to be able to divert the stream thru the box culvert until June 16, 1993, Gamma made plans to start the major roadway embankment on either side of the stream, leaving a gap in the center. Gamma began embankment operations on March 15, 1993, but instead of a single run working from the bottom up, it was limited to two sections. The preliminary embankment was completed on May 15, 1993.

Gamma commenced to close the embankment gap on September 10, 1993, and completed the embankment on October 21, 1993.

Gamma began its paving operations on October 1, 1993, and completed same by December 1, 1993.

Landscaping

Following the winter, the top soil and landscaping were completed by May 15, 1994, well in advance of the specified completion date.

Liquidated Damages

The Department of Public Works withheld $20,800 (104 days ;ca $200 per day) "pending final determination of the extent of liability for liquidated damages" connected with Gamma's failure to complete the concrete box culvert and pumping station foundation on December 1, 1992, as required.

PROBLEMS

Problem Number 1

a.   List each of the excusable delays under the Gamma construction Company contract.

b.   Identify each excusable delay as warranting an extension of time, or as concurrent.

c.   Identify each excusable delay as being compensable or noncompensable.

d.   Where enough information is given in the problem, determine the total time extension, if any, to which Gamma is entitled.

Problem Number 2

a.   Identify all areas where Gamma incurred damages from causes for which the engineer and/or owner are responsible.

b.   For each of the claim areas above, list the kinds of damages for which Gamma is entitled to receive compensation, i.e., labor escalation, loss of production, etc.

   c.   Describe the methods to be used in establishing the various items of damage, and the records that would be required to compute the amount.

SOLUTION

In general, there is insufficient information to make a conclusive determination. One purpose of the exercise is to determine what additional information is required.

Excusable Delays

Excusable Delays on the Critical Path

All delays on the critical path warrant an extension of time. The are additive and determine the job total time extension.

   1.   Lack of Right-of-Way Clearance

       Although the contract stated all right-of-way was available, and although Gamma had scheduled the excavation of the temporary channel to begin on August 12, 1991, Gamma did not get the necessary equipment on site until September 1, 1991. From August 12, 1991, until September 1, 1991, there were concurrent nonexcusable and excusable delays.

       From September 1, to October 3, 1991, the responsibility for delay lies wholly with the owner.

       Proper notice was given.

August 12–September 1	17 days excusable but noncompensable
September 1–October 3	33 days excusable and compensable
TOTAL EXCUSABLE DELAY	50 days

   2.   Additional Time Required to Install Additional Anchor Bars for the HPWL

       Originally, the anchor bars were not on the critical path and therefore not an excusable delay. However, the critical path may have shifted to put the anchor bars on the critical path. If so, any delay is both excusable and compensable. Assuming the critical path did shift, there is not enough information given on the CPM to

determine how much additional time is justified due to the additional bars. You would have to go to the detailed CPM to make such a determination. The main point is the unit price is no longer applicable after the ratio of horizontal to vertical bars is changed.
Excusable and compensable

3. Plan Error in the Top Slab Transverse Reinforcing for the Box Culvert

   The same question here as in 2: Is this activity on the critical path? This is clearly the fault of the owner, and the contractor is entitled to an adjustment in time if on the critical path and damages if they can be proven. One approach on the time adjustment would be the number of days required from the time the error was discovered until the solution is found and the contractor's progress is back to normal. Again, not enough information is given to determine time or damages.

4. Storm: July 8–15, 1992

   (Unusually severe weather conditions beyond the contractor's control)
   Excusable, but noncompensable   7 days

5. Carpenter's Strike: October 16–23, 1992

   (Act by others beyond the contractor's control)
   Excusable, but noncompensable   8 days

6. Ironworker's Strike: July 2–7, 1993

   Although this five-day delay is excusable, it affects work in a noncritical path operation and does not delay the project as a whole. **It is concurrent, noncompensable, and does not warrant time.**

7. Delays Due to Winter Concrete

   This is a complex one to figure. First we have to find out where Gamma was on December 1, 1992. We know there were the following excusable delays, which the owner refused to grant.

Lack of ROW	50 days
Added Anchor Bars	Undeterminable
Rebar Error	Undeterminable
Storm	7
Carpenter's Strike	8
	65 + days

The first thing is to get enough additional information to evaluate the anchor bar and re-bar error delays. Had the owner granted the time extensions, the contractor would not have been forced to place concrete during the winter.

One way to determine how far behind schedule the contractor was at what would have been the winter shutdown is to compare production rates to see if he had more days of concrete placing left than he had excusable delay. If he did, then part of the acceleration costs would be his. Assume that the contractor had more time coming than he was behind schedule and that the owner pays the acceleration costs. The owner has, in effect, bought back the excusable delay and Gamma is no longer entitled to those days of excusable delay.

Now to the question of liquidated damages for missing completion of the culvert on December 1, 1992, prior to the winter shutdown. We have established that Gamma has 65+ days of excusable delay as the winter shutdown approached. Days of time extensions must be given during the construction season; that is, you must replace good working days with good working days. If a time extension had been given, the 48+ days extension would begin at the spring start-up but it would be shortened by those days from December 1, 1992, to the winter shutdown. **Since the culvert was completed on March 15, 1993, no liquidated damages are assessable.**

8. Disruption in the Embankment Operation

By failing to acknowledge excusable delays in a timely manner, the owner forced Gamma into starting embankment before the area was ready. Therefore, the owner is responsible for this acceleration.

DAMAGES

1. Claim No. 1

   a. Damages to Gamma resulting from failure of the Owner to provide entry to the property between September and October 3, 1991, 33 days.

   b. Types of damages involved

      1. Equipment standby costs

      2. Standby labor (labor brought in expressly for this operation and for which no alternate work could be found)

  3. Labor escalation

  4. Indirect costs

  5. Extended period costs

  6. Impact (see claim no. 3)

2. Claim No. 2

 a. Additional costs due to revisions in anchor bars.

 b. Types of damages involved:

  1. Additional costs due to change in the ratio of vertical to horizontal bars

  2. Additional costs arising from loss of production in jury-rig operation (acceleration cost)

  3. Labor escalation

  4. Additional premium time

  5. Indirect costs

  6. Impact

3. Claim No. 3

 a. Plan error in re-steel

 b. Types of damages involved:

  1. Refabrication cost

  2. Resubmittal cost

  3. Equipment and labor standby costs

  4. Indirect costs

  5. Loss of production; inefficiency

  6. Impact

4. Claim No. 7

 a. Wrongful denial of extension of time requiring acceleration in the form of winter concrete.

 b. Types of damages involved:

  1. Winterization costs (i.e., forms, buckets, batch plant, etc.)

  2. Heating and protection costs

3. Concrete additives (more cement, calcium chloride, etc.)
4. Loss of production; inefficiency
5. Labor escalation
6. Additional premium time
7. Indirect costs
8. Impact

8. Owner's Damages for Delays

Most construction contracts contain a provision for liquidated damages. Such clauses amount to a charge against the contractor for not completing the work within the time specified in the contract. A reasonable liquidated damage sum provides added incentive for timely completion and, at the same time, puts a ceiling on the contractor's liability. For the owner, liquidated damages afford protection and reimbursement in the event of late completion.

The great advantage of liquidated damages is that the amount is set beforehand. It is not necessary to calculate actual damages. In many cases, where actual damages would be difficult to prove, this is an expedient method for the owner to recover those damages arising from tardy completion.

The liquidated damages figure sets a reasonable forecast of the harm that will result to the owner from the delay in completion. The courts have held that liquidated damages may not be used as a penalty. Where the damages figure has been proven to be a penalty against the contractor, the courts have held these unenforceable. In other words, liquidated damages should not be a club with which to beat the contractor over the head.

When liquidated damages are challenged, the usual test of fairness hangs on the reasonableness of the estimate of damage at the time the contract was entered into. This is an important concept. If, after the fact, the owner's damages are actually greater or less than that estimate, courts have held that irrelevant. If the amount was reasonable at the time, both parties must abide by those estimates. Naturally, a contractor is not likely to challenge liquidated damages unless he believes the owner has not incurred as much damage as the contractor has to pay.

Liquidated damages are, or can be, an effective tool in construction. The problem is one of fixing on a reasonable figure. This is the owner's problem and one that should be approached carefully. Disputes over liquidated damages are likely to surface after the fact of a delay when a contractor is forced to consider having to pay those damages. At that time, it is not surprising that he should question the amount and the causes of the delay; that is, that he should try to establish that the delay was due to the owner's actions or other excusable causes. So, while we advocate liquidated damages, there is a certain amount of built-in conflict. No one ever wants to pay.

CONTRACT PROVISIONS

The place to begin with any kind of construction problem is with the contract. Here is a liquidated damages clause from the New York City Department of General Services:

> Article 16. Liquidated Damages. In case the contractor shall fail to complete the work within the time fixed for such completion in the General Conditions..., or within the time to which such completion may have been extended, the contractor must pay to the City the sum fixed in the General Conditions..., for each and every calendar day that the time consumed in completing the work exceed the time allowed therefore; which said sum, in view of the difficulty of accurately ascertaining the loss which the City shall suffer by reason of delay in the completion of the work hereunder is hereby fixed and agreed as the liquidated damages that the City will suffer by reason of such delay, and not as a penalty. The Comptroller will deduct and retain out of the moneys which may become due hereunder, the amount of any such liquidated damages; and in case the amount which may become due hereunder shall be less than the amount of liquidated damages suffered by the City, the Contractor shall be liable to pay the difference upon demand by the Comptroller.

The liquidated damages clause of the federal government FAR, para. 52.212–5 states the following:

> If the Contractor fails to complete the work within the time specified in the contract, or any extension, the Contractor shall pay to the Government as liquidated damages, the sum of [Contractor Officer insert amount] for each day of delay.

FAR, para. 52.249-10 states:

> The Contractor's right to proceed shall not be terminated nor the Contractor charged with damages under this clause if—(1) The delay in completing the work arises from unforeseeable causes beyond the control and without the fault or negligence of the Contractor. Examples of such causes include (i) acts of God or of the public enemy (ii) acts of the government in either its sovereign or contractual capacity, (iii) acts of another contractor in the performance of a

contract with the government, (iv) fires, (v) floods, (vi) epidemics, (vii) quarantine restrictions, (viii) strikes, (ix) freight embargoes, (x) unusually severe weather, or (xi) delays of subcontractors or suppliers at any tier arising from unforeseeable causes beyond the control and without the fault or negligence of both the Contractor and the subcontractors or suppliers; (2) The Contractor, within 10 days from the beginning of any delay (unless extended by the Contracting Officer), notifies the contracting officer in writing of the causes of delay.

The contracting officer then must determine whether the delay warrants a time extension. If not, the possibility of assessing liquidated damages looms. The owner must make an analysis of all job delays to determine how many days of delay are excusable, that is, beyond the contractor's control. Liquidated damages are not assessable if there are excusable delays that the owner has not yet granted (see Chapter 6). Sometimes in a complicated project there are overlapping, concurrent delays, some of which may be attributed to either side. These must be sorted out before the owner can be certain that damages are due him. Some courts have held that damages will not be apportioned; that is, if the owner causes some delay, he may not recover any liquidated damages. This is not the modern point of view, however. With increasing frequency, courts are apportioning delay to the guilty party.

One advantage of liquidated damages clauses is that both parties know exactly what a day of delay in completion of a job will cost, assuming the contractor is at fault. For the contractor, this can amount to a sword over his head. When a delay occurs, the first thing both parties should do is to determine who is responsible. Naturally, each side will wish to see the responsibility placed elsewhere. So, as we have stressed continually, accurate records must be kept, schedules must be maintained and updated to reflect the impact of delays, and changes and their impacts must be incorporated in those schedules.

Generally, owners do not assess liquidated damages unless the contractor's delay is flagrant or the contractor asserts a delay claim against the owner. Then the owner is likely to resort to liquidated damages as a defensive tactic. In the public sector, this is less likely to be true. If the job is late and there are no excusable delays, liquidated damages will be assessed.

When disputes reach the courts, most jurisdictions will not grant both liquidated and actual damages. In fact, this is almost unheard of, so an owner should not allow himself to be persuaded that his case would be an exception. One exception, however, is where the liquidated damages clause on its face is limited to certain types of owner damages such as extended engineering, interest, taxes, and where other damages are specifically excluded such as the claims of other contractors affected by this contractor's delay. In such an event, both liquidated and specified actual damages may be recoverable. Moreover, the amount of the liquidated damages previously agreed on can be challenged on the grounds that it constitutes a penalty rather than a fair assessment of the damages owing to the delay. Upon a finding that the liquidated damages set were a penalty, or were unreasonable, then only provable, actual damages are recoverable.

In the absence of a liquidated damages clause, the contractor will be liable for the owner's proven actual damages due to unexcused late completion. These damages may be enormous in relation to the dollar amount of the contract. A contractor should consider what these damages may be before seeking to negotiate a deletion of a liquidated damages clause from his contract. Often contractors will object to liquidated damages clauses without considering the extent of their own exposure to the owner's actual damages absent such provisions.

A prime contractor can hold a subcontractor responsible for liquidated damages assessed by the owner to the extent of subcontractor-caused delay. The prime can, in addition, assess his sub for his actual damages such as extended supervision, overhead, and so on, attributable to the delay, depending, of course, on the provisions of the subcontract.

The period of time for which those liquidated damages must be paid is usually the time between the date specified in the contract for completion and the time by which the job can be "substantially complete" or when the owner takes "beneficial occupancy." Clearly, if the contractor has to pay a sizable sum for each day of delay, it is important, and sometimes controversial, to fix that time period.

SETTING LIQUIDATED DAMAGE FIGURES

How does an owner establish his liquidated damages figure? The following are some considerations:

1. Extra rental of other buildings that might be required because the one being built is not completed.
2. Extra maintenance and utility costs that may be incurred either in the continued use of old high-cost buildings or equipment or in the maintenance of a new area before beneficial use.
3. Interest on the investment or borrowed capital.
4. Extra training required to maintain worker skills pending availability of the building or equipment.

5. Extended supervision cost.

6. Additional operating costs that may result from the continued use of inefficient facility or equipment.

7. Extra costs of split operations resulting from partial occupancy or use of equipment.

8. Loss of revenue; for example, bridge tolls, sale of power from a power plant, building rentals, and so on.

9. Impact costs relating to follow-on contracts.

As we pointed out earlier, the amount of the liquidated damages must be reasonable and not a penalty. In addition to that legal consideration, there is a more practical reason for setting a fair damages figure. Too high an amount may scare contractors away. A high liquidated damages figure could also cause contractors to include a high contingency figure in their bids in order to cover the possibility of paying those damages later on. In such cases, the owner bears that extra cost whether or not the job is delayed. That said, parties don't usually enter into a contract with the idea that they will have to assess or pay liquidated damages. On the other hand, there may be contractors who prefer to gamble on a windfall profit if they finish on time; if not, they feel they can go to court to try to prove that the liquidated damages figure was not reasonable, or that delays were excusable.

REASONABLE COMPLETION DATES

Of course, disputes might be avoided altogether if the completion date were a reasonable one to begin with. If the owner has set an unrealistic date and a high liquidated damages figure, he might be letting himself in for litigation, or at least for claims and requests for time extensions. A contractor might accept a job with that combination of terms because he needs the work or because he believes he can recover through claims and changes. Neither approach is recommended.

Some contracts are written with a specific completion date, some with a specific number of days after the signing of the contract. Parties should be aware of the difference, especially since contracts often take time to process. If the contract includes a liquidated damages clause, all parties should be clear about the date on which the clause may be invoked. Do we need to say more?

During performance of the work, each party should be liable for his own delays. The liquidated damages should not be used to threaten the contractor, nor should the contractor's actions inspire such threats.

The contractor should keep in mind that as unpleasant as it might be to pay liquidated damages, the fact that the figure was set earlier puts a ceiling on his liability. He knows throughout the job exactly what he must pay if he delays the job. Even though no one likes to pay, knowing the amount beforehand eliminates uncertainty.

9. Bonding

TERMINOLOGY

Throughout this chapter we use language specific to the bonding world. *Surety* means the bonding company. The party who furnishes the bond—for example, the contractor—is the *principal*. The party—usually, the owner—benefiting from the bond is the *obligee*. The person who agrees to reimburse, or indemnify, the surety for any money that the surety is required to pay out on the bond is called an *indemnitor*. The principal—for example, the contractor—is always an indemnitor. Frequently, when the surety feels that the principal has insufficient assets, the surety requires others to sign up as indemnitors. These persons might be corporate officers of the contracting firm (and their wives), or a parent corporation may be asked to be an indemnitor. (The surety could also ask for a pledge of stock or a mortgage on real property.)

Most public and many private construction contracts require that the contractor be bonded. For a fee of a dollar or two per hundred dollars of the contract price, a bonding company "guarantees" the contractor's performance and payment by him of subcontractors and suppliers. "Guarantees" appears in quotation marks because the bond provides insurance of a highly qualified sort. In fact, a surety bond is not insurance in the traditional sense. Insurance premiums are calculated actuarially, based on losses and risks. Bond premiums are not. They are calculated as a fee for the extension of credit, much as a bank would charge a fee for issuing a letter of credit if it believes the customer has the resources to repay the bank when the letter is called in. The bank assumes there is no real risk of loss, otherwise, it would not extend credit.

Similarly, a surety will only write a bond if it believes that the contractor has sufficient financial resources and the technical competence to perform the work and pay his bills. The surety assumes there will be no significant risk of loss.

This difference applies to the manner in which bonding companies deal with claims, as compared with insurance companies. When an insurance company receives a claim covered under the policy it has issued, the company expects to pay without recourse against the insured party. When a bonding company receives a claim, it turns to the contractor for a response, since the contractor has agreed to completely indemnify the bonding company against any loss. The bonding company will only consider satisfying the claim once it is convinced that its contractor is unable to do so and that there is no legitimate defense to the claim.

Another aspect of this complicated arrangement is that the contractor is anxious that the bonding company not receive any claims against his performance. Bonds are becoming increasingly more difficult to obtain, and yet they are so often required by owners that a contractor will usually do whatever he can to prevent his bonding capability from being jeopardized.

Practically speaking, then, what does the bond mean? In order to obtain bonding, a financial investigation of the contractor is made. In addition, the surety checks the contractor's technical credentials for performing the work in question.

Issuance of the bond also places the financial resources of the surety behind the contractor. Contractors come and go; sureties go on forever, or at least for a long enough time. However, recovery is not a simple process. If an owner must actually press for recovery under a bond, he should know that the law regarding bonding is extremely complex. If a default situation seems to be in the making, an owner should retain a lawyer knowledgeable in this field, and as soon as possible. Recovery is more likely to result from a lawsuit than from a telephone call or a letter describing the contractor's failings.

The bonding company, as we have stated, will not pay until it is convinced that the contractor will not respond to the claim and that there is no legitimate defense of that claim.

We recommend avoiding litigation as a means of recourse under all but extraordinary circumstances because the legal battle will substantially reduce any benefits and the delay will be more costly than attempting other courses of action.

THE OWNER'S POINT OF VIEW

When should the owner declare the contractor in default? Should the owner permit the contractor to drag the job to completion? Obviously, if the contractor walks off the job, and after adequate notice from the owner he fails to return, there is no choice but to declare him in default.

On the other hand, if it is a question of slowness, continuing delays, or an undermanned project—but work is progressing—there are some points to consider. If the project is 95 percent complete, the default process (including negotiations with the bonding company, obtaining new bids and contracting out, preparation of new contract documents, etc.) in most instances will be more time-consuming than it would be to permit the lagging contractor to proceed at his own pace.

The question is, how far down the percentage scale of completion would it make sense for an owner to default a contractor? As a rule of thumb, it hardly ever pays to default when the job is over 75 percent complete, unless the job is at a complete standstill.

Between 50 percent and 75 percent a judgment must be made. In usual cases of slowness on the part of the contractor, or work stoppage due to lack of financial resources, a discussion with the bonding company is in order. The surety may decide to provide financial support. The owner must judge how long it will take to complete the project at the present rate of progress versus the time to complete if the default, rebid, and award steps are taken. The owner, in making this decision, must also guess whether the contractor on the job will continue at the same slow rate or will get worse or better. Even when the slowdown or other difficulty occurs below the 50 percent completion mark, the owner has to make some serious judgments; but the chance here is better that the default process will in the end gain time.

Generally speaking, if the bonding company learns early enough about the difficulties, its posture as a guarantor rather than an insurer can take a constructive turn. The surety will often lend a helping hand by offering the contractor financial aid or managerial assistance if such actions are thought to prevent default.

In this sense, the owner is protected in an indirect manner. The bonding company will help out to avoid being put in the position of being called on by the owner to complete the work, or to pay.

The contractor will most likely work harder to avoid losing his status with the bonding company. In fact, a contractor who is performing both bonded and unbonded work will most likely put his bonded work first if difficulties arise or if his resources become scarce. The contractor will not want to risk his standing with the surety nor run the risk of being called on to make personal indemnification to the surety. Typically, the surety will seek indemnity to the full extent of its potential exposure under the bonds.

THE PENAL SUM

The amount for which the bond is written, called the *penal sum*, is the maximum amount for which the surety agrees to be liable. Ordinarily, the penal sum is set at 100 percent of the contract price. Sometimes, especially on very large contracts, the penal sum is set at between 50 percent and 100 percent of the contract price (it can also be less).

When default occurs, the surety can either come in and finish the job itself or can pay. How is the amount to be paid determined? The surety is liable for the cost to complete the work minus the unpaid balance on the construction contract, plus liquidated damages, if any are established. These costs cannot exceed the penal sum. If they do, the penal sum is all the surety must pay in all but the most rare instances.

In recent years, there have been "bad faith" assertions against bonding companies that could result in punitive damages above the penal sum. This development rises from state laws relating to bad faith on the part of insurance companies for denying payment of claims without a good faith reason. These statutes have allowed punitive damages to the insured on the basis of their having to seek legal redress and having to go without the benefit the bond would have provided.

In spite of the differences we have pointed out between an insurance company and a surety, bad faith restitution and punitive damages have been sought against sureties. For example, on a performance bond, if the contractor defaults and the surety refuses to respond to the owner's claim on the bond, the owner could lose a great deal of money, more than the penal sum, and might seek to recover punitive damages from the surety. The owner might claim bad faith on the part of the surety to recover such damages. On the other hand, the possibility of frivolous bad faith claims exists. Claimants may institute such suits in order to get greater leverage from a surety or to force the company to capitulate.

Consequently, the response to claims has changed. Sureties seem inclined to respond more promptly and to investigate claims more carefully. The risk of having to pay sums beyond the penal sum should bad faith be alleged is a serious one for bonding companies.

The bad faith claim possibility should be kept in mind by all parties to a bonding arrangement, although in our opinion, the net effect has been alacrity and greater care in claims investigation on the part of bonding companies. At the very least, the result has been greater documentation by bonding companies of their reasons for denying claims. When default occurs, the surety can either come in and finish the job itself or can pay. How is the amount to be paid determined? The surety is liable for the cost to complete the work minus the unpaid balance on the construction contract, plus liquidated damages, if any are established. Those costs cannot exceed the penal sum. If they do, the penal sum is all the surety must pay in all but the most rare instances—for example, where a bad faith claim is successfully asserted.

If the surety decides to finish the work, this is usually accomplished in one of two ways. The surety can hire a completing contractor. In this case, if the final cost exceeds the penal sum, the surety must still pay. The surety may also provide the owner with a lump-sum contract with another contractor acceptable to the owner. The surety pays the owner the amount of that contract less the unpaid amount on the defaulted contract up to the penal sum of the bond. In some cases, the owner lets a completion contract and is reimbursed by the surety on the same basis.

The surety's decision is usually based on what it believes is the least-cost approach. Often, it is not possible to find a contractor willing to complete the project for a lump sum, as the second contractor would have to assume responsibility for unknown conditions in the work begun by another.

The surety, if it is assuming responsibility for the default, generally prefers to make a lump-sum payment in order to put a cap on its loss rather than to have a continuing exposure. That cap is arrived at through negotiations between the owner and the surety.

The owner, naturally, would like to complete the project with the surety's money rather than its own and would also prefer to avoid litigation. If the surety does pay up front, then some sort of release is required from the owner. That release covers the risk of problems with the completing contractor. Consequently, the owner would be well advised to insist that the completing contractor be bonded as a condition of giving the surety the release. It should be obvious that agreement is not reached swiftly in such cases.

COMMON TYPES OF BONDS

Bid Bonds

These guarantee that the contractor will honor his bid price for the work described in the contract documents, will sign the contract if it is awarded to him, and will furnish performance, as well as labor and material payment bonds if required.

A contractor could bid low, be awarded the work, then try to negotiate for more money. The owner would, of course, have the right to take legal action to enforce the bid. However, the time and money involved might be considered so great that he would accede to the contractor's demands. So the bid bond, in the main, is a sort of insurance when owners are dealing with unknown contractors. The bid bond requirement tends to exclude all but serious qualified bidders. If a private sector owner is dealing with a pool of bidders he knows well, the bid bond might seem superfluous. In any case, the guarantee is not unqualified. The bonding company is relieved of liability if the owner makes any modification or addition in going from the bid documents to the contract documents; for example, changes the design, time of performance, general conditions, and so on. The same applies if the owner fails to live up to any of his obligations under the bid documents, such as the time for making an award.

The amount of the bid bond (called the *penal sum*) is specified in the bid. It is usually in the range of 5% to 20 percent of the bid price, depending on the size of the project.

There are two common types of bid bonds: (1) liquidated damages and (2) difference-in-cost. The liquidated damages type requires forfeiture of the entire penal sum as "liquidated damages." This is to cover the owner's cost of rebidding, delay in the start of the project, and any difference in cost upon rebidding. The difference-in-cost type requires payment, up to the penal sum of the bond, of the owner's actual rebidding cost and the actual differences in cost on rebidding, assuming it is higher. If on rebid, the low bid is sufficiently lower to cover the owner's rebid costs, then, under this type of bond, there would be no liability. But this rarely happens.

Performance Bonds

As the name suggests, the performance bond guarantees that the contractor will live up to his obligations under the contract. The bonding company's liability is, however, conditional on the owner living up to his contract obligations. Many owners fail to appreciate the significance of this latter condition.

This point illustrates clearly the distinction between a bond and an insurance policy. In most cases, it is understood that the insured did not start the fire or cause the flood, or otherwise contribute to the cause that gave rise to the claim. On the other hand, with the bond, the question becomes a matter of degree on both sides. Were the contractor's breaches so substantial as to warrant default by the owner? Were the owner's breaches so substantial as to relieve the surety of liability?

Since the surety was not on the job, it does not have first-hand knowledge of the facts. The surety investigates the situation to assess the performance of contract obligations by both parties. That situation is rarely as clear as asking, "Was there a fire?" as in the case of an insurance claim.

Even when the surety listens to the contractor's version and the owner's version of the events that led to the slowdown or financial problems, the surety cannot be a totally impartial judge. The surety, after all, wishes to be indemnified by those indemnitors mentioned earlier. Therefore, the surety is more than likely to side with the contractor unless it appears that the contractor's contentions are substantially without merit. In most cases, out of duty to those indemnitors, the surety will let the contractor have his day in court.

Most owners fail to understand this point and are incensed when the surety sides with the contractor. In paying for the bond, directly or indirectly, what the owner buys is the security of knowing that if he is right, and can prove it, he will ultimately be compensated for his damages.

This is a far cry from getting a check from one's insurance company. However, bond premiums would be much higher were they to provide protection comparable to that of an insurance policy.

A Joint-Obligee Bond

A joint-obligee performance bond is one where there are two obligees, typically the owner and a lending institution. This form of bond makes the bank, which has been providing the construction financing, the beneficiary of the bond along with the owner.

Labor and Material Payment Bonds

This category of bonds covers those people who supply labor and material for the project (including subcontractors, suppliers, renters of equipment, etc.). A performance bond may also cover these items; however, a performance bond does not give direct rights to such persons. If the owner pays for labor and materials and obtains reimbursement under the performance bond, this cuts into the penal sum. An owner may not wish to diminish that sum. Subs and suppliers cannot claim and recover under the performance bond. They can do so only under a labor and material payment bond.

Therefore, it makes sense for an owner to require both kinds of bonds. The penal sum of the performance bond is in addition to the penal sum of the payment bond. The owner should consider these points:

1. The premium for the two bonds is generally the same as for one.
2. The penal sums are greater: they are the sum of the two bonds. But the premium, remember, is the same.
3. Subcontractors and suppliers may not wish to rely on the contractor's credit, or may do so only at a premium price. The owner may believe that the payment bond will result in savings in sub and supplier prices that will be passed along to the owner by the contractor.
4. Under the lien laws in some states, the owner can be directly liable to subs and suppliers unless a payment bond has been posted by the general contractor.

5. For public relations reasons, some owners may prefer to have payment bonds. They can then satisfy a moral obligation (even if there is no legal obligation) to make sure that subs and suppliers are paid. If the subs or suppliers contact them, they can merely send a copy of the payment bond and recommend that a remedy under the bond conditions be pursued.

The Federal Government Miller Act

The federal government, through the Miller Act, requires payment bonds for all construction undertaken by federal agencies. There are two principal reasons for this:

1. There is no right to file a lien on a federal project.
2. It is an avowed public policy to insure that subs and suppliers on federal projects are paid.

Many states have enacted "Little Miller Acts" that apply to state public works projects. These do not extend to private construction within the state. In general, the owner has no contract with the subs and suppliers. Unless the jurisdiction in which the project is located has a statute (for example, a lien law) providing otherwise, the owner has no obligation to the subs and suppliers, since they are not in privity of contract, that is, a party to the contract. If they are not paid by the contractor, they cannot recover from the owner.

Single Instrument Payment and Performance Bonds

Typically, the performance bond and the payment bond are two separate instruments; their penal sums are cumulative. Sometimes, the performance bond and the payment bond are in a single instrument. In such a case, there is a single penal sum for both obligations. Some courts have so held. In such an event, the owner is better protected if he gets two separate bonds, especially if the premium charged is the same in both cases.

Completion Bonds

This is a variation of the performance bond. In general, sureties do not like to write them because they (a) guarantee completion regardless of the cost (without the limitation of a penal sum), and (b) they require the surety to complete the project, without the option of merely paying money or providing a satisfactory completing contractor.

Guarantee Bonds

Often, these are guarantees in the contract documents that extend beyond the life of the performance bond (generally one to two years after acceptance of the work). These typically apply to guarantees on paving, roofing, or specialized equipment. It can be desirable to require specific bonds to secure these guarantees.

As an alternative, owners will often hold a sum of money to secure these obligations for their duration. This must be provided for in the contract documents. Sometimes the contract will allow a choice: the contractor will allow $Y for X years to secure the paving guarantee, or he will provide the guarantee bond. Many of these bonds, especially those offered by manufacturers, are very restrictive in their language, and recovery is not easy. Wording should be reviewed carefully to see if the owner is getting anything of value.

Claims under these bonds are difficult to establish. Once the contractor has left and the facility is operating, defects can be laid to poor maintenance, improper usage, and other causes beyond the contractor's control or responsibility, none of which would be covered by such a bond.

Lien Discharge Bonds

Under the lien laws of some states, the owner may not make further payments to a general contractor after a lien has been filed without running the risk of direct liability to the lienors—that is, having to pay a second time. This bond, when provided by the general contractor, relieves the owner from that liability.

Subcontractor Performance and Payment Bonds

Often, the contractor may require bonds of his subs. Sometimes the owner will insist on this for certain subs. This may be the case even if the contractor has himself furnished bonds to the owner. In doing so, the owner wishes to avoid potential trouble with key trades, such as the steel erector or the concrete sub. Default by a key sub on the critical path of a project can have a disastrous effect on the job. The owner or contractor will have to pay a little more to obtain this additional bond, but in some cases may find this requirement worth that expense.

PERFORMANCE BOND CLAIMS

Contractor defaults don't occur overnight. A vigilant owner should be aware of difficulties long before a contractor goes under and should try to remedy the situation. With the owner's cooperation, temporary problems can be overcome. In this book, we have outlined good administrative practices. An adversary stance is to be avoided whenever possible.

However, if the owner has reason to think that the contractor is in real trouble, he should ask for explanations first, then and if necessary, he should tell the contractor that the bonding company will be informed.

Some common symptoms of trouble are these: payment inquiries from vendors and subcontractors, requests for accelerated payments, shortages of labor on the job, and inability to obtain responses from the principals of the contracting firm. As we said earlier, the contractor will probably make every effort to avoid notice to the surety.

Three standard lines of defense are available to the surety when a claim is made. These are:

- That the owner did not give the surety adequate notice;
- That the owner caused or contributed to the contractor's problems; and
- That the owner made cardinal, or major, changes in the contract provisions.

Any of these may obviate the surety's obligations under the bond.

The Owner's Position

The owner should know that these are standard defenses and should conduct himself accordingly. For example, the owner must give adequate notice. Most performance bonds have provisions regarding notice. This gives the surety the opportunity to head off a default. If the bond requires prior notice to the surety, an owner must not take the matter into his own hands and complete the project without giving such notice. If he does, he can forfeit recovery from the surety.

Unless the bond so states such notice requirements, the owner has no legal obligation to inform the surety. Sureties often make routine requests for information regarding project status and the adequacy of the contractor's performance. The owner should avoid a perfunctory sign-off if he does submit this information. If these reports do not indicate anything out of the ordinary, when in fact there is, the surety can claim that the owner misled it. Owners would be better off not responding at all than to do so incorrectly or incompletely.

Once the bonding company is informed of the problem, investigation may take many weeks or even several months. (As we have observed, the development of bad faith claims seems to have accelerated these investigations.) During this time, work on the project is slowed down or even stopped. So the owner must choose his timing carefully. He should not use the bonding company as a club to threaten the contractor. Nor should he wait too long to inform the surety of a major problem.

For information purposes, the owner can send the surety duplicates of letters stating concern about lack of progress. The owner might refer in such letters to the contract provisions relating to default, stating that this is a remedy if progress is not improved. Certified mail (return receipt requested) protects the owner by providing proof of receipt.

The Surety's Defenses

Under the legal theory of subrogation, the surety has available to it all the defenses the contractor has against the owner. In other words, the owner has no greater rights against the surety than he does against the contractor.

This means that the owner, guilty of delays and interferences to the contractor, cannot default the contractor and then look to the surety to complete the work and to pay liquidated damages.

The surety will pursue the same defenses as the contractor when dealing with delays and other construction problems. The bonding company will say that the owner prevented site access, delayed delivery of equipment, suspended work, or in some other way interfered with the work, and consequently caused the contractor hardship.

As elsewhere, diligent documentation, progress schedules, and other records are essential to the owner. It should be obvious that the contractor would have nothing to lose and everything to gain by assisting the surety in building his case. The surety also has other defenses available against the owner. It may claim that cardinal or material changes to the construction contract have been made, changes that have not been consented to by the surety. The contractor may have asked for these changes or the owner may have consented to them. This doesn't change things for the surety. If the bonding company did not consent to the change, the bonding agreement could be voided.

Most construction contracts give the owner the right to make changes, and most bonds waive notice of these changes to the surety. However, such provisions do not extend to major changes. When in doubt as to whether a change is "major," the owner should get the surety's consent to a change to the construction contract. In most cases, that consent is routinely granted when requested before the fact.

Changes in Payment Procedures

Changes are not limited to physical changes in the scope of the work but can also include changes in payment procedures. An owner may agree to a contractor's request for accelerated payments. The owner may do this out of a desire to help the contractor out of temporary difficulty. If this assistance doesn't prevent a default, the surety will claim that this accelerated payment was a departure from the contract and that the owner consequently should forfeit part or all recovery.

DEALING WITH PERFORMANCE BOND SURETIES

The following is some practical advice for dealing with sureties when a claim is made by an owner.

Establish Credibility Early

When the notice of default is given to the surety by the owner, the surety will turn first to the contractor and other indemnitors to hear their position.

The contractor and those other indemnitors naturally have a vested interest in preventing the surety from paying since they are obligated to repay the surety for every dollar it pays out. Thus, the surety is likely to hear two stories.

The surety will be inclined to believe the contractor's version. If the surety pays over the contractor's objections, the contractor will not indemnify the surety voluntarily. The surety will most likely have to sue to recover. Thus, the owner's principal objective should be to establish credibility with the surety early, to convince the surety that the contractor's defenses are not valid. This is achieved through candor and a manifest desire to give the surety the full story. It starts with early notice to the surety.

Be Reasonable

Since the surety invariably has the option of letting the owner complete the project at his own expense and suing for recovery later, the owner should provide the surety a positive incentive to pay at the time the default is declared.

As we have mentioned, the surety will be under pressure from the indemnitors not to pay. The owner has a considerable burden here. He must convince the surety that it will be less costly to pay now than suffer a subsequent lawsuit when the project is completed. Exorbitant or unreasonable claims by the owner will only serve to support the surety's inclination to defend rather than to pay.

On the other hand, if the surety is offered a bargain, or has the opportunity to put a cap on an otherwise indeterminate claim, the surety may leap at the chance. Also, if for no other reason than good public relations, sureties like to settle amicably when they are shown to have no valid defense. If they can settle for a reasonable amount, they probably will.

Do It Right

The laws and procedures affecting construction bonds are complex and not well understood. Owners entering a prospective default situation are well advised to seek counsel and guidance from persons experienced in dealing with sureties.

A surety is well within its legal rights to assert technical defenses and will often do so. Competent advice will minimize the likelihood of such defenses and will impress on the surety the owner's ability to successfully pursue the case should the surety decide to defend.

Consider Alternatives to Default

Just as a divorce is not always the best solution to matrimonial problems and is usually the last resort, a default is not to be taken lightly. The owner must be aware above all that recovery from the surety is difficult at best. That final resort may be more time-consuming and costly than helping the contractor to avoid default. Initial proper election procedures and good administration practices can help avoid this situation.

What follows are sample bond forms published by the EJCDC. Note the relative simplicity of these forms in contrast to the complexity of effecting recovery as described in this chapter.

Performance Bond

Any singular reference to Contractor, Surety, Owner or other party shall be considered plural where applicable.

CONTRACTOR (Name and Address):

SURETY (Name and Address of Principal Place of Business):

OWNER (Name and Address):

CONTRACT
Date:
Amount:
Description (Name and Location):

BOND
Date (Not earlier than Contract Date):
Amount:
Modifications to this Bond Form:

Surety and Contractor, intending to be legally bound hereby, subject to the terms printed on the reverse side hereof, do each cause this Performance Bond to be duly executed on its behalf by its authorized officer, agent or representative.

CONTRACTOR AS PRINCIPAL
Company: (Corp. Seal)

Signature: _____
Name and Title:

SURETY
Company: (Corp. Seal)

Signature: _____
Name and Title:
(Attach Power of Attorney)

(Space is provided below for signatures of additional parties, if required.)

CONTRACTOR AS PRINCIPAL
Company: (Corp. Seal)

Signature: _____
Name and Title:

SURETY
Company: (Corp. Seal)

Signature: _____
Name and Title:

EJCDC No. 1910-28-A (1996 Edition)
Originally prepared through the joint efforts of the Surety Association of America, Engineers Joint Contract Documents Committee, the Associated General Contractors of America, and the American Institute of Architects.

1. The CONTRACTOR and the Surety, jointly and severally, bind themselves, their heirs, executors, administrators, successors and assigns to the Owner for the performance of the Contract, which is incorporated herein by reference.

2. If the CONTRACTOR performs the Contract, the Surety and the CONTRACTOR have no obligation under this Bond, except to participate in conferences as provided in paragraph 3.1.

3. If there is no OWNER Default, the Surety's obligation under this Bond shall arise after:

  3.1. The OWNER has notified the CONTRACTOR and the Surety at the addresses described in paragraph 10 below, that the OWNER is considering declaring a CONTRACTOR Default and has requested and attempted to arrange a conference with the CONTRACTOR and the Surety to be held not later than fifteen days after receipt of such notice to discuss methods of performing the Contract. If the OWNER, the CONTRACTOR and the Surety agree, the CONTRACTOR shall be allowed a reasonable time to perform the Contract, but such an agreement shall not waive the OWNER's right, if any, subsequently to declare a CONTRACTOR Default; and

  3.2. The OWNER has declared a CONTRACTOR Default and formally terminated the CONTRACTOR's right to complete the Contract. Such CONTRACTOR Default shall not be declared earlier than twenty days after the CONTRACTOR and the Surety have received notice as provided in paragraph 3.1; and

  3.3. The OWNER has agreed to pay the Balance of the Contract Price to:

    3.3.1. The Surety in accordance with the terms of the Contract;

    3.3.2. Another contractor selected pursuant to paragraph 4.3 to perform the Contract.

4. When the OWNER has satisfied the conditions of paragraph 3, the Surety shall promptly and at the Surety's expense take one of the following actions:

  4.1. Arrange for the CONTRACTOR, with consent of the OWNER, to perform and complete the Contract; or

  4.2. Undertake to perform and complete the Contract itself, through its agents or through independent contractors; or

  4.3. Obtain bids or negotiated proposals from qualified contractors acceptable to the OWNER for a contract for performance and completion of the Contract, arrange for a contract to be prepared for execution by the OWNER and the contractor selected with the OWNER's concurrence, to be secured with performance and payment bonds executed by a qualified surety equivalent to the Bonds issued on the Contract, and pay to the OWNER the amount of damages as described in paragraph 6 in excess of the Balance of the Contract Price incurred by the OWNER resulting from the CONTRACTOR Default; or

  4.4. Waive its right to perform and complete, arrange for completion, or obtain a new contractor and with reasonable promptness under the circumstances;

    4.4.1 After investigation, determine the amount for which it may be liable to the OWNER and, as soon as practicable after the amount is determined, tender payment therefor to the OWNER; or

    4.4.2 Deny liability in whole or in part and notify the OWNER citing reasons therefor.

5. If the Surety does not proceed as provided in paragraph 4 with reasonable promptness, the Surety shall be deemed to be in default on this Bond fifteen days after receipt of an additional written notice from the OWNER to the Surety demanding that the Surety perform its obligations under this Bond, and the OWNER shall be entitled to enforce any remedy available to the OWNER. If the Surety proceeds as provided in paragraph 4.4, and the OWNER refuses the payment tendered or the Surety has denied

liability, in whole or in part, without further notice the OWNER shall be entitled to enforce any remedy available to the OWNER.

6. After the OWNER has terminated the CONTRACTOR's right to complete the Contract, and if the Surety elects to act under paragraph 4.1, 4.2, or 4.3 above, then the responsibilities of the Surety to the OWNER shall not be greater than those of the CONTRACTOR under the Contract, and the responsibilities of the OWNER to the Surety shall not be greater than those of the OWNER under the Contract. To a limit of the amount of this Bond, but subject to commitment by the OWNER of the Balance of the Contract Price to mitigation of costs and damages on the Contract, the Surety is obligated without duplication for:

  6.1. The responsibilities of the CONTRACTOR for correction of defective Work and completion of the Contract;

  6.2. Additional legal, design professional and delay costs resulting from the CONTRACTOR's Default, and resulting from the actions or failure to act of the Surety under paragraph 4; and

  6.3. Liquidated damages, or if no liquidated damages are specified in the Contract, actual damages caused by delayed performance or non-performance of the CONTRACTOR.

7. The Surety shall not be liable to the OWNER or others for obligations of the CONTRACTOR that are unrelated to the Contract, and the Balance of the Contract Price shall not be reduced or set off on account of any such unrelated obligations. No right of action shall accrue on this Bond to any person or entity other than the OWNER or its heirs, executors, administrators, or successors.

8. The Surety hereby waives notice of any change, including changes of time, to the Contract or to related subcontracts, purchase orders and other obligations.

9. Any proceeding, legal or equitable, under this Bond may be instituted in any court of competent jurisdiction in the location in which the Work or part of the Work is located and shall be instituted within two years after CONTRACTOR Default or within two years after the CONTRACTOR ceased working or within two years after the Surety refuses or fails to perform its obligations under this Bond, whichever occurs first. If the provisions of this paragraph are void or prohibited by law, the minimum period of limitation available to sureties as a defense in the jurisdiction of the suit shall be applicable.

10. Notice to the Surety, the OWNER or the CONTRACTOR shall be mailed or delivered to the address shown on the signature page.

11. When this Bond has been furnished to comply with a statutory or other legal requirement in the location where the Contract was be performed, any provision in this Bond conflicting with said statutory or legal requirement shall be deemed deleted here from and provisions conforming to such statutory or other legal requirement shall be deemed incorporated herein. The intent is that this Bond shall be construed as a statutory bond and not as a common law bond.

12. Definitions.

  12.1 Balance of the Contract Price: The total amount payable by the OWNER to the CONTRACTOR under the Contract after all proper adjustments have been made, including allowance to the CONTRACTOR of any amounts received or to be received by the OWNER in settlement of insurance or other Claims for damages to which the CONTRACTOR is entitled, reduced by all valid and proper payments made to or on behalf of the CONTRACTOR under the Contract.

  12.2. Contract: The agreement between the OWNER and the CONTRACTOR identified on the signature page, including all Contract Documents and changes thereto.

  12.3. CONTRACTOR Default: Failure of the CONTRACTOR, which has neither been remedied nor waived, to perform or otherwise to comply with the terms of the Contract.

  12.4. OWNER Default: Failure of the OWNER, which has neither been remedied nor waived, to pay the CONTRACTOR as required by the Contract or to perform and complete or comply with the other terms thereof.

(FOR INFORMATION ONLY--Name, Address and Telephone)
AGENT or BROKER: OWNER'S REPRESENTATIVE (Engineer or other party):

Payment Bond

Any singular reference to Contractor, Surety, Owner or other party shall be considered plural where applicable.

CONTRACTOR (Name and Address):   SURETY (Name and Address of Principal Place
                                                  of Business):

OWNER (Name and Address):

CONTRACT
   Date:
   Amount:
   Description (Name and Location):

BOND
   Date (Not earlier than Contract Date):
   Amount:
   Modifications to this Bond Form:

Surety and Contractor, intending to be legally bound hereby, subject to the terms printed on the reverse side hereof, do each cause this Payment Bond to be duly executed on its behalf by its authorized officer, agent, or representative.

CONTRACTOR AS PRINCIPAL   SURETY
   Company: (Corp. Seal)   Company: (Corp. Seal)

   Signature: _____   Signature: _____
   Name and Title: Name and Title:
                                                      (Attach Power of Attorney)

(Space is provided below for signatures of additional parties, if required.)

CONTRACTOR AS PRINCIPAL   SURETY
   Company: (Corp. Seal)   Company: (Corp. Seal)

   Signature: _____   Signature: _____
   Name and Title: Name and Title:

EJCDC No. 1910-28-B (1996 Edition)
Originally prepared through the joint efforts of the Surety Association of America, Engineers Joint Contract Documents Committee, the Associated General Contractors of America, the American Institute of Architects, the American Subcontractors Association, and the Associated Specialty Contractors.

The CONTRACTOR and the Surety, jointly and severally, bind themselves, their :irs, executors, administrators, successors and assigns to the OWNER to pay for bor, materials and equipment furnished for use in the performance of the Contract, hich is incorporated herein by reference.

With respect to the OWNER, this obligation shall be null and void if the ONTRACTOR:

2.1. Promptly makes payment, directly or indirectly, for all sums due Claimants, and

2.2. Defends, indemnifies and holds harmless the OWNER from all claims, demands, liens or suits by any person or entity who furnished labor, materials or equipment for use in the performance of the Contract, provided the OWNER has promptly notified the CONTRACTOR and the Surety (at the addresses described in paragraph 12) of any claims, demands, liens or suits and tendered defense of such claims, demands, liens or suits to the CONTRACTOR and the Surety, and provided there is no OWNER Default.

With respect to Claimants, this obligation shall be null and void if the ONTRACTOR promptly makes payment, directly or indirectly, for all sums due.

The Surety shall have no obligation to Claimants under this Bond until:

4.1. Claimants who are employed by or have a direct contract with the CONTRACTOR have given notice to the Surety (at the addresses described in paragraph 12) and sent a copy, or notice thereof, to the OWNER, stating that a claim is being made under this Bond and, with substantial accuracy, the amount of the claim.

4.2. Claimants who do not have a direct contract with the CONTRACTOR:

1. Have furnished written notice to the CONTRACTOR and sent a copy, or notice thereof, to the OWNER, within 90 days after having last performed labor or last furnished materials or equipment included in the claim stating, with substantial accuracy, the amount of the claim and the name of the party to whom the materials were furnished or supplied or for whom the labor was done or performed; and

2. Have either received a rejection in whole or in part from the CONTRACTOR, or not received within 30 days of furnishing the above notice any communication from the CONTRACTOR by which the CONTRACTOR had indicated the claim will be paid directly or indirectly; and

3. Not having been paid within the above 30 days, have sent a written notice to the Surety and sent a copy, or notice thereof, to the OWNER, stating that a claim is being made under this Bond and enclosing a copy of the previous written notice furnished to the CONTRACTOR.

If a notice required by paragraph 4 is given by the OWNER to the ONTRACTOR or to the Surety, that is sufficient compliance.

When the Claimant has satisfied the conditions of paragraph 4, the Surety shall omptly and at the Surety's expense take the following actions:

6.1. Send an answer to the Claimant, with a copy to the OWNER, within 45 days after receipt of the claim, stating the amounts that are undisputed and the basis for challenging any amounts that are disputed.

6.2. Pay or arrange for payment of any undisputed amounts.

The Surety's total obligation shall not exceed the amount of this Bond, and the nount of this Bond shall be credited for any payments made in good faith by the irety.

8. Amounts owed by the OWNER to the CONTRACTOR under the Contract shall be used for the performance of the Contract and to satisfy claims, if any, under any Performance Bond. By the CONTRACTOR furnishing and the OWNER accepting this Bond, they agree that all funds earned by the CONTRACTOR in the performance of the Contract are dedicated to satisfy obligations of the CONTRACTOR and the Surety under this Bond, subject to the OWNER's priority to use the funds for the completion of the Work.

9. The Surety shall not be liable to the OWNER, Claimants or others for obligations of the CONTRACTOR that are unrelated to the Contract. The OWNER shall not be liable for payment of any costs or expenses of any Claimant under this Bond, and shall have under this Bond no obligations to make payments to, give notices on behalf of, or otherwise have obligations to Claimants under this Bond.

10. The Surety hereby waives notice of any change, including changes of time, to the Contract or to related Subcontracts, purchase orders and other obligations.

11. No suit or action shall be commenced by a Claimant under this Bond other than in a court of competent jurisdiction in the location in which the Work or part of the Work is located or after the expiration of one year from the date (1) on which the Claimant gave the notice required by paragraph 4.1 or paragraph 4.2.3, or (2) on which the last labor or service was performed by anyone or the last materials or equipment were furnished by anyone under the Construction Contract, whichever of (1) or (2) first occurs. If the provisions of this paragraph are void or prohibited by law, the minimum period of limitation available to sureties as a defense in the jurisdiction of the suit shall be applicable.

12. Notice to the Surety, the OWNER or the CONTRACTOR shall be mailed or delivered to the addresses shown on the signature page. Actual receipt of notice by Surety, the OWNER or the CONTRACTOR, however accomplished, shall be sufficient compliance as of the date received at the address shown on the signature page.

13. When this Bond has been furnished to comply with a statutory or other legal requirement in the location where the Contract was to be performed, any provision in this Bond conflicting with said statutory or legal requirement shall be deemed deleted herefrom and provisions conforming to such statutory or other legal requirement shall be deemed incorporated herein. The intent is, that this Bond shall be construed as a statutory Bond and not as a common law bond.

14. Upon request of any person or entity appearing to be a potential beneficiary of this Bond, the CONTRACTOR shall promptly furnish a copy of this Bond or shall permit a copy to be made.

15. DEFINITIONS

15.1. Claimant: An individual or entity having a direct contract with the CONTRACTOR or with a Subcontractor of the CONTRACTOR to furnish labor, materials or equipment for use in the performance of the Contract. The intent of this Bond shall be to include without limitation in the terms "labor, materials or equipment" that part of water, gas, power, light, heat, oil, gasoline, telephone service or rental equipment used in the Contract, architectural and engineering services required for performance of the Work of the CONTRACTOR and the CONTRACTOR's Subcontractors, and all other items for which a mechanic's lien may be asserted in the jurisdiction where the labor, materials or equipment were furnished.

15.2. Contract: The agreement between the OWNER and the CONTRACTOR identified on the signature page, including all Contract Documents and changes pthereto.

15.3. OWNER Default: Failure of the OWNER, which has neither been remedied nor waived, to pay the CONTRACTOR as required by the Contract or to perform and complete or comply with the other terms thereof.

(FOR INFORMATION ONLY--Name, Address and Telephone)
AGENCY or BROKER: OWNER'S REPRESENTATIVE (Engineer or other party):

Part II: Prosecuting and Defending Claims

10. Documentation and Record Keeping

The second half of this book carries certain legal overtones, the sense of documenting one's case in court. But if the first task, that of record keeping, is conscientiously conducted, many construction differences are likely never to reach the courtroom.

Ultimately, what is at stake in any claim is money. But as a matter of sound administrative practice on any project, records must be kept, filed in an orderly fashion, and referred to as the occasion demands. Memories become faulty with time, verbal agreements entered into in good faith cannot be recalled in detail, setting the stage for a potential dispute or claim.

Documentation should be maintained whether or not either party foresees a change or a claim. In particular, anything that might conceivably affect the cost of the project should be documented. Any changes, however inconsequential they may seem at the time, must be recorded. They may be typed or written memoranda; they may be photographs with notes; or they may be entries in the contract plans or specifications. This latter system provides an excellent method of recording technical changes. For example, a note on the footing plans might state that on a specific date the owner's representative, by name, directed that footing E-8 be enlarged 1 foot in a northerly direction with appropriate increases in the length of reinforcement. This note should be initialed and dated. It is not unusual on large and complex projects to have more than 500 changes. Obviously, these can have a synergistic effect; it's a lot easier to plot the effects of one change or another and their total impact on the job if every change is clearly recorded.

No one wishes to go to court or to resort to other methods of formally settling disputes. Both sides incur risks in doing so, not the least of which is a reputation for litigiousness or claims consciousness. In the long run, court cases are expensive. In addition to the obvious legal fees, there are less visible expenses, for example, diverting personnel from other more productive activities to those related to the lawsuit, such as time spent in court, with attorneys, in checking documents, and so forth.

But disputes do reach litigation. When they do, the sheer weight of paper can overwhelm the opposition, bolster one's case, and prove one's point. What follows is a discussion of the types of things that need to be recorded, the significance of the paper paraphernalia that attaches itself to any construction project, and suggestions about the storing and retrieval of that paperwork.

First, some general admonitions. With the broad freedom-of-information laws and broader interpretation of "discovery," just about any scrap of paper is likely to be accessible to the parties in a lawsuit, and could become part of the legal proceedings. Double entendre, wisecracks, or personal remarks in the margin or text of job records could become acutely embarrassing years later in the judicial setting. Still, embarrassing records are better than no records. It has been argued that a certain amount of scribbling in the corners adds authenticity to such records, particularly if it clarifies or ties to another document. But, be aware, anything on paper could be subject to further scrutiny. (See the inspector's report at the end of this chapter. The handwritten version is preferable to a subsequent typed version when a dispute is being settled if it is legible. Both, of course, may be submitted.)

STORAGE AND RETRIEVAL

We are talking about a considerable amount of paper in most cases. The volume of paper is in proportion to the size of the project and of the dispute. Before it piles up, a system for filing should be established, as records may be kept for years on a long-term project. Workable categories should be sorted out; which categories these might be is to some extent a personal matter. The important thing is accessibility and some degree of orderliness. It is imperative that at least one complete record of all paper related to the project be kept intact.

Cross-files according to subject areas, (i.e., concrete, payroll, borings, progress meeting minutes, etc.) can be maintained according to preference. Copies of documents and correspondence having to do with specific segments of the work or with particular individuals may be kept separately by the appropriate persons who, in certain cases, may be in widely separated locations.

Originals of field records generally are kept in the project office until job completion. Originals of contract documents are kept either in a regional office or a headquarters office. Originals of other documents such as weather conditions and strikes may be kept in still another office, such as a division office.

The important thing is that all top-level personnel associated with a project must know where the originals are and, at the appropriate time, make them available for negotiation purposes or for court appearances. If documents are sent to warehouse storage, responsible project officers must know what has been sent, how it is filed, and who controls access.

The contract documents must always remain intact. These, signed by both parties, will include the plans, which in many cases are also signed or at least initialed by both parties. They should also contain the performance bonds. This point should be clear to all on the job: No one should remove a relevant page for his particular use, or be in a position to lose or misplace any part of those documents while using it for reference. Conformed documents can be made available and will serve all purposes, except that final appearance in court.

With projects now increasing in size, and with claims following in number, complexity, and dollar volume, computers are now being used for cross-referencing and information storage. There are no rules of thumb about what is the optimum project size where the use of a computer in retrieval is effective. The claim size must certainly be in the multimillion dollar range, and the numbers of documents in the tens of thousands of pages.

In the absence of documentation, when both sides are working in a professional manner to negotiate and settle claims, both will attempt to reconstruct events and establish the validity of the claims. This is, unfortunately, a most unsatisfactory way to proceed because everything relies on memory, and a very valid claim can be lost because neither party can substantiate what should be a proper settlement. Further, if the documentation was done but couldn't be found, all those hours of careful record keeping were wasted.

CORRESPONDENCE

The importance of sending letters cannot be overemphasized. Unlike the other categories of records—notes, diaries, invoices, meeting minutes, and so on—letters convey the sense of "shared" information. The recipient cannot say at a later date that he was unaware of a troublesome situation. In some cases, unacknowledged receipt can be construed as acquiescence or as lack of cooperation. Where matters of top importance are being discussed through correspondence, the use of return receipt certified mail or the inclusion of a duplicate copy with a place for the recipient to note receipt is recommended.

Letters should be written about everything that could impact the project. Even in the most harmonious of working situations, verbal suggestions, changes, promises, or complaints should be put in letter form and copies of such letters sent to enough people to spread that knowledge around. This can be done in the briefest of letters, but all correspondence should be dated. A serial number can be assigned to each letter. Each letter can then be logged into a central register.

Keep the principles of good letter writing in mind. Inform the recipient why you are writing the letter. Is it for his information? Is it a direction to him? Does it call for a reply? When do you expect a reply? If you are referring to a continuing situation, use the same basic descriptive terms in each letter and refer to previous correspondence on this subject by date. A good system to keep track of original correspondence is to number each sequentially. The numbered letters can then be entered in a computer and quickly accessed when questions occur.

Receipt of letters should always be acknowledged, even if in a cursory and noncommittal fashion. For example, "We received your letter of xxx about yyyy and are reviewing the situation. We expect to give you a complete reply by..." Again, send copies of these letters to project principals. The more copies distributed, the less likely the problem will disappear through a crack in the floor only to reappear when it has grown into a catastrophe.

PHOTOGRAPHS

Photographs of the work should be taken at every stage by both the owner and the contractor. This starts with prebid visits to the site. Photographs should be taken on the date of the start of work. Specific progress photographs on a predetermined time basis should also be included. Any unusual conditions or situations must be photographed. Finally, photographs should be taken of the completed project. Several staff members can be assigned this task. It is imperative that all photographs be dated and that the location and the photographer be noted. It is also helpful to include a person or some object of known dimensions in the photo to give a sense of scale. Without this identification, such photos are of little use in documenting a claim or change. Obvious candidates for this photographic record are changed conditions, damaged equipment or materials, blocked access to the site, flooding, and heavy snow. General work progress should also be photographed. At some future time, a dated photograph indicating some state of construction—the extent of a concrete pour in a certain area—can belie or back up someone's sketchy notes or memory. Random shots are also essential. Some detail in the background may buttress a point in contention years later, for example, that the steel re-bars are placed at a particular date during the work. A photograph may clarify a condition of the work during construction that is not observable during a claim unless test holes are cut into the concrete.

For example, in one relevant case, a reinforced concrete building was being analyzed after construction as to adequacy under a new seismic code. The design drawings were not clear as to the penetration of the column reinforcing into the floor slab, and the shop drawings could not be located ten years afterward. Fortunately, a set of photographs were found indicating that the column reinforcing projected into the slab.

Sophisticated cameras and professional photographers are not necessary; the most pedestrian of pictures might still be worth more than a witness's thousand words. A 35mm camera using fine-grain film in the hands of an amateur photographer can provide inexpensive photographs, which can be blown up into clear 8" X 10" glossies that will clearly define the status of the work when the photograph was taken. Polaroid cameras do not have similar benefits; nevertheless, a Polaroid shot of a hole in the pavement or disintegrated pavement or a crack in a brick masonry wall may very well tell the story. Polaroids also have the advantage of giving the photographer immediate knowledge that he has a usable photograph. He doesn't have to wait until the film is developed, by which time a condition may have changed.

Aerial photography can be useful on large projects, such as dams, that spread over large areas. The use of time-lapse and video photography has become widespread as a means of documenting progress. These methods are excellent: the time lapse shows progress over a long period of time; and the sound portion of videos provide verbal commentary made at the time of the filming.

RECORDS CHECKLIST

The preceding points on the storage of information, the need for correspondence on sensitive items and photos, and/or the caveat on making personal comments in job records are all general precepts to be followed by both the owner and the contractor. What follows is a checklist of points to be covered in a typical job sequence. In all cases, dates, time of day, location if appropriate, and people in attendance, should be noted.

Prebid

Records should be kept for the following: the Invitation to Bid (newspaper, Commerce Business Daily, or letter), Dodge Reports, Browns Letters, site visit notes and photographs, minutes of any prebid meetings, proposed schedules, estimates including work sheets, project log, and all addenda and logs of telephone conversations.

Some owners have found prebid meetings of value in clearing up ambiguities or unclear sections of the plans and specifications. They can also be used to clarify site conditions and provide information concerning other contracts in the general vicinity. At the meeting, the knowledgeable owner will state that no information given at the meeting is binding, nor a part of the contract, until a written set of the prebid meeting minutes is distributed with questions and answers included. These should be made an addendum to the contract.

Bid

The contractor should have copies of all papers submitted, together with all supporting bid calculations, quantity takeoffs, subcontractors' and suppliers' quotes, estimated productivity of labor and equipment, and proposed project schedule in detail. Where subcontractors' prices and supplier's prices were received by telephone, there should be a written memorandum of each quotation together with date, time, and names of all parties involved. There should be a record of the receipt of all addenda received and any consultant reports. The names of the other bidders and the amounts bid by each should be recorded and kept in the file.

Requirements that original bid papers be held in a depository are increasingly included in contracts. The idea is that these can be retrieved in the case of dispute. At this point, there is some controversy and disagreement as to the value of this requirement.

The owner should record the names of each organization submitting a bid, the amount of the bid, and whether or not bonds and other required material are attached, together with any protests or statements that were made by bidders or attendees. The names of the owner's representatives present should be recorded by both the owner and the contractor.

Precontract

Records should be kept for the following items: minutes of negotiation meetings, minutes of qualification hearings where applicable, job schedules, and record of any changes in costs that were negotiated. Any decisions agreed upon by the owner/contractor/engineer should be included. As part of any change, all calculations should be maintained as a permanent job file. Photographs and notes of any site visits during this period should also be stored.

Contract

Keep the original contract documents in a secure file (not to be used as working copies); conformed contracts should be made available for the job.

Construction

Minutes of a preconstruction conference should definitely be included. Such conferences are generally held between the award of the contract and the start of work. (See Appendix 3 for an example of such minutes.) Copies of job schedules (including revisions) should be stored. Records should be made of all verbal instructions and field orders.

Job diaries, including those of foremen, field superintendents, and management are also part of the record. These form a record of what happened on the job each day—visitors, weather, materials delivered, trades working, names of all men working, and so on. All diaries, including personal diaries, should be readily accessible. Parties maintaining these should understand that under the freedom-of-information laws and discovery rules in litigation, the contents of those diaries are all available to the other side.

The fact that a diary is kept at home does not preclude its being entered into evidence during a pretrial questioning, court proceeding, or arbitration hearing. Essentially, any record, in whatever form, kept in connection with the project, by anyone, is available to all parties. These can even include tape-recorded job "diaries." If it can be established that such were a part of a systematic procedure on the job and not a random instance, it is possible to introduce such records under the doctrines of "business records" or otherwise as "past recollection recorded."

Shop Drawings

The shop drawing log should be maintained accurately, with any delays in approval duly noted and, where possible, explained. These drawings can be one of the more critical aspects of a project; and if improperly handled, they can create major problems and delay on the project.

Shop drawing reviews are subject to misuse. Contractors have been known to make changes to the work and the intent of the design without highlighting the point, or possibly, without recognizing that they have changed the design. In turn, architects and engineers have been known to effect design changes with resultant changes in construction costs in the process of approving shop drawings without acknowledging (or again, recognizing) that they have done so. Neither intentional action is professional unless it is spelled out to the other party and the change noted. If one party believes the other has made a change, it should be discussed at once.

The engineer should be aware of the legal significance of his signature on shop drawings. In spite of exculpatory language on the approval

stamp—that is, "approval for general conformance with design requirements, not as to details and dimensions"—courts and administrative boards
are loathe to enforce exculpatory language that protects only the author
of a document.

As-built drawings should be kept up to date; seemingly minor field
changes should not be filled in later when the memory is dim. These
might prove to be particularly important in the claim situation that involves changed conditions, increased quantities, or a failure.

Job Schedules

Job schedules and actual progress can be charted in several ways, such as
S-curves that plot money against time, simple bar graphs that indicate
progress in various separate categories of work, and the critical path
method (CPM). Progress may also be recorded with the aid of a computer on large complex projects.

Particular effort should be made to maintain accurate files of these
charts and of any revisions made. In the final preparation for defense of a
claim, these charts can prove to be the most telling piece of evidence to
prove or disprove a case. They are essentially the final product of all
documentation. Actual work completed can be overlaid against the original schedule and the causal relationship between the segments analyzed.
In this manner, ripple or impact effects can be plotted, and their effect
upon costs down the line can be measured. For example, a steel strike
may delay one segment upon which another depends. This in time causes
another portion of the work to be performed during the winter, which in
time required different equipment than that originally considered. CPM
charts are a sophisticated analysis method and deserve careful scrutiny
and review. (See Chapter 4 on delays and Chapter 6 on costs of delays.)

Owner's Records

While the basic records kept by an owner or a contractor are essentially
the same (except for fixed-price contracts where the owner doesn't have
much incentive to keep cost records), the owner has additional records
that he keeps as a matter of form. These include inspectors' reports, concrete form approvals, borrow pit approval, field direction or clarification, safety observations and direction, and administrative decisions (such
as permits, easements, transmittal of shop drawings, approval of material, and payment calculations).

Change Order Files

For each change that occurs or is believed to have occurred during the course of the work, a separate self-contained file should be maintained that includes the order, relevant correspondence, a record of all negotiations concerning the change, estimates of cost, and the calculations on which the estimates are based. Where possible, photocopies of the appropriate sections of the specifications and the contract drawings will make the file more complete and simplify work during negotiations. At the end of the chapter, we have included an auditor's report on the costs of a change. This kind of document should always be kept on file.

Daily Reports

Records should be kept for daily staffing of the project, divided into trades and subcontractors, part of which should include the time cards kept on the project. Time cards are one of the basic sources of data concerning a project.

When there is a claim situation, it is desirable to have both the owner and the contractor agree as the work progresses upon the classification of labor, the man hours, the equipment used, and any material involved in the item in dispute.

Payment Requisitions

The methods of payment are covered by the contract, and while many methods of payment are provided for, the two most common are:

1. Payment at predetermined intervals of time, such as monthly, based on percentage of completion.
2. Payment at completion of a certain phase of the project or of a certain dollar amount of work, such as $100,000.

The payment is made on the basis of a requisition submitted by the contractor to the owner. Or the owner can make his own estimate of work and pay based on that estimate. In either case, both the contractor and the owner ought to agree upon the work performed during the payment period. Calculations by which these estimates are obtained should be kept by the party who prepared them. They are one source of a future reconstruction of the job and the preparation of an as-built CPM chart.

Minutes of All Meetings

These should include time and place where meeting was held and those in attendance. The minutes should indicate whom each participant represents. If they are continuing meetings, it is desirable to assign a number to each set of minutes to avoid the possibility of a set being overlooked or mislaid. These minutes should be kept in the same form for each meeting.

Prior to the meeting, an agenda should be prepared and distributed. (A sample of job meeting minutes appears at the end of this chapter.) A job meeting should cover such items as progress, any delays, effect of strikes or of material delivery delays, staffing of project, start of a new operation, status of change orders, status of shop drawing review, information required from the owner, information required from the contractor, and the need for clarification of drawings or specifications.

Change orders or agreements can start at such job meetings. Their insertion in the minutes is the beginning of documentation on a particular change order or agreement. It is desirable, however, that change orders or agreements be followed up by letters or written orders. The final item in the minutes should be the date, time, and location of the next meeting.

Several methods may be used to obtain concurrence with the minutes by those in attendance. At the following meeting, the opening item on the agenda can be the approval or changes to the prior minutes; or both parties can sign the minutes when they have been approved or concurrence has been reached. If the minutes are distributed to all parties attending, and no objections are voiced, it can be assumed that the parties concur with what the minutes contain. Hence, if a party receives minutes that are incorrect or incomplete, he should make sure his objections are recorded in writing somewhere. Otherwise, years later in a claims setting, it will be difficult to overcome the presumed correctness of the minutes.

Field Reports

Daily records should be kept of such items as cubic yards of concrete delivered, truckloads of earth or rock removed, truckloads of sand fill, deliveries of structural steel, shift progress in a tunnel, and the inspector's reports on the work—including the location that he inspected each day. Daily reports should also include areas where work was performed, estimate of production, problems, visitors, weather, lost time or reports on inefficiency, and similar remarks. Records of estimated production are invaluable in a claims situation, especially where comparisons can be made between production before, during, and after the period when problems arose. (See samples of Inspector's Daily Report at the end of this chapter.)

Weather

Daily weather records should be kept. In cases of delays attributed to weather conditions, the National Oceanic and Atmospheric Administration can provide reference point data. The general contract usually contains time extension clauses excusing delay for unusually severe weather. (This subject is dealt with in Chapter 4.) The effect of the weather upon the job should be indicated. A series of rainy days can have a severe effect upon an earth-moving operation, while it would have no effect upon interior work in a building.

Material and Equipment Records

Delivery dates for all materials should be kept. Where there are delays in the delivery of such material, they should be indicated. All quotations, purchase orders, invoices, and similar documents should be kept on file. A record should be kept of damaged material or unsuitable supplies. Photographs may be used to document such damage. Any storage required for materials or equipment should be noted, especially if related to a change or delay of work. Unavailability of material and the reason for it should be documented. Cost of equipment on the project should be recorded.

Idle or extra time for expensive equipment can be a major cost in a claim. Ripple effects of changes involving equipment can be a significant factor. Storage, removing such equipment from another job, work under adverse conditions (winter months), or simply the delay caused by moving bulky equipment to a different location on a site should be recorded. (Chapter 6 on the costs of delays deals with this subject in more detail.)

A record of the types of equipment used each day and the number of hours these are used should be kept. A full description of every piece of equipment should be available in an equipment register or a similar record. When equipment is being repaired, the nature of the breakdown and the repair work should be recorded, as well as the time and materials used for repair.

Accident Records

The circumstances and the details of each accident that occurs on a job should be described. The amount of detail in connection with each accident should be in direct proportion to the seriousness of the accident. In addition, a record should be made of measures taken to insure safety. These can include regular meetings, use of films, use of lectures.

Generally, accidents are entirely an internal problem of the contractor, but there can be situations where accidents or potential accidents involving employee safety can have a major effect upon job progress, and thus lay the groundwork for potential delays and claims. For example, if a tunnel becomes unsafe to the extent that miners will not go into it, special rehabilitation work may provide the basis for a claim. This may involve the installation of special temporary steel support and cause not only a dispute as to the method of payment, but also a further delay due to the time required to set such steel. During such a time, a sizable work crew, together with equipment, is not productively engaged in proceeding on the tunnel.

Delay Records

Any delays due to strikes, weather, lack of access to the site, late delivery of materials, unavailability of materials, accidents, or any other reason should be kept in a delay file. Specifics should be listed.

For instance, in the case of a strike, the dates when work stopped, when the strike was settled, and when the work was resumed all should be spelled out in detail. Add to this what the effect of the strike on the project; Did it affect the entire project? Did it stop an operation? If it did stop an operation, what was the effect of this on the overall progress? Strikes elsewhere that affect the work should be recorded. A national steelworkers strike or a national teamsters strike could have an effect upon the project. (See Chapter 4 for more detail.)

Subcontractors

The owner and the contractor can require of major subcontractors that similar records be kept on labor, equipment, delays, and similar matters. Naturally, if the subcontractor is a party to the claim, his records would be part of the claim, and he would be responsible for his own record keeping.

Overhead

Records of overhead costs on the job over a period of time are necessary if a claim is filed. Therefore, they should be kept as a matter of routine. This includes the time of all personnel, whether they are at the job site or in a headquarters office when they are doing work related to the project. It will include the costs of trailers, light, heat, telephone, supplies, and postage. The items that can be considered to be overhead are numerous. Overhead can include any costs connected with the project that are not specifically chargeable to an item of work. Whether or not an item is overhead is usually dependent upon the manner in which a contractor keeps his books.

The Eichleay Formula is one of the methods used to allocate job overhead. This formula and other methods of computing costs are discussed in Chapter 7.

Time Cards

The time card on the average project is coded by the labor classification of the workman and by the type of work done. Each worker has a particular code; for example, every carpenter will be in the 700 series, with the carpenter's helpers in the 800 series and the laborers in the 900 series. (See sample time cards at the end of this chapter.)

Every time card should include the date, the shift, the code number, and name of every workman in that particular crew; the number of hours worked; the items worked on; the location; and the equipment and materials used. The time card should provide a space for the foreman to make his estimate of production, and to list any problems, visitors, weather, lost time, or inefficient time. As a cross-check, the timekeeper will certify that these men were on the project at that particular time and were performing that particular operation, along with the number of hours they worked. The payroll clerk will also check numbers; and depending upon the project, a member of a management will also check the work items.

The time card process starts when the employee is hired. The worker is assigned a craft-related employee number at that time. For example, all carpenters could be in a 5000 series. Firms that have many projects, or those with more than 200 employees are advised to use computers. In a computerized payroll and cost accounting system, basic data such as the employee's social security number, craft, employee number, pay rate, and so on, are entered either into a computer or a CD used with the computer. Anything that affects pay rates, overtime rates, tax deductions, union

FOREMAN'S TIME CARD

SHIFT NO.

DATE

FOREMAN

PROJECT CODE

LOCATION CODE

FOREMAN'S SIGNATURE

APPROVED _____ SUPERINTENDENT

JOB DESCRIP. CODE

BADGE NO.	EMPLOYEE NAME	CODE	RATE	HOURS		COMMENTS (ACCIDENTS, ETC.)
				OT		
				RT		
				OT		
				RT		
				OT		
				RT		
				OT		
				RT		
				OT		
				RT		
				OT		
				RT		
				OT		
				RT		
				OT		
				RT		

This time card is generally used on computerized payrolls and cost accounting. Foreman enters data including employees' names, badge number, work description, work code, and craft code. Computer does the rest.

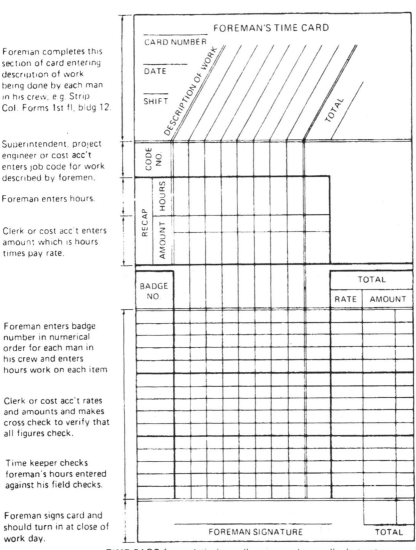

Foreman completes this section of card entering description of work being done by each man in his crew, e.g. Strip Col. Forms 1st fl, bldg 12.

Superintendent, project engineer or cost acc't enters job code for work described by foremen.

Foreman enters hours.

Clerk or cost acc't enters amount which is hours times pay rate.

Foreman enters badge number in numerical order for each man in his crew and enters hours work on each item

Clerk or cost acc't rates and amounts and makes cross check to verify that all figures check.

Time keeper checks foreman's hours entered against his field checks.

Foreman signs card and should turn in at close of work day.

TIME CARD for a relatively small project and generally, but not necessarily for a non-computerized payroll and cost accounting system.

207

deductions, or other payments by the employer is entered into the computer or CD.

Similarly, every operation on which the contractor wishes to keep a record of costs (labor in this case, but the system is used to record other costs as well) is assigned a code number. For example, forms for concrete may be subdivided into classifications of footings and slabs on earth, then into walls, beams, soffits for slabs, and so on. In sum, these subdivisions can be subdivided into prefabricated forms, erect, strip, and clean, and oil. So, concrete forms may have a 6000 designation. Within the series, we could have 6010 (footing forms), 6011 (erect footing forms), 6014 (strip footing forms), 6015 (clean and oil footing forms), 6020 (wall forms), and so forth.

These codes generally require fine-tuning for each particular project. The foreman must know the codes and also the details of the system so that he can enter proper numbers and descriptions. When a card is completed for a shift and turned in, it must be checked by a responsible member of the project administrative staff for accuracy. Depending on job size this can be the project manager, the superintendent, or the cost engineer.

The information on the time card, along with the basic information stored on a CD is fed into the computer. This includes the employee's hours worked as a total as well as the hours worked that are designated under a particular coded item or items.

The computer then is used to output payroll data on a weekly basis. The output can be mailed to the job site; or if the computer is on site, a printout can be provided there. Checks are made out mechanically at either location.

Cost data may be printed on a weekly or a monthly basis. The responsibility rests with the estimator or the cost engineer to feed the necessary quantities of work prepared during the period into the computer. Thus, total stored labor costs can be distributed into total and costs per item of work. On smaller projects, all these calculations can be performed manually either on site or at the main office. Some sample time cards follow.

SUMMARY AND SAMPLES

Records are kept for many purposes on a construction project. A contractor requires cost records and productivity records for use in the preparation of item costs to be used in the preparation of future bids. As a matter of good business, as the job progresses, the contractor should determine how his actual expenses compare with his estimated expenses. He should be calculating what his anticipated profit is on a continuing basis compared to the profit he estimated when his bid was prepared. Payroll records must be kept so that the proper payroll taxes, such as federal, state, and local income taxes can be deducted, together with social security and similar taxes and charges applicable to payroll.

The owner is generally interested in whether or not the contractor is paying the prevailing wage rate and, at times, may be interested in keeping his own costs on particular items of work for reference in future projects. There are also records required by federal, state, or local regulations. For example, those relating to affirmative action, equal opportunity, minority hiring goals, and so on.

Most important, in terms of this book, records are kept to document claims or to defend against claims or extra costs caused by a wide spectrum of delays or changes.

All parties to a project should keep records. Once it has been determined that a delay has occurred or that there is a change to the contract documents, the records should provide an equitable means of determining whatever additional costs are incurred. If records are not maintained or have been lost, one party or the other may suffer unjustly. Claimants may be forced to rely on experts' reasonable estimates, for example, or be forced to accept the other party's calculations of costs, or go to court.

The contractual documents are the obvious reference point for the start of a claim. It is the claimant's responsibility to prove not only that there was a deviation from the contract documents, but that the incurred additional expenses came either directly or indirectly as a result of this deviation. The proof of the claimant's position must rest with the documentation that he maintained and the documentation available under freedom-of-information laws, pretrial disclosure, or through the goodwill of the party from whom he is requesting recompense.

Sample Minutes of Job Meeting

A typical heading for a set of minutes for a job meeting, followed by the outline of business conducted during that meeting, is on page 210. In this fictional case, the project is being performed under a construction manager (CM), who is responsible for keeping the minutes.

ABC CONSTRUCTION MANAGEMENT CO.

Minutes of Job Meeting
Owner: Fairfax Developers
Project: Timon Housing

Meeting No.   _____
Page   _____
Date   _____
Contract No.   _____

Phase II Grading and Road Paving
          DEF Construction Company

Present: Mssrs.

T. E. Smith	DEF Construction Co.	B. C. White	Timon Housing
R. A. Jones	DEF Construction Co.	D. F. Right	ABC Construction Management Co.
A. C. Black	Timon Housing		

Date: _____   Location: _____

Outline of basic entries:

1. General and Special Announcements
2. Review of Minutes of Last Meeting
3. Payment Requests—status, past and present.
4. Job progress by specific items with emphasis on critical path items
5. Job Administration
     Shop drawing status
     Change order status
     Owner furnished materials and equipment
     Quality control reports
     As-built drawings
     Progress photos
     Site control (access, parking, etc.)
6. County restrictions
7. Contract Item List—list of materials needed such as roadway base stone, cast iron inlets, road signs, manhole casting.
8. Safety Reports
9. Labor Conditions
10. Equal employment opportunity reports
11. Weekly payroll reports.
12. Date of next meeting, time and location.

Distribution list

     All present
     Others

Signed by _____
Title

Sample Inspector's Daily Reports

Two samples of an Inspector's Daily Report during construction of the Teton Dam appear on pages 211-226. The formula still holds, although computers have expedited the process. The first is handwritten; the second is the same report, but typed. Although the typed version may be more legible, the handwritten copy lends more authenticity to the report.

SNCO-4

UNITED STATES
DEPARTMENT OF THE INTERIOR
BUREAU OF RECLAMATION
SOUTHERN NEVADA CONSTRUCTION OFFICE

INSPECTOR DAILY REPORT

.507

ROUTING | INT
Chief Inspector
Field Engineer
Const. Engr.
Office Engr.
Contract Adm.
240

PROJECT _____ S-N-C-O _____

Feature _PUMPING PLANTS_ Date _THURS. APRIL 10 1990_ Shift _DAY 6 AM - 3 PM_

Spec. No. _DC7347_ Weather _CLEAR COOL TO WARM_

Contractor _____ Contractor Rep. _____

Sub Contractor _____ Location of Work _HACIENDA PUMPING PLANT_

Safety _SATISFACTORY - 5 MIN. TAILGATE SAFETY MEETING CHANGED TO MON. OF EACH WEEK._

Description of Work _6:15 AM - 3" & 4" SUMP PUMPS OPERATED_

WHAT _DURING PAST 24 HOURS. 6" DISCHARGE - 252 GPM._

WHERE _9 AM - DISCHARGE MANIFOLD ENC. N. 35 E PORTION OF_

WHO _NORTH WALL ELEV. 1846 TO ELEV. 1853 LINES 'B' & 'C'_

WHY _AND LINES 1 TO 5 WAS PLACED BY PLACEMENT_

WHEN _CREW. CONCRETE WAS REASONABLY CONSISTENT_
AS TO WORKABILITY. BATCH 70 Y³ USED 69¾ Y³ ¼ Y³
WASTE. 12:30 AM PLACEMENT COMPLETE.
2 CARPENTER CREWS CONTINUE WORK ON LEVEL
II SOUTHWALL N° 5 FORMS. GETTING FORMS
READY FOR WPRS SURVEY CREW CHECKOUT.
WPRS SURVEY CREW CHECKED OUT INSIDE
WALL FORM FOR LINE & PLUMB. WILL
COMPLETE CHECKOUT @ 7:30 AM FRIDAY 4/11/90
PLACEMENT OF 100 Y³ SCHEDULED FOR 9:00 AM
FRIDAY, 4/11/90.
2 CARPENTER CREWS CONTINUE WORKING
ON 2ND DECK FORMS & BEAM FORMS.
IRONWORKERS COMPLETED PLACING & TYING
REBAR FOR BEAMS & TOP & BOTTOM MATS IN II
DECK.

(OVER)

ELECTRICIANS CONTINUE CONDUIT & LIGHT
BOX INSTALLATIONS IN 2ND DECK
HOURS WORKED CRAIG 6AM - 7PM
              HART 7PM - 6AM

PLUMBERS COMPLETED INSTALLATION OF
DRAIN SYSTEM FOR THIS AREA.
LABORERS CONTINUE · CLEANUP &
BLOWDOWN OF FORMS AND MATERIALS
DIRT SPREAD - AIR CHAMBER TEE ENC. AREA.
  3-10 WHEEL DUMP TRUCKS & 2 SEMI
TRAILERS (END DUMPS) CONTINUE HAULING
BACKFILL MATERIAL FROM WASTE AREA
FOR USE AS COMPACTED EMBANKMENT
AROUND & ABOUT STRUCTURE.
SPREADING BY JD 450 DOZER & SPREADING
& COMPACTION BY CAT 825B' WHEEL
ROLLER. WACKER BEING USED FOR
COMPACTION AROUND AND AT STRUCTURE
WALLS
RETEST No 1- COMPACTION BY WACKER
ELEV 1854.00 - LINE 14 & LINE 'a'
RESULTS 93.3% & .4 DRY- WILL REWORK &
RETEST.
TEST No 2- COMPACTION BY WHEEL ROLLER
  ELEV 1864.00   LINE 'a' 15' NORTH & LINE 22 50' WEST
RESULTS · 99.87% - .14 WET ACCEPT
TEST No 3. COMPACTION BY WACKER
  ELEV 1864.00   LINE 'a' 5 FT. NORTH & LINE 22 40' WEST
RESULTS - 96.8% - 1.2 DRY ACCEPT
3 PM
OPER & OILER w/ HC 138A CRANE; MASTER MECHANIC
w/ TRUCK, WELDER & TOOLS & 2 LABORERS STARTED
ADDING A 30' JT OF GCM PIPE TO SUMP OF
3" PUMP. WILL ALSO CHANGE PIPING &
ADD HOSE FOR ADDITIONAL DEPTH
6 PM JOB NOT COMPLETE - 3" SUMP PUMP NOT BACK IN
OPERATION. WILL START AT 6:30 AM FRIDAY 4/11/90

SNCO 9

UNITED STATES
DEPARTMENT OF THE INTERIOR
BUREAU OF RECLAMATION
Southern Nevada Construction Office
CONCRETE INSPECTORS PLACEMENT REPORT

ROUTING	INIT.
CHIEF INSPECTOR	
FIELD ENGINEER	
LEAD ENGINEER	
OFFICE ENGINEER	
CONTRACT ADMIN	

HACIENDA PUMPING PLANT FEATURE _____ CONTRACTOR PETER KIEWIT SONS SPEC NO DC7347
FOREMAN H. ARGABRIGHT WEATHER CLEAR-WARM DATE 4/10/80 SHIFT 7am-3pm DAY

PLM NO	STRUCTURE	STATION FROM	TO	ELEVATION FROM	TO	TIME START	END
1	DISCH. MAN. ENC #5	LINES A TO C	LINES 1 TO 5	1846-1/2	1853-33	9:10am	12pm
2	NORTH WALL						
3							
4							

PLM NO	CY CONCRETE PLACED GROUT	3/4"	1-1/2	3"	GVMNT	CY CONCRETE WASTED CONTR	MAX SIZE	REASON	PLACING TEMP NO TAK	MAX	MIN
1			69 3/4			1/4		OVERORDER		75°F	70°F
2											
3											
4											

PLM NO	METHOD OF PLACING	CONSOLIDATING	PROTECTING	CURING
1	CRANE w/2Y3 BUCKET	INTERNAL VIBRATION	CLEAR & WHITE CURING COMPOUND	
2				
3				
4				

SAFETY & REMARKS REBAR AS DETAILED AND AS PER SPEC. DWGS

RELIEVED _____ RELIEVED BY _____ INSPECTOR

SNCO-7

UNITED   STATES
DEPARTMENT OF THE INTERIOR
BUREAU OF RECLAMATION
SOUTHERN NEVADA CONST. OFFICE
INSPECTORS   DAILY   REPORT   FOR   *THURS* 4/10/80
DATE

LABOR ~AND~ EQUIPMENT

SPEC'S. NO. *DC934-7*   INSPECTOR *Jul. M. Chang*

LABOR

Name of Employee	Class	Rate	Hour	Description of Today's Work	Item
M. ROBERSON GENCARPFMN			11½		
~L. BRISON~ ~CARP.~				VACATION	
D. NERO CARPFMN			10¼	SUPERVISION	
G. GRAHAM CARPFMN			10½		
C.P. JOHNSON CARPFMN			8		
C.R. LEWIS CARPFMN			11		
H. LOOSBROOCK "			10½	BUILD FORMS	
J. GRAHAM "			10½	SET FORMS	
D. COURTNEY "			10½	STRIP FORMS	
G. STERRETT "			10½	SET FALSEWORK	
J. MᶜCULLAR "			10½		
R. DOZIER "			10½		
A. MAESTAS "			8		
A. ANDERSON "			8		

EQUIPMENT

Equipment	Model	Make	Hour	Capacity	Today's Work	Item

SNCO-7

UNITED   STATES
DEPARTMENT OF THE INTERIOR
BUREAU OF RECLAMATION
SOUTHERN NEVADA CONST. OFFICE
INSPECTORS DAILY REPORT FOR   *THURS* DATE *4/10/90*

LABOR   AND   EQUIPMENT

INSPECTOR *Jules M. Craig*

SPEC'S. NO. *OC7347*

LABOR

Name of Employee	Class	Rate	Hour	Description of Today's Work	Item
J. SALAZAR	CARP.		10½		
				BUILD FORMS	
D. BROD AWAN	CARP APP		10½	SET FORMS	
W. HUGHES	" "		10½	STRIP FORMS	
K. LESTER	" "		8	BUILD SAFETY HANDRAILS,	
R. HOUSKA	" "		10½	STAIRS, ETC.	
R. KEEVER	" "		8	SET FALSEWORK	
AL WARD	CEM FIN FMN		8	SUPERVISON	
C. DU SHANE	C/F APP.		8	CONCRETE REPAIRS	
M. DELEO	CEM. FIN		8		

EQUIPMENT

Equipment	Model	Make	Hour	Capacity	Today's Work	Item

UNITED STATES
DEPARTMENT OF THE INTERIOR
BUREAU OF RECLAMATION

SNCO-7

SOUTHERN NEVADA CONST. OFFICE

INSPECTORS DAILY REPORT FOR

LABOR ⟶ EQUIPMENT

THURS DATE _4/10/90_

SPEC'S. NO. _OC7347_

INSPECTOR _Jules M. Craig_

LABOR

Name of Employee	Class	Rate	Hour	Description of Today's Work	Item
L. SMITH	LABOR FMN		11½	SUPERVISION	
E. SACHETTI	LABORER		8		
J. WATERS	"		8		
A. DELAROSA	"		8	SAND BLASTING	
W.H. WILSON	"		10¼	BLOW DOWN	
J. LOGAN	"		11'	CLEAN UP	
M. BOBB	"		10½	HELP CARPENTERS	
M. KEITH	"		8		
A. CHAVEZ	"		8		
P. SILVA	TMSTR		8	WATER TRUCK	
PHIL JUAREZ	MSTR. MECH.		6	WELD GC MP EXTENSION	

EQUIPMENT

Equipment	Model	Make	Hour	Capacity	Today's Work	Item

UNITED STATES
DEPARTMENT OF THE INTERIOR
BUREAU OF RECLAMATION

SNCO-7

SOUTHERN NEVADA CONST. OFFICE

INSPECTORS DAILY REPORT FOR THURS 4/10/90 DATE

LABOR *AND* EQUIPMENT

SPEC'S. NO. DC 7347   INSPECTOR _Jules M. Craig_

LABOR

Name of Employee	Class	Rate	Hour	Description of Today's Work	Item
R. ALLISON	OPER. FMN		8		
J. MATTHEWS	OPER		8	LOAD TRUCKS	
L. ABRAMS	"		9	HAUL BACKFILL MATERIAL	
A. PAYNE				SPREAD BACKFILL MATERIAL	
J. SANDCANTI	OPER		8	COMPACT EMBANKMENT	
R. GABEL	TMSTR		8	HAUL & PLACE ROCK	
G. CHADWICK	"		8		
R. JAMES	"		8		
J. PERDON	"		8		
C. BAIRD	"		8		
J. LANCASTER	OPER.		12	CRANE WORK W/ FORMS	
J. DREW	OILER		11½	UNLOAD MATERIALS	
A. MATTHEWS	LABOR		8	COMPACT EMBANKMENT	
R. WAGNER	"		8	FLAGMAN - HANDWORK - DIRT	

EQUIPMENT

Equipment	Model	Make	Hour	Capacity	Today's Work	Item
CRANE	HC138A	LINKBELT	12 / 8	25 T	FORMS & MATERIALS	
DOZER	JD450C	JOHN DEERE			SPREAD BACKFILL	
VIBRATORY ROLLER (SHEEP FOOT		ESSICK)			COMPACTION	
WACKERS	STE 300	MIKAYSA	8		"	
WATER TRUCK		FORD	6	1200 GAL	DUST CONTROL	
LOADER	966C	CAT	8	$3/5 y^3$	LOAD TRUCKS	
BACKHOE	780	CASE	4	$2 y^3$	HAUL ROCK	
TRUCKS - 3 (10 WHEEL END DUMPS, 2 - TRAILER					HAUL BACKFILL	
END DUMPS)			24		MATERIAL	

SNCO-7

UNITED STATES
DEPARTMENT OF THE INTERIOR
BUREAU OF RECLAMATION
SOUTHERN NEVADA CONST. OFFICE
INSPECTORS DAILY REPORT FOR   THURS
DATE 4/10/90
LABOR ~AND~ EQUIPMENT

SPEC'S. NO. DC7347   INSPECTOR Jules M. Craig

LABOR

Name of Employee	Class	Rate	Hour	Description of Today's Work	Item
STEEL ENG'R'G.					
J. HUFFSTEDLER	I/W FMN		8		
R. CAMPBELL	I/W FMN		8		
DALE COWAN	I/W		8	PLACE & TIE REBAR	
JIM NELSON	I/W		8		
L. LAUNCIEAUX	I/W		8		
BOB PIPER	I/W		8		
ROSENDIN ELEC					
J. HEISHMAN	ELEC		8	PLACE & SECURE CONDUIT	
J. BUETTNER	ELEC		8		
B. P. WILLIAMS					
P. BRODHY	PLBR FMN		4	DRAINS	
M. VELENCE	"	"	4		

EQUIPMENT

Equipment	Model	Make	Hour	Capacity	Today's Work	Item

SNCO-4

UNITED STATES
DEPARTMENT OF THE INTERIOR
WATER AND POWER RESOURCES SERVICE

SOUTHERN NEVADA CONSTRUCTION OFFICE

INSPECTOR DAILY REPORT

ROUTING	INT
Chief Inspector	
Field Engineer	
Const. Engr.	
Office Engr.	
Contract Adm.	

PROJECT _____ SNCO _____

Feature Pumping Plants _____ **Date** Thurs. 4/10/90 __ **Shift** 6:15 A.M. - 3:30 P.M.

Spec. No. DC-7347 _____ **Weather** Clear, Cool to Warm _____

Contractor _____ **Contractor Rep.** _____

Sub Contractor _____ **Location of Work** _____

Safety Satisfactory - 5 min. tailgate safety meeting changed to Monday of
each week.

WHAT
WHERE
WHO
WHY
WHEN

Description of Work 6:15 A.M. - 3" and 4" sump pumps operated during past 24 hours.
6" discharge - 252 gmp+. 9:10 A.M. - discharge manifold enc. No. 5 and portion
of north wall elev. 1846.42 to elev. 1853.33 lines 'a' and 'c' and lines 1 to 5
was placed by placement crew. Concrete was reasonably consistent as to
workability. Batch 70 Y^3 used 69 3/4 Y^3 1/4 Y^3 waste. 12:30 P.M. placement
complete.

2 carpenter crews continue work on level II southwall No. 5 forms. Getting
forms ready for WPRS survey crew checkout. WPRS survey crew checked out inside
wall form for line and plumb. Will complete checkout at 7:30 A.M. Friday,
4/11/90. Placement of 100 Y^3 scheduled for 9:00 A.M. Friday, 4/11/90.

2 carpenter crews continue working on 2nd deck forms and beam forms. Iron
workers completed placing and tying rebar for beams and top and bottom mats in II
deck.

Electricians continue conduit and light box installations in 2nd deck.

Plumbers completed installation of drain system for this area. Laborers continue
clean-up and blow down of forms and materials dirt spread - air chamber tee enc.
area. 3 - 10 wheel dump trucks and 2 semi trailers (end dumps) continue hauling
backfill material from waste area for use as compacted embankment around and about
structure.

Spreading by JD 450 Dozer and spreading and compaction by Cat 825B wheel roller.

Wacker being used for compaction around and at structure walls.

Retest No. 1 - compaction by wacker.

Elev. 1854.00 - line 14 and line 'a'.

Results: 93.3% and .4 dry - will rework and retest.

(OVER)   **INSPECTOR**

Test No. 2 - Compaction by wheel roller.

Elev. 1864.00 - line 'a' 15' north and line 22 50' west

Results - 99.8% - .4 wet accept

Test No. 3 - compaction by wacker

Elev. 1864.00 - line 'a' 5 feet north and line 22 40' west

Results - 96.8% - 1.2 dry accept

3:30 P.M. - oper and oiler w/HC138A crane; master mechanic w/truck, welder and tools and 2 laborers started adding a 30' jt of GCM pipe to sump of 3" pump. Will also change piping and add hose for additional depth.

6:30 P.M. - job not complete - 3" sump pump not back in operation. Will start at 6:30 A.M., Friday 4/11/90.

SNCO 9

UNITED STATES
DEPARTMENT OF THE INTERIOR
BUREAU OF RECLAMATION

Southern Nevada Construction Office

CONCRETE INSPECTORS PLACEMENT REPORT

ROUTING	INIT.
CHIEF INSPECTOR	
FIELD ENGINEER	
CONST ENGINEER	
OFFICE ENGINEER	
CONTRACT ADMIN	

FEATURE ___Pumping Plant___   CONTRACTOR _____   SPEC. NO ___DC-7347___

FOOTAGE _____   WEATHER ___clear, warm___   DATE ___4/11/90___   SHIFT ___7 - 3:30___ Day

PLM NO	STRUCTURE	STATION		ELEVATION		TIME	
		FROM	TO	FROM	TO	START	END
1	Disch. Mani. Enc. #5 &	Lines	Lines	1846.42	1853.33	9:10 A	12:30 P
2	North wall	'a' to	1 to 5				
3		'c'					
4							

PLM NO	CY CONCRETE PLACED				CY CONCRETE WASTED				PLACING TEMP		
	GROUT	3/4"	1-1/2"	3"	GVMNT	CONTR	MAX SIZE	REASON	NO TAK	MAX	MIN
1			69 3/4			1/4		Overorder		75 F	70F
2											
3											
4											

PLM NO	METHOD OF:			
	PLACING	CONSOLIDATING	PROTECTING	CURING
1	Crane w/2 Y³ Bucket	Internal Vibration	Clear & White Curing Compound	
2				
3				
4				

SAFETY & REMARKS   ___Rebar as detailed and as per spec. drawings.___

CONTINUE REMARKS ON BACK OF SHEET

RELIEVED _____   RELIEVED BY _____   INSPECTOR _____

SNCO-7

UNITED   STATES
DEPARTMENT OF THE INTERIOR
WATER & POWER RESOURCES SERVICE

SOUTHERN NEVADA CONST. OFFICE

INSPECTORS   DAILY   REPORT   FOR

LABOR   AND   EQUIPMENT

Thurs.
DATE 4/10/90

SPEC'S. NO. DC-7347

INSPECTOR

LABOR

Name of Employee	Class	Rate	Hour	Description of Today's Work	Item
M. (	Gen Carp Fmn		11½		
D. X	Carp Fmn		10½	Supervision	
G. X	Carp Fmn		10½		
C. X	Carp Fmn		8		
C. X	Carp Fmn		11		
H. X	Carp Fmn		10½	Build Forms	
J. X	Carp Fmn		10½	Set Forms	
D. X	Carp Fmn		10½	Strip Forms	
G. X	Carp Fmn		10½	Set False Work	
J. X	Carp Fmn		10½		
R. X	Carp Fmn		10½		
A. A	Carp Fmn		8		
A. A	Carp Fmn		8		

EQUIPMENT

Equipment	Model	Make.	Hour	Capacity	Today's Work	Item

SNCO-7

UNITED STATES
DEPARTMENT OF THE INTERIOR
WATER & POWER RESOURCES SERVICE

SOUTHERN NEVADA CONST. OFFICE

INSPECTORS DAILY REPORT FOR

LABOR ⟶ EQUIPMENT

DATE 4/10/90   Thurs.

SPEC'S. NO. DC-7347   INSPECTOR _____

LABOR

Name of Employee	Class	Rate	Hour	Description of Today's Work	Item
J. S	Carp.		10½		
				Build Forms	
D. B	Carp App.		10½	Set Forms	
W. H	Carp App.		10½	Strip Forms	
K. L	Carp App.		8	Build Safety Hand Rails,	
R. H	Carp App.		10½	Stairs, etc.	
R. K	Carp App.		8	Set False Work	
Al	Cem Fin Fmn		8	Supervision	
C. D	C/F App.		8	Concrete Repairs	
M. D	Cem Fin		8		

EQUIPMENT

Equipment	Model	Make	Hour	Capacity	Today's Work	Item

SNCO-7

UNITED   STATES
DEPARTMENT OF THE INTERIOR
WATER & POWER RESOURCES SERVICE

SOUTHERN NEVADA CONST. OFFICE

INSPECTORS DAILY REPORT FOR

LABOR   AND   EQUIPMENT

Thurs.
DATE 4/10/90

SPEC'S. NO. DC-7347

INSPECTOR

LABOR

Name of Employee	Class	Rate	Hour	Description of Today's Work	Item
L. SS	Labor Fmn		11½	Supervision	
E. S S	Laborer		8		
J. W S	Laborer		8		
A. D S	Laborer		8	Sand Blasting	
W. H S	Laborer		10½	Blow Down	
I. L S	Laborer		11	Clean-up	
M. I S	Laborer		10½	Help Carpenters	
M. K S	Laborer		8		
A. S S	Laborer		8		
P. S S	TMSTR		8	Water Truck	
Phil I S	MSTR. MECH.		6	Weld GCMP Extension	

EQUIPMENT

Equipment	Model	Make	Hour	Capacity	Today's Work	Item

SNCO-7

UNITED STATES
DEPARTMENT OF THE INTERIOR
WATER & POWER RESOURCES SERVICE

SOUTHERN NEVADA CONST. OFFICE

INSPECTORS DAILY REPORT FOR

LABOR ~AND~ EQUIPMENT

Thurs.
DATE 4/10/90

SPEC'S. NO. DC-7347   INSPECTOR _____

LABOR

Name of Employee	Class	Rate	Hour	Description of Today's Work	Item
AX	Oper Fmn		8		
MX	Oper.		8	Load Trucks	
AX	Oper.		9	Haul Backfill Material	
SX	Oper.		8	Spread Backfill Material	
CX	TMSTR.		8	Compact Embankment	
CX	TMSTR.		8	Haul & Place Rock	
JX	TMSTR.		8		
PX	TMSTR.		8		
BX	TMSTR.		8		
LX	Oper.		12	Crane work w/forms	
DX	Oiler		11½	Unload Materials	
MX	Laborer		8	Compact Embankment	
WX	Laborer		8	Flagman - Hand work - Dirt	

EQUIPMENT

Equipment	Model	Make	Hour	Capacity	Today's Work	Item
Crane	HC138A	Linkbelt	12	75T	Forms & Materials	
Dozer	JO450C.	John Deere	8		Spread Backfill	
Vibratory Roller	(Sheep Foot Essick)				Compaction	
Wackers	Ser. 300	Mikaysa	8		Compaction	
Water Truck		Ford	6	1200 GAL	Dust Control	
Loader	966C	Cat	8	3/5 Y^3	Load Trucks	
Backhoe	780	Case	4	24.3	Haul Rock	
Trucks - 3 (10 wheel end dumps and 2- trailer end dumps)			24		Haul Backfill Material	

SNCO-7

UNITED STATES
DEPARTMENT OF THE INTERIOR
WATER & POWER RESOURCES SERVICE
SOUTHERN NEVADA CONST. OFFICE
INSPECTORS DAILY REPORT FOR
LABOR *AND* EQUIPMENT

Thurs.
DATE 4/10/90

SPEC'S. NO. DC-7347

INSPECTOR _____

LABOR

Name of Employee	Class	Rate	Hour	Description of Today's Work	Item
Steel Engr.					
J. HX	I/W Fmn		8	Place and Tie Rebar	
R. CX	I/W Fmn		8		
Dale X	L/W		8		
Jim X	I/W		8		
L. X	I/W		8		
Bob X	I/W		8		
Elec.					
J. X.	Elec		8	Place and Secure Conduit	
J. X	Elec		8		
B.PX					
P. X	Plbr Fmn		4	Drains	
M. X	Plbr Fmn		4		

EQUIPMENT

Equipment	Model	Make	Hour	Capacity	Today's Work	Item

AUDIT REPORT

What follows is an audit report on a federal project.

UNITED STATES DEPARTMENT of the INTERIOR
OFFICE OF THE SECRETARY
OFFICE OF AUDIT AND INVESTIGATION
CENTRAL REGION

1841 Wadsworth
Lakewood, Colorado 80215

MEMORANDUM AUDIT REPORT

Memorandum

To: Commissioner, Bureau of Reclamation

From: Regional Audit Manager

Subject: Audit of Change Order No. 2 (Powerhouse), Specifications No. DC-6910, Teton Dam, Power and Pumping Plant, Teton Basin Project

We have completed our audit of costs incurred under Order for Changes No. 2 (Powerhouse), Specifications No. DC-6910, Teton Dam, Power and Pumping Plant, Teton Basin Project. Field work was completed at the Contractor's construction site at Newdale, Idaho, on December 13, 1974.

Your request was for an audit of costs incurred under Order for Changes No. 2 (Powerhouse) and Order for Changes No. 4 (Riprap). Order for Changes No. 2 resulted from not encountering rock in the northwest corner of the foundation of the power and pumping plant at elevation 4980 as anticipated in the specifications. The change order, as revised by the Contractor and submitted to the Bureau of Reclamation (LBR) per Contractor's letter dated June 26, 1974, claimed costs (and profit) totaling $177,848. Costs claimed under this change order were incurred during October and November 1972.

Order for Changes No. 4 resulted from increased costs associated with a change in the designated riprap source; the original quarry failed to provide the larger sizes of rock required by specifications. The change in quarry sites resulted in an increased haul distance and nonproductive costs incurred in attempting to utilize the original quarry site.

The original contract price for supplying and placing a ton of riprap on the dam was $2.85 per ton. In the Contractor's letter dated July 2, 1974, a proposed price of $6.80 per ton to supply and place riprap on the dam was made. This figure was revised in a subsequent letter from the Contractor dated November 26, 1974. The revised figure was set by the Contractor at $10.31 per ton.

Following a discussion of the increased costs of riprap with the Contractor's representatives, including the assistant comptroller from the home office, were informed by the assistant comptroller that he and his office felt that Order for Changes No. 4 was not submitted with due consideration to appropriate contract provisions and relative correspondence and that resubmittal was in order. He stated that a resubmittal would be made at a later date, possibly by the latter part of December 1974. Consequently, our audit was limited to costs claimed under Order for Changes No. 2.

AUDITED COSTS

The Contractor claimed costs totaling $177,848 under Order for Changes No. 2 (see attached exhibit). LBR has reimbursed the Contractor $7,900 under this change order. As a result of our audit we have questioned $16,336 of the 177,848 as follows:

Direct costs	$10,568
Indirect costs	5,768
	$16,336

Questioned costs are discussed in the remainder of this report under the captions direct costs and indirect costs.

DIRECT COSTS

Direct costs claimed under Order for Changes No. 2 and related questioned costs follow:

Items	Claimed costs	Questioned costs	Approved costs
Direct labor	$36,278	($4,718)	$31,560
Equipment ownership	20,151	(2,062)	18,089
Equipment operating expense	19,018	(3,044)	15,975
Materials	29,557	300	29,857
Escalation	15,277	15,277	
Small tools	2,123	(242)	1,881
Plant operations	6,028	(688)	5,340
Overtime premium	2,166	(114)	2,051
Totals	$130,598	($10,568)	$120,030

Direct costs adjustments shown above resulted from a review made by LBR personnel (with minor audit adjustments) of hourly rates and utilization of labor and equipment. LBR has already discussed these adjustments with the Contractor's personnel.

INDIRECT COSTS

Indirect costs applicable to the change order were computed as follows:

	Per contractor	Per audit	Questioned costs
Direct costs	$130,598	$120,030	
Indirect rate	23.79%	21.09%	
Indirect costs	$31,082	$25,314	$5,768

The indirect cost rate per Contractor and per audit were computed as follows:

	Per contractor	Per audit
Total direct costs	$6,404,651	$6,404,651
Total indirect costs	1,523,464	1,350,471
Indirect rate	23.79%	21.09%

Audit adjustments to the Contractor's indirect cost accounts were as follows:

Accounts	Accounts balances as of 12/31/72	Audit adjustments	Accounts adjusted balances
Travel expense - job connected	$9,943	($77)	$9,866
Entertainment, donations and contributions	678	(678)	
Insurance - other than payroll	90,828	(90,828)	
Taxes - other than payroll	35,165	(35,165)	
Mobilization expenses paid:			
General expenses (excluding items listed above)	1,297,131	(1,018)	1,296,113
General plant	89,719	(45,227)	44,492
Totals	$1,523,464	($172,993)	$ 1,350,471

Following are explanations of indirect costs questioned.

Travel expense - job connected

We have questioned $77 in travel costs. Sixty-two dollars in airfare was incurred by an employee's wife to attend a convention and $15 for entertainment. Only costs incurred by an employee for official business are allowable per 41 CFR 1-15.205-46(a), and entertainment costs are unallowable per 41 CFR 1-15.205-11.

Entertainment, donations and contributions

We questioned the $678 balance in the entertainment, donations and contributions account. Entertainment costs are unallowable costs according to 41 CFR 1-15.205-11. Donations and contributions are unallowable costs according to 41 CFR 1-15.205.8.

Insurance and taxes

Insurance and tax expenses applicable to equipment were recorded in the Contractor's general expense accounts ($90,828 and $35,165) and were used by the Contractor in the computation of his indirect cost rate. In his claim as a direct charge, the Contractor computed equipment ownership costs using percentage rates published by the Associated General Contractors of America (AGC). The AGC rates include a factor for equipment insurance and tax expenses, thus resulting in a duplication of costs.

Since the Contractor was unable to furnish us with data showing the breakdown of insurance and tax expenses, we have deducted total insurance costs of $90,828 and total tax costs of $35,165 in computing the overhead rate. Information available to us indicates that substantially all of the Contractor's recorded costs for insurance and taxes are related to equipment in use at the site.

Mobilization expenses paid

In accordance with the advertised contract, the Contractor received payments for mobilization and preparatory work for plant and equipment in a lump sum of $3 million. The mobilization costs were recorded in deferred asset accounts and are being amortized over the contract period. A portion of the mobilization costs appeared in the general expenses and general plant expense accounts (indirect costs accounts) for the period under consideration and in turn were included in the Contractor's computation of the indirect rate. Since these costs are applicable to the advertised contract and not the negotiated change orders, we have identified and deducted such costs ($1,018 and $45,227) in our computation of the indirect cost rate. Order for Changes No. 2 did not involve any mobilization or preparatory work.

While the indirect cost rate of 21.09 percent is applicable to this change order, it should not be considered applicable to change orders involving costs incurred after January 1974, because the Company recently underwent a reorganization. The reorganization resulted in a redistribution of overhead costs to the various divisions within the organization. It is our belief that this redistribution of overhead costs to the division sponsoring the Teton Dam project will differ from the costs distributed to the division prior to the reorganization.

UNSUPPORTED COSTS

The 10 percent profit rate (margin) applied to this claim is a negotiable amount to be determined by agreement between the Contractor and LBR. We have therefore classified the entire amount of profit proposed by the Contractor ($16,168) as unsupported.

We have discussed the audit findings with appropriate representatives of LBR and the Contractor at the Newdale, Idaho site. In the event there is need for additional information, you should contact me at the above address. Copies of correspondence pertaining to the negotiation of this change order should be forwarded to this office.

cc: F. X. Smith, Chief
Division of Construction
Bureau of Reclamation

SUMMARY OF COSTS APPLICABLE TO
CHANGE ORDER NO. 2 (POWERHOUSE)
SPECIFICATION NO. DC-6910
TETON BASIN PROJECT

	Revised claim	Questioned costs	Unsupported costs	Approved by audit
Direct labor	$36,278	($4,718)		$31,560
Equipment ownership	20,151	(2,062)		18,089
Equipment operations expense	19,018	(3,044)		29,857
Materials	29,557	300		29,857
Escalation	15,277	15,277		
Small tools	2,123	(242)		1,881
Plant operations	6,028	(688)		5,340
Overtime premium	2,166	(114)		2,051
Indirect costs	31,082	(5,768)		25,314
Profit	16,168		($16,168)	
Totals	$177,848	($16,168)	($16,168)	$145,344

11. Claims Presentation

Claims have a tendency to escalate with time. A principal point of this text has been that enlightened contracting practices and project management can avert time-consuming and costly claims. The choice of contract type is the first step in establishing the environment in which claims are likely (or not) to be filed. As we have discussed, fixed-price contracts in which risk is placed on the contractor are more likely to result in claims than any other kind of contract. However, various provisions in those contracts can mitigate the claims potential. We have discussed such provisions as variation in quantity clauses, differing site conditions clauses, and others that work to spread risk equitably. Prompt attention to requests for extensions of time and other actions on the part of the owner are sound project management practices that can also help prevent claims from escalating into disputes.

But assuming that a claim does arise, here is a typical sequence of events.

THE CLAIMS PROCEDURE

First, a letter is written by the contractor to the owner, pointing out a problem or a potential problem on the project. This might be a temporary impediment to site access or some other violation of the owner's obligation under the contract. The owner can immediately face the problem by holding a meeting with the contractor at which the parties discuss whether the contractor is due an extension of time and possibly additional payment because of this situation. Or the owner can ignore the situation and the letter of notice and allow the minor problem to escalate into a major one. Obviously, there are possibilities between these two extremes. At this stage, letter writing can also become a weapon. One side might send a barrage of letters designed to keep the other off guard and to build up the record. Knowing that a responsible owner or contractor will set up a claims file on the receipt of such a letter, the other side can count on triggering this record-keeping process. A lot of letters can keep the opponent so busy that the extent of real problems can become obfuscated. This is not an effective tactic, nor is it an effective way to solve problems and finish a project.

The proper approach is for the party who receives the letter to accept a potential claims notice in good faith. He should start his own file and then prepare and accumulate pertinent documentation.

The contractor, in making a claim, should give details and refer to the contract documents to show why his claim is legitimate. This need not be done in the first letter, but it should be done early in the proceedings.

The contractor should keep records of time, materials, and potential delay to the overall project as part of his documentation of the expense caused by the situation for which a claim is made, and furnish copies of these to the owner.

To sum up the contractor's position at this point, he wants to establish two things in these letters and records: (1) that his claim is legitimate under the contract, and (2) that expenses have been incurred because of the claim situation.

MEETINGS

At this stage, meetings between the owner's representative—the architect or engineer—and the contractor can help achieve an early resolution of differences. Unfortunately, too many owners tend to postpone any corrective action while they study the claim. In some cases, this takes months and even years. Thus, owners often contribute to the escalation process.

When claims escalate for whatever reason, attorneys will frequently attend meetings and take over the presentation of the parties case or defense. If resolution is not achieved, the case passes out of the hands of those original parties and goes to arbitration, the courtroom, an appeals board (in the case of some public bodies), or some other formal dispute forum. Let's examine the sequence of events in more detail.

1. Notice

The first step is for the contractor to recognize that a claims situation exists. He should acknowledge the problem and realize that slippage at this stage could lead to further damage down the line. He will recognize and calculate those effects, both time and money, on the eventual completion of the contracted work. At this point those calculations are estimates based on the contractor's experience.

The contractor should send the owner a concise letter describing the situation and his opinion of its effect on the project. The letter should refer to that section of the contract concerning "notice."

The contractor should be certain that the notice is timely under that clause and that it is presented to the proper individual. If all facts are not available when that first letter is written, others can be added in a follow-up letter. It is more important to get timely notice out than to include all relevant material by that deadline. The omission of this timely notice could preclude recovery by the contractor.

It is the owner's responsibility to acknowledge such letters promptly. Even if the reply indicates only that the matter is being looked into, or some other form of acknowledgment, the intent to cooperate has been established. This can be very important later on should the situation become more complicated.

All parties should keep separate files for each claim, and if possible, agree on a numbering or other identification system to facilitate correspondence and access to information in the future.

2.   Preparation

The next step in this long process is the preparation of the claim itself by the contractor. The contractor must accumulate the justification for the claim under the contract terms, along with information on its material costs, labor costs, and all other costs, plus the effect of the claims situation on other phases of the project. There may be a considerable time gap between the first step and the second. If the contractor can assemble his justification quickly, the owner sometimes agrees to make partial payments while the claims paperwork progresses. This is not unusual on major claims where a major delay in payment to a contractor could lead him to financial difficulties. Clearly, when this happens, the owner is acknowledging to some extent that the claim is justified.

As the work progresses for which the contractor is claiming additional payment, copies of the appropriate labor time cards, additional materials, equipment, tools and supplies, overtime costs, and other items outlined in Chapter 10, should be kept in the file. This information should be furnished to the owner through the appropriate party as provided in the contract.

At the same time, the contractor must continue buttressing the legal arguments for his claim.

3.   Assembling the Claim

Who puts the claim together? As we have seen, there are two elements to claims: the legal justification and the cost data. If the contractor has a sizable staff that handles large dollar volume construction work, much of the cost data can be assembled in-house. For smaller firms, or for complex claims, the contractor may employ a claims consultant.

On small claims, the staff can generally also provide the justification. But it is advisable that attorneys work on claims as they increase in complexity and size. These may be in-house counsel or outside attorneys who specialize in construction claims.

Assembling a claim is a team effort. If the claim is complex and an attorney is involved, he should assemble the final document, or at the very minimum, review the final work assembled by others.

The expense of having an attorney document and prepare a claim can be high. It is generally more economical to retain a claims consultant to assemble the data and then retain an attorney for the final product. As an alternative, the attorney may retain the claims consultant.

The level of documentation and justification assembled for a claim depends on several factors. The complexity of the claim and the dollars involved are most important. The needs and requirements of the owner are also important. Certain public agencies may demand more documentation than others.

The ability and resources of the owner are another consideration. Some owners don't have the manpower to analyze a massive amount of data. Others have the staff to adequately review a claim. For example, a large utility with major on going construction and a knowledgeable staff may require limited justification together with reasonable estimates of costs. The utility's own staff can verify this material. A municipality, for public accountability, may require extremely detailed justification and cost documentation.

Many contractors can assess the needs of the owner; and claims consultants and attorneys who have processed claims for a particular owner in the past can be invaluable in this area. Certain agencies, both municipal and federal, have better reputations than others in claims processing. A knowledgeable contractor will be aware of those reputations when he bids. He will, to a degree, provide for the cost of financing extra work when the agency has a poor reputation. In prosperous times, contractors may even avoid such agencies, and as a result, those agencies can have difficulty attracting bids.

Ordinarily, the processing of a claim, timely or untimely, does not affect the progress of the work. There are exceptional cases where the lack of payment is a hardship to the contractor and can force him to slow down or into default or even bankruptcy. The basic contract agreement provides that a contractor must proceed with the work, invoking the contractually specified disputes resolution procedure as his sole means of relief.

To sum up, a well-prepared and complete claim is always a good idea. Here are some other points to keep in mind concerning the claims process.

1. *The contract should be thoroughly understood.* Too often contract provisions are regarded as a lot of boilerplate. The contract should be read carefully, and when a potential claims situation arises, it should be reread before embarking on the claims course. The boilerplate may decide the case before it begins.

2. *The owner can reject the claim as unacceptable.* This may happen whether or not the claim document is unsatisfactory, but grounds for such action should be avoided. Make the claim clear, concise, and thorough. The appearance of a claim document is not merely cosmetic. The presentation should indicate that the claimant means business, that he hasn't thrown together the equivalent of some scribbling on the back of an envelope. If the owner wants to send back a claim as a tactic in his battle campaign, he will; the contractor needn't provide ammunition in the form of careless documentation and presentation that will be used against him.

3. *The opponent should be sized up carefully.* We have already discussed this point in relation to the capability of the owner to analyze the claims justification and the cost data. To carry the analysis one step further, some owners may have a reputation for emphasizing certain aspects of claims. For example, an owner may be a stickler for exhaustive legal justification. In discussions with the owner's job- and home-office staff, determine the kind of detail the owner requires on quantities, costs, productivity, labor, and equipment rates. What kind of documentation does he want? Does he need receipted invoices, or notarized or certified payrolls? How does he pay for fuels and repairs, for tools and supplies? Is he willing to accept some percentage of the equipment rates cited in manuals prepared by equipment associations?

    Does the owner have a litigious reputation? If so, prepare the claim with the thought that the judge and jury may be the final audience for your case.

    This advice works for all parties. Each can, and will be, gearing up to deal on the basis of the other side's reputation.

4. *Remember that the claim will likely be judged by someone who has not been on the job.* That could be the mayor of the municipality or the head of a federal agency. If possible, have that person visit the job site.

Prepare an executive summary that is brief and persuasive. Depending on the case, anything from two to ten pages should do it. Put time and effort into this; it is the broad brush stroke description of the claim. It should be convincing to that ultimate arbiter who, as we have said, may never have been on the job. The nonengineer, the person unaccustomed to dealing with contracts, will need all the assistance available when he reviews and judges the claim. The executive summary is the first impression of the claim; it should give the reader a clear concise picture of what the claimant wants and why he is entitled to it.

A good executive summary also makes a favorable impression on the top official who is familiar with the procedures. The claim may be settled expeditiously on the basis of a convincing summary. On the other hand, that top official may set it aside for a long period of time if it is confusing, disjointed, or contradictory. The next section of the claim can be an enlargement of the summary, as much as ten times as long, with considerable detail but without all the minutiae.

Finally, the last section of the claim should include all the calculations, backup material, correspondence, and other documentation in an appendix. This may even be a separate volume. There are no hard rules here. The claim structure should be based on one side's estimate of who will read it and on who will make the final decision.

5. *The claim presentation should include photographs, charts, copies of relevant correspondence, and any other appropriate visual evidence.* These can be more persuasive than pages of text and will assist reviewers who have not been to the job site.

6. *Avoid holding back any information with the idea of saving a compelling bit of evidence for the "perfect moment."* If you believe that you have overwhelming proof of the merits of your case, present it as early as possible. The other party may concede immediately and save both sides a lot of time and money. It is, of course, rare that a single item can tip the scales so radically. It is also an understandable temptation to want to hold a trump card. But there is no point in marching everyone to the courtroom for a show-stopper presentation, when the case could have been resolved at an earlier stage.

A considerable amount of judgment about the parties and the case should go into the decision about when and where to present particular evidence.

7. *The claims presentation should include statements, and if possible, documentation, of the steps taken to mitigate the condition that led to the claim.* Telephone calls, letters sent, and meetings held should be outlined. Good project management and administration assumes that such steps were taken; therefore, these are a key part of the claims presentation.

8. *Contractors should avoid the trap of thinking that certain agencies or owners are likely to pay so many cents on the dollar.* This leads to situations in which contractors overstate their claims to protect themselves. Unfortunately, however, there are some owners who operate this way. We reiterate: Contractors should know as much as possible about the owner and his policies. Cost documentation must be credible, calculations must be logical and available to the owner. Pages of formulas may be impressive, but the result must be reasonable. Much can be lost by having inflated figures exposed as being just that. In the case of federal contracts, if it can be shown that the contractor intentionally misrepresented the costs, severe penalties may be imposed.

In each of the steps discussed here, there should be continued communications between the parties: telephone calls, letters, and meetings. This communication provides opportunities for early settlement. In a situation in which the dispute drags on, such communication provides opportunities for documenting that each side knew or was informed of the other side's actions. And at the very least, it provides the vehicle for both sides to check their positions on the facts and figures so that misunderstandings don't occur. The owner's file on the claim should be growing at the same rate as the contractor's. When the final decision and settlement are being debated, discussion should focus on the legitimacy of the claim, not on those facts and figures.

CALCULATING THE CLAIM

There are many approaches to calculating damages. Basically, what the contractor wants to establish is how much additional cost he incurred. Put another way, he wants to establish the difference between what the job would have cost without the condition that gave rise to the claim and what cost was due to the claims situation.

A contractor knows what the job actually cost if he kept basic cost records, and he should be able to document every amount spent. The problem is in determining what it would have cost without the claim situation and in convincing the owner that these figures are accurate.

The difference between the bid price and the actual costs is not a sound basis for establishing damages. The bid price will rarely be accepted as gospel. The variations in bids presented by competing contractors for the same project show that there is a definite difference in opinion as to what a particular portion of the work will cost.

The contractor keeps cost records as a matter of routine. He compares his actual costs with the estimated costs with which he originally prepared the bid. Then, when these figures differ notably, he tries to determine why they do. The contractor attempts to determine if the difference is his own fault, either through inaccurate estimates, poor productivity, job delays that are not the responsibility of the owner, or other problems that are not under the owner's control. If, however, the problem does seem to be the owner's responsibility, then the contractor begins the claim process.

Even without cost records, the sophisticated contractor can recognize a condition that will increase his expense. Even so, to provide documentation, he must go through the cost calculations and quantify the cost that can be attributed to the claim situation. More detail on these calculations appears in Chapter 7.

An important point to remember is that no one can know what the actual cost of the project would have been without the changes. There can be estimates, and in rare instances a contractor may have done the same kind of work elsewhere. If there were no changes on that other project, we have a comparison. However, it would be extremely rare that there would be an exact "control group" set of circumstances for another project. The weather and the season of year may have been different even if the work were the same; the level of expertise and the productivity of the workers might have been different.

Thus, to a certain extent, there is a subjective aspect to the financial calculations in any claim. Many owners will make a judgmental decision as to cost. They will even state that their decision is subjective in the final report. This is particularly true in cases where loss of efficiency and the resulting costs are discussed.

Ultimately, the acceptance and the settlement of a claim without recourse to the courts (or to other formal dispute resolution forums) may depend on, and be determined by, subjective factors. The skill of tacticians, the credibility of witnesses and experts, and the effect of these parties on the representative of the owner may count as much as do the facts in the case. But those facts must be assembled with care. The caveat about subjectivity is merely that: preparation should be as factual and as thorough as possible.

The precedents in the case, either legal or those that can be used to calculate costs, should be made known. We discuss costs of delays, acceleration, and changes in other chapters. Now in attributing costs to these items, the contractor has to find out which costs are allowable. His claims consultant and attorney can provide such information. Here are some important points to consider:

1. There are objective aspects of the added costs that should be easy to quantify if good records have been kept: labor costs, supplies, materials, equipment costs, field office expenses, and so on.

2. There is a second category of costs that is less objective but that has been granted by owners for a sufficient number of years so as to provide precedent. These might include allocation of jobsite and main-office overhead to the costs of the work change, the inefficiencies of work forced into winter months or rainy seasons, limitations placed on access to the site, and occasionally the interest on capital borrowed by the contractor to perform this change in the work. (The federal government pays interest costs only in unusual circumstances.)

3. Finally, there are items for which calculations must be subjective, or to put it another way, for which there are no precise calculations. Examples might be the determination of the cost of change in the quality of supervision, of lowered morale on the job, of a change in the learning curve of workers due to the change in the work, and of the need for increased manpower or equipment on the site due to the claims situation.

Another difficult area to quantify is the inability of the contractor to take on other profitable work. While there are precedents for recovery of such damages, they are usually held to be too speculative to be allowed. A convincing and imaginative analysis is essential if a significant portion of the claim depends on such items. The services of either a claims consultant or an attorney with experience in this field would be advisable. In some cases, the services of both might be needed. They would know which areas have been allowable and which methods of calculation should be acceptable. The contractor's staff and counsel should be capable of judging the validity of such approaches.

Thus, actual labor costs, payroll records, and material expenses are only the beginning of the work needed to document a complex claim. The chapter on the contractor's costs of delays includes more on this subject, as well as a lengthy example of the calculations involved in claims.

RESPONSE TO A CLAIM

The owner must respond to the claim. The letter of acknowledgment can be brief; the response itself is a detailed reply to the claim that can take considerable time, regardless of the amounts of money involved. Lack of acknowledgment can be interpreted as lack of cooperation on the part of the owner.

That owner, having had previous notification of a potential claims situation, should have long since begun to accumulate his own file on the claim. He should also have met with the contractor. Even so, the analysis and the detailed response can take six months or longer.

A slow reaction on the part of the owner can be a tactic, albeit unethical. An owner may wish to gain leverage by dragging his feet with the idea that when the contractor reaches the negotiating table, he will be so eager for settlement that he will accept less. On the other hand, the owner risks arousing the ire of the contractor, who may then push the claim as a matter of principle to the bitter end, which could be a prolonged lawsuit. Ultimately, though, both parties will sit down at the negotiating table. There are a minority in the construction industry who relish these skirmishes. The industry, however, would be better served if this expenditure of energy and intelligence were directed elsewhere. One such direction would be an expeditious settlement of problems along the way, before they reach the crisis stage, with top contractor and owner personnel involved.

THE INITIAL MEETING

The first meeting during the negotiating stage is usually an informal one. Each side will try to evaluate the position of the other and the degree of seriousness that the other attributes to the particular claim.

The principal purpose of this first meeting is to establish procedures, state positions, and work toward a meeting of minds. Major areas of agreement should be determined, major areas of disagreement should be highlighted, and a middle ground where neither side is convinced should be ascertained.

Coming to the negotiating table may be the beginning of an extended process where experts with different purposes and viewpoints act in good faith in order to arrive at an equitable and legal settlement. Parties should avoid playing games and trying to escalate the discussion into a battle of wits. If the negotiations fail and the claim does escalate to another level, the ultimate judge in the case will be less interested in personalities and tactics than in the facts.

At this initial meeting, both sides walk through the claim. There should be a serious attempt to agree on the facts. If the facts cannot be agreed upon, an acceptable method of determining those facts should be established. For example, if weather is an important factor, there are several sources such as job records and the National Weather Service that can provide documentation. The parties should agree on which data they will accept. If the type and use of equipment on the site at a specific time is important, parties could agree on a method of determination acceptable to all. They can also agree to have that information available for the next meeting.

A second and important function of the first meeting is that it allows each group to evaluate the other. Parties can attempt to analyze the other side's strategy. Here are some points to keep in mind:

1. *Be prepared.* Even at the first meeting, it is important to be thoroughly prepared. Have a game plan. Decide beforehand who is going to speak and when. Select a team leader. Make sure everyone on the team knows who that person is. Prepare an outline, listing the important items that should be covered at this meeting. As we have stressed, each side is evaluating the other; appearing to be unprepared can leave an unfavorable impression.

2. *Who comes to the meeting?* It is courteous to inform the other side of the approximate number of people that will attend the meeting. It is also appropriate to indicate who they are. If one side brings an attorney without telling the other side, or if the top brass descends on the meeting for one side, while the other has sent only lower-rank people, the impact is obvious. It is well worth the effort to send a list of the people who plan to attend the meeting. If the owner will be represented by a vice president, then the contractor should certainly have the opportunity to have representation of equal rank.

   Two people are essential. One is the person who can speak for the firm with authority. The other is someone who has actually been on the job site. Failing the latter, there must be someone who is intimately involved with the project and knowledgeable about the problems.

3. *Know what you will settle for.* Both sides should know before that first meeting how much or how little they will settle for. Have a high and a low figure in mind from the start.

4. *Be flexible.* As the meeting progresses, be capable of adjusting the presentation according to your evaluation and perception of

the other group. Do they seem to want a fast, concise overview, or do they plan on an all-day meeting to discuss details?

5. *Is there a personality conflict?* If there is, regroup forces as quickly as possible—definitely before the next meeting, and if possible, during this one. A private meeting between the two principals can also resolve this sort of conflict by removing the sources of friction, or by other mutually agreeable means.

6. *Be tactful.* Strong words, accusations, and finger-pointing will not be helpful. In fact, such actions are counterproductive. The personal approach with recriminations frequently forces the parties into court. A more amicable—yet still factual—approach may resolve the matter in less time and at less expense.

FORMAL NEGOTIATIONS

The closing item of the initial meeting should be to establish the time, place, and the agenda for the next meeting. There should also be some agreement about who will attend that next meeting, and an agenda should be at least outlined.

Parties should aim at resolving the claim in that second meeting. At the very least, they should aim at deciding at that time whether agreement can or cannot be reached. If it cannot, the claim should move to the next level of decision making.

In big-money and more complex claims categories, it is not unusual for a series of meetings to be held. Many of these will be at staff level for the purpose of agreeing on data and matters of fact. The final negotiations will be in the hands of the principals.

These formal negotiations represent the final effort to come to agreement before submitting to a higher, outside authority. In the case of the federal government, this usually means going to the agency's board of contract appeals, after which the claim can go to the claims court, and ultimately, though rarely, to the U.S. Supreme Court. With other owners, the case may go to arbitration or to local or state courts. All are time-consuming processes, and ones in which the final judgment is out of the parties' hands. Thus, the stakes are high. The participants need to be skillful tacticians operating under some prearranged strategy. Ideally, project management practices and enlightened contract arrangements should prevent disputes from reaching this "United Nations Security Council" stage. However, having reached that stage, parties should be prepared.

Each has had an opportunity to size up the other at that initial meeting. But in addition to the first impression that the principal participants have of one another, it's a good idea to do some research. What are the reputations and backgrounds of these people? If the owner is a public agency, is the principal present in a position to make the final decision, or is he subject to being overruled by a higher authority? Could his yielding on the claim jeopardize an opportunity he might have for promotion? Could it lead to a transfer? Is there an election coming up that could influence his actions? Could this case become a political issue that might be blown out of proportion by the incumbent's political opponents?

Both sides should investigate the financial aspects of the claim. Will the settlement of this claim force the owner to find new financing? Will the owner have to return to a political body, such as a city council, for approval to spend the additional money? If the claim is settled at a lower figure, does the owner have the financing immediately available to pay the claim? In the case of the contractor, does he need the cash for immediate financing, or does the claim in any way affect his ability to obtain a bond on new work?

Find out if the other party has retained outside counsel and a claims consultant. If so, check out the reputations of these individuals, too. Neither side knows what facts the other has in its arsenal. Parties to the negotiations, including counsel and consultants, may have written papers themselves on similar subjects that could be relevant to the claim. This would be so if the information in such papers (in technical journals, etc.) appears to contradict statements made by those individuals during negotiations.

With freedom-of-information laws, the contractor can research the files of the public owner in matters pertaining to the case. Similarly, contract clauses may give the owner the right to see much of a contractor's documentation. In addition, an owner has a certain leverage. If, for example, he calls up a contractor and says that he needs to see certain documents in order to evaluate the contractor's claim, few contractors would be likely to deny the owner access to those documents.

In negotiation, advantage is a psychological matter, and a great deal depends on one's perspective. Contractors think the owner has the advantage at the beginning of the negotiations. The owner's staff have read and carefully studied the contractor's claim. The contractor has no idea what discrepancies or flaws have been found. Yet owners think the contractor has the advantage. He is on the offensive; he can call the shots. Probably both points of view are valid under certain circumstances.

The following are principles that apply to the later stages of negotiation:

1. Much of what has been said during the initial meeting applies to subsequent meetings. Have an agenda and a game plan. Designate a team leader. Preparation is very important.

2.   At the meeting, a spokesman for either side may run through what was agreed on at prior meetings. This gives the participants a chance to start out on an amicable basis, agreeing on something. Of course, in some cases there may be a different interpretation of what happened previously.

3.   Minutes should be kept of everything that is agreed on at these meetings. Thus, in addition to that verbal run-through, there is a document to which the parties have agreed or which they may amend. All parties present at the meeting should read and sign off on those minutes, particularly if agreement was reached on any subitems. Parties may have a great reluctance to signing off on anything at this point, feeling that such action will close off options later on in the bargaining. However, parties negotiating with candor and in good faith should be willing to stand by agreements that they have reached.

If final agreement is not reached at this meeting, the time, date, and site should be set for the next meeting.

Time and timing are very important. While holding frequent meetings does not guarantee that the dispute will be settled quickly, if they are not held at regular intervals, the dispute is bound to last longer. Months and even years have been required to reach settlement in many cases. Obviously, however, a definite period of time is required between meetings in order for parties to mull things over and to prepare factual analyses in response to points made at the previous meeting.

The passage of time apparently inspires some confidence in the justice of the decision. That optimum period of time "tells" each party that reasonable thought, analysis, and careful consideration were given to the problem. While this view of the time it takes for proper judgment is far from exact, it seems to be a fact of negotiating life.

What is that optimum period? It varies from case to case. But time means different things to different groups in the industry. A large federal agency, with staff to match and respectable resources including engineers and attorneys, has almost infinite time compared to a contractor who, in the worst of cases, could be on the brink of bankruptcy. Agencies will be there "forever" and have their own regular pace of accomplishing business. The Court Reorganization Act of 1982 has several provisions for hastening that pace; that law is discussed in Chapter 12.

Contractors, unless they are giants, are often working with borrowed money and have cash flow problems. It can be assumed that some will settle as quickly as possible and for less than what they have asked for, in order to obtain fluid capital.

Time, in most cases, is on the side of the owner. For example, he is rarely required to pay realistic interest charges on the money due the contractor. Government at all levels borrows money at a lower rate of interest than that of inflation. So with or without interest, the contractor is paid with dollars of lesser value when a dispute takes a long time to resolve. There is, therefore, the likelihood that the owner is saving money if the dispute goes on for years.

Unless there is a complete stoppage of work as a result of the claim, and these are rare, inflation works to the advantage of the owner. Most construction contracts provide that the work must continue while claims are being negotiated or adjudicated. For a contractor to threaten work stoppage amounts to an idle threat. If the contractor should carry out such a threat, he risks default and loss of his bonding.

4. Negotiating strategy is a personal matter. There are more or less standard approaches, and these may work for one individual and not for another. One member of the group is tough and firm, while another member of the team talks about compromise and argues the other side's case. If he doesn't actually argue, he may appear to be sympathetic to certain points made by the other side. The object is to make the opponents think they had better accept the "white hat's" offer or they'll be stuck with the "black hat's" harsher demands.

Playing "dumb" is another approach that may lure the adversary team into revealing more information than they intended.

Scornful accusations sometimes can be effective with some personality types. In other cases, such an approach can create hostilities and alienate parties so that settlement is effectively blocked. The use of such tactics relies to some extent on intuition and subjective judgments.

More often than not, there is no correlation between the sophistication of the negotiators and the size and complexity of the claim. The more sophisticated participants in a negotiation proceeding, regardless of the size of the claim, have heard everything and will probably dismiss the tactics we have just described as time-wasters. The skilled and seasoned negotiator is more impressed by facts and the quality of the presentation, written and oral, than by role-playing. (We point out, in the next chapter on courtroom resolution of disputes, that most judges are equally unimpressed by dramatic stratagems.)

On the other hand, a city engineer who hears one claim a year, may react in quite a different way to such negotiation tactics. And the claim may be enormous. In addition, at the local level, no matter what the size of the claim, newspaper publicity, the interference of public officials and the public will all be factors. The same factors will affect the contractor. His reputation could well suffer as the result of adverse publicity.

If the contractor wants repeat business with any owner, he should avoid the stance of the tough, unyielding claims negotiator. Nor should he ever leave the impression that he is excessively claims-conscious. In the private sector, an owner may have no choice but to accept his bid, if it is the lowest, but his projects will be monitored very carefully.

To sum up, negotiating can be something of an art. The stakes are important, but so are the players. It's relatively easy to make a mistake about the personality or temperament of the adversaries. In any case, thorough preparation and documentation are the best tools to begin with.

5. Prolonged negotiations have ramifications other than the financial ones mentioned earlier. After long delays, all parties need to be reeducated as to the facts of the claim. Attorneys, witnesses, the staff, the principals all have to take the time to reimmerse themselves into the case. Some parties lose interest, others become discouraged. The staffs of the parties can change; people are transferred, or they retire, leave for jobs elsewhere, or even die. Continuity becomes a problem, even when copious minutes are maintained on all negotiations and other meetings.

6. Visual aids are most effective and persuasive. Scale models of a dam, a cross-section of a tunnel, CPMs, schedules, graphs, blown-up photographs—anything that depicts the claim situation will be convincing to both the expert negotiator and the tyro.

These can get the point across more effectively than pages of text, no matter how well expressed. A large four-color chart done by a skilled draftsman or graphic artist, can be a blockbuster. Such a chart can show, for example, how a site was blocked and precisely where and how men and machines had to be deployed around the impediment. In another example, such illustrations can be used to show rock conditions in tunnels. Photographs and a series of cross-sections showing locations of previously unknown fault planes can be very convincing.

If the written text of the claim closely explains the same facts, the illustrations can be more dramatic and convincing at the negotiating table.

The critical path chart is one of the basics of the delay claim display arsenal. Use it to establish and dramatize the effect of delay on the job. But be aware that the chart can also be used against the maker if flaws in logic or errors in input are discovered.

7. Outside parties may be brought in to bolster one's case during the negotiation phase. (These parties will ultimately be called in as witnesses if the case reaches an appeals board or other formal dispute resolution forum.) Choose these persons with care. An expert in a field may be a poor speaker; he may wander off the point or open doors one would not wish not open. The persons used as expert witnesses should be reliable and available when they are needed. Such witnesses should be prepared: they should know how the case is being argued, what the strategy is, and what points are being made that relate to him. There may be regional prejudices to be considered. For example, a witness from the North may arouse certain undesirable reactions in a negotiating session in the South.

Prior to the meeting at which he appears, there should be a meeting between the members of the negotiating team and the witness so that no contradictory or harmful statements are likely to be made. If the credibility of a witness is damaged at the negotiating table, the damage to the case can be considerable.

SUMMARY

To sum up, it is clear that a great deal of time, energy, and money go into these negotiation proceedings. The process has certain dramatic aspects, complete with role-playing and rehearsals and visual aids.

There can be many styles of negotiating as there are negotiations. Similarly, there are many tactics that can be used. We have tried to describe those styles and tactics that we have found to be the most successful over the long run.

As we pointed out earlier, it would be better for the industry if this energy were transferred to avoiding large-scale conflicts. But the realities of the industry dictate that parties be prepared to negotiate effectively. That being the case, the ideas in this chapter aim at effective, informed, and ethical strategies.

12. Formal Dispute Resolution

Once negotiations have failed, the next step is to bring the dispute to an outside party of avowed impartiality. The principal forums for resolution of construction disputes are the courts, arbitration, and in public work, administrative boards of appeal. More recently, Dispute Review Boards (DRBs) and other notable variations on arbitration have evolved; they are discussed in this chapter.

It must be remembered that once a claim or a dispute is lodged under one of these formal methods, control passes out of the hands of those involved. It is an axiom that a good settlement during negotiation is better than a good lawsuit.

Construction industry participants have more frequently recognized the advantages of settling disputes outside the courtroom. For example, the 1990 revisions of the EJCDC general conditions included the phrase:

> It is contemplated that the dispute resolution agreed upon by the owner and contractor will be set forth in an exhibit captioned "dispute resolution agreement." Absent such agreement, the owner and contractor would be left to the courts or to consensual forums.

Alternate dispute resolutions were officially recognized by the federal government in the November 1990 Administrative Dispute Resolution Act. This law strongly recommends alternate dispute resolution forums, stating that "Congress finds that administrative proceedings have become increasingly formal, costly, and lengthy, resulting in unnecessary expenditures of time and in a decreased likelihood of achieving consensual resolution." The act encourages federal agencies to take advantage of the "wide range of dispute resolution procedures" and even encourages agencies to take the lead in further development and refinement of such techniques. Under the new law, each federal agency is required to designate a senior official as a dispute resolution specialist.

The use of alternative modes of dispute resolution have grown tremendously. The 1996 EJCDC revisions to the general conditions go beyond contemplation and formally endorse mediation, followed by binding arbitration. The documents note the changes and improvements in the American Association of Arbitration Construction Rules and provide suggested language for mandatory mediation and arbitration clauses.

In 1996, the Dispute Avoidance and Resolution Task Force (DART), an independent program of the American Arbitration Association (AAA) endorsed the Declaration of Principles for the Prevention and Resolution of Disputes. The industrywide coalition heralds a new standard of behavior in the construction industry and promotes the use of private cooperative dispute resolution and prevention techniques at the earliest stages of construction planning.

CONSTRUCTION INDUSTRY ARBITRATION

In recent years, parties to disputes of all sorts have resolved them without resorting to the courtroom. Arbitration, which is quasilegal in nature, is the most frequently used of several such procedures. An impartial panel of arbitrators hears both sides of the dispute. Parties may be represented by counsel; witnesses may be called in to testify; and other formal courtroom procedures can be incorporated. However, many of the time-consuming legal niceties are bypassed. While this is intended to save precious time, the significant protection afforded by the formal judicial apparatus is missing because the forum is not a court of law, nor is the arbitrator a judge or even necessarily a lawyer. Yet the arbitrators' decision is in most states final and legally binding, not subject to an appeal in court.

The AAA has established procedures for selection of arbitrators experienced in appropriate areas. Record keeping and coordination are also performed by the AAA, as well as training for arbitrators. Fees go to the association for these services and to the arbitrators for their time and travel expenses. These costs can be substantial, in many instances more than for a prolonged court trial. Though most cases are settled in two days of hearings, these may be widely spaced in time. Some complicated cases have had in excess of 100 hearings and gone on for years.

Arbitration has been criticized as being just as costly and time-consuming as the court trials parties had hoped to avoid. As a result, the American Arbitration Association has developed streamlined new rules and procedures. A fifty-five–member AAA Construction ADR Task Force composed of construction professionals, engineers, lawyers, insurers, and academics undertook a comprehensive review of dispute avoidance and resolution services. The task force recommendations were unanimously endorsed in October 1996.

Under the new rules, disputes are broken down into Fast Track, Standard Track and Large Complex Case Track cases. Smaller, fast-track cases involving claims less than $50,000 are to be resolved in sixty days. Discovery is eliminated and arbitrators are selected from a special prequalified pool who serve on an expedited basis. The smallest of claims are to be decided on the basis of documentation alone, thereby completely avoiding the use of witness testimony and complicated legal maneuvering.

Standard-track cases, involving claims between $50,000 and $1 million, include liberal amendment of claims and counterclaims and arbitrator authority to control the discovery process. A written breakdown of awards is prepared only if requested by the parties in a timely manner.

Complex cases involving disputes in excess of $1 million are heard by a team of three highly trained arbitrators. These procedures include a mandatory preliminary hearing and broad arbitrator authority over the proceedings. In addition, the hearings are presumed to be held on a consecutive or block basis.

The task force also developed standard qualifications and minimum training requirements for arbitrators, including ten years of construction industry experience and mandatory retraining every three years. The task force also recommended that the AAA actively promote nonbinding mediation, partnering, and dispute review boards, and develop associated services.

Binding arbitration is different from the other dispute resolution mechanisms in that it is final and can only be reviewed by a court where procedural infirmity, fraud, or conflict of interest on the part of an arbitrator can be proved.

In the private sector, enabling legislation exists in most states. This gives the forum some considerable force, and makes it incumbent on the parties to approach arbitration very seriously. The proceeding is not just one skirmish in a long battle. The decision is final. No one can be compelled to arbitrate unless he has agreed to do so, but once he has, he must abide by the finding. To be realistic, when contracts are drawn up, contractors may agree to all sorts of things including arbitration, because they want the work or because some other aspect of the contract seems more important at the time. The important thing here, as we have discussed throughout this book, is that parties should thoroughly understand the meaning and implications of that contract. In the case of binding arbitration, parties are surrendering the protection of the formal legal machinery.

There is one out. If both sides agree when the dispute arises that they do not want to resolve the difference through arbitration, they can resort to other means. But both must agree. Action cannot be taken unilaterally if arbitration is provided for in the contract.

An advantage of arbitration over other methods of dispute resolution is the expertise of the individuals who act as arbitrators. As just stated, the new AAA procedures require that arbitrators have ten years of construction industry experience. Attorney arbitrators must devote half of their practice to construction matters. All arbitrators are required to undergo ongoing training and are approved and monitored by regional construction advisory committees. The AAA provides disputants with a list of qualified arbitrators and their biographical information from which the parties make their selection. If the parties cannot agree, the AAA appoints an arbitrator from the names not stricken from the list.

Knowledgeable arbitrators should be able to come to just decisions. In many courts of law, the judge may have heard a divorce suit one day, an accident case the next, and a construction dispute the next. The judge can't be expected to be familiar with the intricacies of the industry, nor with what is acknowledged common practice. Much time and explanations are thus saved in arbitration; arbitrators don't need to have elementary construction procedures and practices explained to them. As industry leaders, which most would be, their sympathies lie with their fellow practitioners (on both sides of the disputes). They are less likely to render a punitive award than are many juries who feel the need to find a villain in the piece. However, there is a growing acknowledgment of the right of arbitrators to award such damages, and such awards could become more common in the future. The CPR Institute for Dispute Resolution suggests that parties who wish to preclude arbitrators from awarding punitive damages should include a provision to that effect in the contract. Even in public agency boards of contract appeal, the board may not have as full a knowledge of industry practices as many arbitrators have. What's more, arbitration is private and it is not conducted in the public arena as is a court proceeding.

The arbitrator has a vested interest in an equitable resolution: he is part of the construction industry establishment; he wants both parties to go away satisfied if that is at all possible. At the very least, arbitrators are likely to want both parties to feel that they got a fair shake.

A 1985 survey by the construction attorneys group within the American Bar Association reflects client attitudes toward arbitration. The survey was sent to 3,000 attorneys with experience in arbitration. Out of the 576 attorneys who responded, 62 percent indicated their clients would probably elect to arbitrate again. These results were for cases involving claims under $250,000. Clients in cases where the claims were over $250,000 were slightly less willing; 46 percent would arbitrate again. More complete survey results appear in Appendix 2 of this book.

Deciding on Arbitration

The decision to go the courtroom route has to be mutual if an arbitration clause has been included in the contract, as we have pointed out. When disputes occur, it's human nature for each side to seek a means to gain the upper hand. The party who thinks he has the most equitable case might prefer to go to arbitration where the fairness of his position will be recognized. On the other hand, a party who thinks he has the law on his side, no matter how unfair it may seem, would probably want to go to court.

However these decisions are made, it is clear that there is a construction industry trend toward arbitration, or to variations thereof. The growing numbers of industry cases that go to arbitration are proof of acceptance. The majority of respondents in the 1994 Multidisciplinary Survey on Dispute Avoidance and Resolution in the Construction Industry indicate that they will encourage the use of binding arbitration in the future (see Appendix 8). More telling are the positive responses to predispute techniques, such as design review boards and Partnering and less formal mediation techniques. (See Appendix 11 for standard AAA demand for arbitration and answering forms.)

Arbitration among Parties Themselves

Parties can elect to arbitrate a dispute themselves and proceed along AAA or similar guidelines without working through the association. There is undoubtedly less control of the situation in this case. The presence of an experienced outside group such as the AAA increases the likelihood of orderliness in the proceedings. Should the parties want to go it alone, the CPR Institute for Dispute Resolution offers guidelines for nonadministered arbitration. The organization also provides lists of potential arbitrators who take on significant administrative duties as well as decision-making authority.

International Arbitration

The FIDIC general conditions provide for arbitration to settle disputes. The AAA has recently made efforts to streamline arbitration between United States and foreign parties. The association has revised its international arbitration rules, including a waiver of punitive damages, in an effort to make U.S. arbitration more predictable and attractive to foreign firms. The AAA rules are modeled after the UNCITRAL Model Arbitration Rules, an international arbitration standard. Additionally, the AAA has entered into cooperative agreements with arbitral institutions in fifty-three other nations.

Alternate Dispute Resolution

In the last decade, variations on arbitration procedures have increased. These attempt to provide more expedient dispute resolution and avoidance. A description of several follows.

1.   Mediation.

Mediation is the process of settling disputes through conferences conducted by a neutral third party who facilitates negotiation. It is often used after negotiations between parties have reached an impasse and lines of communication need to be reestablished. Unlike Dispute Review Boards, which meet with all parties simultaneously and render decisions on the issues before them, mediators hold ex parte caucuses with the individuals involved in the dispute and gather necessary information to promote settlement without presenting decisions. The mediation process is typically as follows:

1.   The parties meet jointly with the mediator to exchange facts and briefly summarize their positions; if desired, witnesses can appear at this time,

2.   The mediator caucuses privately with each party, and proceeds to shuttle back and forth between the parties, presenting offers and counteroffers as authorized.

3.   If a settlement is reached, the parties are brought together again to confirm its terms and sign an agreement.

The mediator's role is primarily to cultivate empathy and understanding between the parties and to identify and address barriers to the effective resolution of disputes. A good mediator will serve as a medium for the cultivation of a more productive relationship between the opposing parties, and will act alternately as a communication channel, an impartial confidant, a tension breaker, and a voice of sanity.

Because a mediator is not a decision maker, and because all discussions are kept confidential, a party is more likely to disclose information that it might otherwise keep hidden regarding its position and the facts surrounding the dispute. The privacy inherent in mediation is one of its chief advantages. The American Arbitration Association's Construction Mediation Rules assure that "confidential information disclosed to a mediator by the parties or witnesses in the course of the mediation shall not be divulged by the mediator...nor shall the mediator be compelled to divulge such records or to testify in regard to the mediation in any adversary proceeding or judicial forum." Parties to a typical mediation agreement are similarly prohibited from introducing into evidence, either directly or indirectly, views, proposals, or admissions made during the mediation. This assures all participants that they will not be prejudiced by their candor if they choose to stop mediation and return to the status quo. In addition, a number of states having enacted specific legislation to preserve the confidentiality and nondiscoverability of mediation proceedings.

Although the mediator must maintain confidentiality at all times and cannot disclose information received during the ex parte caucuses, the knowledge he gathers may facilitate more realistic negotiation. Hidden interests and settlement alternatives may be revealed as a result of this process and which direct negotiation between the parties would not have uncovered. In his role as negotiator, a mediator may offer his or her interpretation of a party's legal or factual position in order to give an objective and accurate assessment of the party's position. There may be an advantage to choosing an attorney to mediate a dispute since the advice offered may be more extensive and insightful. With an understanding of the strengths and weaknesses of his position, a party may be more willing to reach a realistic settlement rather than risk an expensive loss in binding adjudication that might follow. For this reason, it is critical that the parties to the mediation have settlement authority, and the involvement of senior management representatives is strongly encouraged. Where insurance is an issue, insurers should be made part of the process as well.

When should the parties mediate? Mediation is most appropriate when: (1) the parties are familiar enough with the case to negotiate intelligently; (2) the parties share a good faith interest in settling; (3) the parties want to maintain an ongoing business relationship; and (4) privacy and confidentiality is a primary concern. Parties should probably avoid mediation when: (1) the dispute turns on a novel question of law; (2) credibility is a major issue; (3) the opposing party is considered untrustworthy or unlikely to compromise. If these "conditions" are met, mediation should be tried. Nearly 95 percent of all civil litigation settles before reaching trial, and roughly 85 percent of all cases submitted for mediation result in settlement. Mediation is certainly an attractive and effective option.

2.   Dispute Review Boards.

Previously known also as an Independent Board of Consultants, the Dispute Review Board nomenclature has won out. The principal advantage is that the DRBs address construction problems before they reach the adversarial stage. The procedure was first used at the Eisenhower tunnel in Colorado in 1975, where it achieved its objective with notable success. The use has grown significantly in the last decade.

The Technical Committee on Contracting Practices of the Underground Technology Research Council, American Society of Civil Engineers, has developed a widely used model for Dispute Review Boards. The board consists of three members: one chosen by the contractor and approved by the owner; one chosen by the owner and approved by the contractor; and the third selected by the other two and approved by the owner and the contractor, and usually serving as the board chairperson. None of the board members should have a conflict of interest with any project participant, and all parties should feel confident that the board is impartial and objective. If the impartiality of a prospective board member is called into doubt, a thorough inquiry should be conducted, and another selection made if a finding acceptable to all seems unlikely. Trust is an important element in the effectiveness of Dispute Review Boards and should be cultivated whenever possible.

Another criterion for selection is construction experience and technical expertise. Board members should be knowledgeable on the particular subject matter involved in the construction project, but it is not necessary that members have legal skills or experience. An experienced peer can be effective in promoting the resolution of a dispute by creating a more trusting and less formal environment conducive to the settlement of job-site disputes. It is helpful, however, if at least one member has served on another board and has experience with the operational aspects of the process.

Sound judgment, leadership, and an understanding of the technical details and contractual significance of problems that may arise in the course of the project are important qualities, and add greatly to the prospects of success. Experience in the type of construction involved in the project, in the engineering principles relevant to it, in cost and scheduling procedures, and in claims analysis are all qualifications that should be represented on the board and that should guide the selection process. The fees and expenses incurred in setting up and operating the board should be shared by the contractor and owner.

Including DRB provisions in bid documents is believed to result in the submission of lower bids. Since inclusion of a DRB clause may indicate that an owner is fair and willing to cooperate, a contractor might not feel the need to insert a high contingency into the bid to cover the prospect of unresolved claims. The dispute can be resolved quickly, before it becomes unmanageable and its disruptive effects on the job accumulate. Construction can continue and shutdowns and delays can be avoided. DRBs encourage parties to identify problems early, deal with them promptly, and realistically evaluate their positions.

The board is familiar with the project, having received progress reports throughout the duration of the job, and has had the opportunity to observe the problems presented to it as they occur. DRBs also have a better understanding of the issues brought before them due to their construction expertise, an expertise often not found in the judges and juries who traditionally decide disputes. The parties are likely to accept a DRB recommendation since it is the product of an impartial, mutually acceptable panel's decision-making process. Such a decision is hard to reject in good faith. DRBs give an owner's representatives a basis for settling disputes and making a concession that might otherwise be perceived as an unnecessary capitulation to a contractor's position.

The very existence of a DRB on a job increases the likelihood that problems will be resolved between the parties themselves, without recourse to a neutral. Parties do not want to appear uncooperative or foolish in front of the board. Minor disputes are often settled independently because parties do not want a board, which may ultimately be deciding a significant dispute, to perceive them as petty or unreasonable. The total cost of setting up and maintaining a Dispute Review Board has generally ranged from .05 to .50 percent of the total contract cost, a fraction of the expense of full litigation or arbitration. This may even be entirely offset by the contractor's lower bid price.

In short, Dispute Review Boards work. To date, there have been relatively few court challenges to any DRB recommendations. If litigation or arbitration were to occur, however, admissible DRB records and findings could reduce the time-consuming and costly process of discovery and facilitate adjudication.

There are differing schools of thought concerning the admissibility of DRB recommendations in subsequent proceedings. Proponents believe that parties will take the process more seriously if they know in advance that the DRB's findings may be reviewed by a trial court or arbitration panel. Opponents argue that DRB recommendations are properly categorized as hearsay and therefore should not be admissible in court. The specifications should reflect the parties' wishes on these matters.

The method has been strikingly successful. In the previous edition of this book, we reported that boards have been used or are being used on about one hundred projects, valued at about $6 billion. The 1996 edition of *The Construction Dispute Review Board Manual* reported that as of January 1994, the number of projects in progress or in the planning stages had risen to 250 at a total cost of $19 billion; 68 DRB projects had been completed, 211 disputes had been successfully settled, and no claims had gone to the litigation stage.

DRBs are now being used for fairly complex buildings, transportation projects, and other work. Many state transit authorities routinely use DRBs on complex projects with multimillion-dollar price tags.

One specific example is the Mt. Baker Ridge Tunnel, a large and innovative tunnel project in Seattle, owned by the Washington State DOT. The cost of maintaining the board was $134,000 on a $36 million project, or .003 percent. Fees were split evenly between the state and the contractor. The proceedings were admissible in court and the recommendations were not binding.

The board made recommendations on three matters. In the first case, they found that the term "one-inch settlement" in the tunnel was ambiguous, and recommended that the contractor's interpretation of this clause was reasonable. The other two disputes involved payments relating to decisions already made by the state and the contractor. The board collected the information and tried to determine what payment would have been reasonable at a given time. They often analyzed the data differently than either the contractor or the state had done. In one case, the board accorded responsibility and cost to the parties on a 60/40 basis, with the state bearing the 60 percent share. In the other case, the contractor bore an 80 percent responsibility and, consequently, the same share of the cost. In the end, this state-of-the-art tunnel was completed ahead of schedule and under bid.

The cost of the DRBs is typically far less than 1 percent of the project cost, as it was at Mt. Baker Ridge. (A complete list of projects and a more exhaustive discussion of the method and its track record appear in *Avoiding and Settling Disputes in the Construction Industry*, published in 1991, by the American Society of Civil Engineers.) (Appendix 5 of this book contains some of this information and a model contract specification for a DRB.)

DRBs are not perfect, however. In the case of the Los Angeles Metro, for example, a DRB member was dismissed because of the appearance of impropriety. The comparatively relaxed culture of DRBs and other proactive dispute resolution mechanisms can lead parties to abandon strict formality. As board members are in a position to make important, expensive, and sometimes unpopular decisions, it is important that they avoid even the appearance of bias. Careful board selection and a timely and open response to concerns over impartiality should eliminate such breakdowns in the DRB process.

International Dispute Review Boards   In recent years, there has been some international experimentation with the use of DRBs. Ordinarily, under the FIDIC general conditions, the engineer is given the authority to resolve disputes between the contractor and owner. If the decision is unacceptable to either party, they may move to arbitration. The insertion of DRBs before or after the engineer's decision has resulted in additional delay. The World Bank has responded by eliminating the engineer from the dispute resolution process altogether. The bank's sample documents assign disputes to a DRB, after which either party may reject the recommendation and move on to arbitration. The bank's adoption of DRBs will likely bolster the growing interest in the use of neutrals overseas. Whether DRBs will flourish as they have in the United States, however, remains to be seen.

3. Minitrials.

The minitrial is in effect a structured mediation proceeding, usually set in motion by a submission agreement between the parties. Like mediation, it is a private and nonbinding settlement procedure, but unlike mediation it gives the parties the psychological satisfaction of having their "day in court," without the attendant cost, delay, and aggravation. Although each minitrial agreement sets forth the ground rules for its operation, the minitrial process can be divided into three phases: preparation for the hearing, conduct of the hearing, and negotiation among management representatives with settlement authority.

The preparatory phase is usually brief, with limited discovery permitted. The exchange of documents is voluntary. Prior to the hearing, the parties can prepare 10- to 15-page position papers for simultaneous exchange. It is recommended that the minitrial hearing not contain any surprise evidence.

The second phase of a minitrial, the hearing, is also known as the "information exchange." Here management representatives can interact, unfettered by the rigid structure imposed by a typical court. Comfortable and socializing surroundings promote cooperation and the resolution of differences. The "us" versus "them" sentiment is downplayed, with the parties meeting over lunch between presentations. Presentations are made to the minitrial panel, which consists of key executives from each party with full authority to settle, and, in most cases, a mutually agreed upon neutral advisor. The presentations, which generally last two to three hours, are allowed to proceed uninterrupted and give the parties an opportunity to weigh the strengths and weaknesses of the other side's position. Members of the panel may ask clarification questions, and at the end of each presentation a short rebuttal period is allowed.

The minitrial "information exchange" typically lasts one or two days and is immediately followed by the third phase: a meeting of the panel executives to negotiate the dispute. If the executives are unable to reach a settlement agreement, the neutral advisor may then attempt to mediate the negotiation and ultimately may write an opinion outlining the probable outcome of adjudication. If an impasse still exists, a "cooling off" period is recommended prior to the initiation of legal proceedings. This often affords the parties an opportunity to digest the information gathered at the hearing and make a less impassioned decision.

4. Other Variations.

Baseball arbitration is just one of many newer approaches to settling disputes. This method borrows from the sports world, where it has been used to negotiate the amount of player's salaries. Each party puts its last best offer in writing. The arbitrator cannot make a compromise award, but instead must choose one of the two offers. This puts the onus on the parties to put the most realistic figure possible forward. Increasingly, the courtroom seems the last resort in resolving construction industry disputes. There are several other alternate dispute resolution methods in the early stages of development, and we have no doubt that many more will surface.

A 1990 survey by the construction attorneys group within the American Bar Association reports on the use of mediation, minitrials, and other ADR devices. A report on that survey appears in Appendix 8 of the book.

5. Private Judging.

State and federal court congestion has brought about a the creation of private judging firms such as JAMS/Endispute, an ADR provider, Washington, DC. The industry hires retired judges (and lures working judges with high salaries). These "real" judges hear claims under quasijudicial conditions. For a fee, parties can avoid extensive backlogs and get a timely "trial" date.

The general criticism is that final judgment may follow the law but will have little to do with the realities of the construction industry. If you are going to pay a fact-finder, why not get one who is familiar with technical construction processes?

FEDERAL GOVERNMENT DISPUTE RESOLUTION

The federal government has had a formal mechanism for resolving contract disputes for decades. Contract appeals boards have been set up within the various federal agencies engaging in construction to hear contract disputes. These have evolved in response to need, and have been codified into law over the years. The procedures were updated and refined in the Contract Disputes Act of 1978 and subsequently modified in the Court Reorganization Act of 1982.

The contract appeals boards are made up of appointed officials who are considered knowledgeable and impartial judges, capable of ruling on contract disputes. They are supposed to be experts in construction practices as well, although the boards hear other kinds of contract disputes. The main idea has been to reduce the number of cases going to the courts. The boards are able to hear disputes on a more timely basis than are the courts. The system also has the advantage of having specialized judges. The 1978 law allowed claimants to bypass the boards and go directly to the then existing U.S. Court of Claims if they chose. The effect of this change was insignificant. The caseload of agency boards was only minimally affected, and few cases went directly to court. The 1982 statute eliminated the Court of Claims and created the U.S. Claims Court as the court of original jurisdiction for contract claims against the government. The act also created the Court of Appeals for the federal circuit. Today, claimants may bring cases before either the agency boards or the Claims Court. Appeals relating to decisions of either may be appealed to the appellate court.

The boards have generally worked well over the years. However, the direct access to the courts has not resulted in a diminution of appeals; in fact, the opposite has occurred. A criticism of the boards that has not been addressed, and possibly can never be addressed, is that the judges may be partial to the agency that appointed them. They certainly know the practices of their agency and the personalities of the agency personnel involved. However, in practice this issue has rarely been raised in a contract dispute. The judges are more conversant with construction practices than are most courtroom judges. One consideration may outweigh the other.

The federal boards have great advantages over most state and local government forums. One is that the procedures are fairly uniform. The rules are the same, too, though practices in different agencies may vary somewhat. As we point out in the next section, the methods at state and local levels of government vary greatly. Some are a good deal more effective and free from local influence than are others. The federal boards conduct hearings in a quasilegal atmosphere. Many of the formalities and the rights that one would have in the courts remain, such as the right to subpoena. Another legalistic aspect is that the findings of the boards may be appealed.

Contractors who deal with the federal government should be familiar with the workings of these boards or should have staff or consultants who are. They should also know that rules may differ between agencies. The federal agencies naturally have staff schooled in these procedures.

Some important points about the federal board procedures follow:

1. Each agency that lets contracts must set up an appeals procedure. Each must have a caseload that justifies three full-time board members. Those agencies that do not have a sufficient caseload refer disputes to another agency board.

2. The boards must have a minimum of three full-time members. Each member must have five years experience in public contract law. The board members are appointed by the head of the agency.

3. The boards have jurisdiction over contracts that deal with the procurement of property and services. Services include construction, alteration, repair and maintenance of real property, and the

disposal of personal property. Disputes over leases of property by or from the federal government are also heard by the boards.

4. All disputes relating to contracts may be heard before an agency board. The claim may be under the provisions of the contract, or it may arise from an alleged breach of contract.

5. A government contract may include a clause—and virtually all do—that states that the contractor must proceed with work while disputes are pending. In other words, the contractor has no grounds on which to stop work while the dispute is being resolved, except under very rare circumstances where the contractor can show that continuing would be senseless and result in a useless product.

6. Agency boards have procedures similar to those of the courts. They can authorize depositions, administer oaths, and subpoena witnesses and books and papers relevant to the case. If a witness refuses to obey a subpoena, an appropriate U.S. District Court can order the appearance of that person before the board. Failure to comply can result in a citation for contempt of court.

7. A contractor is permitted to bypass the appeals board. He may take the case directly to the Claims Court. An attorney is the best adviser in such cases.

8. Both parties may appeal the decision of the appeals board. Conclusions of law may be appealed as a matter of right. Findings of fact can be appealed only if they are not supported by sufficient evidence. The case would then go to the Court of Appeals for the federal circuit and possibly to the U.S. Supreme Court as the ultimate appeal.

9. The 1978 law includes antifraud provisions. At the time the claim is submitted, the contractor or the authorized representative most knowledgeable about the project or who has overall responsibility for the contractor's operations must certify that his information is accurate and complete and accept the consequences. The government thus places the liability for error or misstatement on the contractor.

   The government is entitled to recovery of the amount of the claim proved to be unsupported if misrepresentation can be shown. The government can also recover the cost of reviewing the claim.

10. The payment of interest on damages has been a controversial issue for years. As we have noted in this book, it is often to the benefit of the owner *not* to settle a claim promptly. The owner

sometimes pays no interest on damages or pays so little compared to prevailing rates that he has little incentive to act quickly. The interest rate is established by the Secretary of the Treasury and is revised every six months. That interest rate has thus provided some compensation for claimants and some incentive for the government to act expeditiously. However, the new rates cannot be said to be quite comparable to prevailing rates.

The 1978 law governing the federal boards and their procedures states that interest will be paid from the date that the contracting officer receives the claim until payment, if the board finds money due the claimant. The law, however, is not specific as to what constitutes a claim from which date the interest accrues, and thus has resulted in much litigation. A 1995 decision, however, calls for a broad and plain reading of the term claim as simply "a written demand seeking, as a matter of right, the payment of money in a sum certain." This definition results in interest accruing from the moment any disputed claim is submitted.

The claim must set forth an amount sought, which if it exceeds $50,000, must be certified with appropriate language. If all this is not done, the contracting officer may not act; and even if he does, neither the boards nor the court will take jurisdiction of the case. The contractor will be forced to begin again, and if there is a statute of limitations, he could lose his right to claim.

To sum up, board cases are conducted in much the same manner as a courtroom proceeding and with many of the same rules. The decisions of the boards can be cited as legal precedent and they can also be appealed. The boards may be bypassed, and cases can go directly to the Claims Court. While many board hearings are conducted in Washington, DC, there is no requirement that this be the case. In the interest of saving travel money for witnesses and others involved, hearings are occasionally conducted outside the capital.

PROCEDURES—TIME LIMITATIONS

Once a contractor has made a claim, certain procedures must be observed, and there are time limits for subsequent actions.

- The contractor must submit his claim in writing to the contracting officer. That officer is identified in the contract documents.
- The contracting officer must respond within 60 days of receiving the claim. At that time, if the claim is for less than $50,000, he must make a decision in the case. If the claim is for more than

$50,000, the contracting officer has two choices: he can rule on the claim within 60 days or he can inform the contractor of the time within which he will make a decision. If the contractor finds the time period unreasonable, he can appeal to the contract board to direct the officer to rule in more timely fashion. Frequently, the board will take the appeal and deem the failure of the contracting officer to decide as a denial of the claim. Of course, the board may agree with the contracting officer and leave the matter as it was.

- When the contracting officer makes his decision, he must state that he is issuing a final decision and the basis upon which it was made, and include a description of the administrative appeals process. If he does not respond to the claim at all within the 60-day limit, it is tantamount to a denial of the claim. The appeals process then commences.

Appealing a Decision

The contractor must file his appeal with the agency board within ninety days of the receipt of the contracting officer's decision. At this point, the contractor may decide to go directly to the U.S. Claims Court. Should sixty days pass since the contracting officer received the claim without his issuing a final decision or telling the contractor in writing how much additional time is needed, the contractor may appeal. In this case, he has twelve months in which to act.

If the dispute goes to the appeal board, the decision of that body may be appealed. Either party can do this. The contractor would seek the advice of his attorney before appealing the board's decision. The federal agencies would, as we have noted, seek the concurrence of both the agency head and the U.S. Attorney General before filing an appeal of its own board's ruling. Such an appeal has rarely been sought and occurs only when the government believes that to allow the board or Claims Court decision to stand would set a bad precedent. Appeals from the Claims Court to the Court of Appeals must be filed within ninety days of the Claims Court decision. Appeals from board decisions must be filed within 120 days.

Under the law, claims under $10,000 can be settled by a single member of the board and within 120 days. The 1978 law also provides that each agency can establish its own rules for achieving this end.

Most federal agencies have developed such procedures. The U.S. Postal Service, for example, has rules governing claims under $10,000, those under $25,000, and those under $50,000. All, of course, comply with the provisions of the federal laws governing dispute resolution.

STATE AND LOCAL DISPUTE RESOLUTION

Sovereign immunity is the principle that the entity that makes the laws cannot be sued under those laws. The idea goes back to early Anglo-Saxon times and to the concept that the king is above the law. The principle is still honored in that the federal government and states now "consent" to be sued. The limited grounds on which they may be sued are usually set forth in statutes. The federal government in consenting to be sued has, as we have seen, set forth an orderly method for resolving construction claims.

The fifty states, however, have developed many different approaches to the recovery process. A wise contractor will find out what the law is in his own state and in others where he might be involved in a public works project.

In most instances, the first step at the state level is the same as that at the federal level. The contracting officer, or whoever is named in the contract, makes the first-line decision. The next step, if that initial ruling proves unsatisfactory, can vary considerably.

In some states, the claim must be appealed to the state comptroller; in others, the claim is filed with a local court or with the state court of claims. A description of some other variations follows.

Administrative Boards

A few states have instituted administrative boards of appeals, and these vary from state to state where they do exist. Most are patterned after the federal boards of contract appeals, but the test of time has not yet been applied. One criticism is that state boards are likely to be made up of political appointees and therefore to have political overtones. Another criticism is that the individuals appointed have little or no experience in the field of construction and contract law. Rulings of these boards may be final by statute in some cases. In other states, judicial appeal may be permitted or required. Procedures may be more or less formal in the legal sense. The boards meet at intervals and may have jurisdiction only over certain kinds of disputes.

Review Boards

In some cases, a review panel is appointed to oversee certain categories of public expenditures, highways and roads, for example, or water and sewage waste disposal. These bodies may be the first line of appeal when a construction claim is likely.

In short, it is important to know what kind of relief can be obtained in a particular state. When a claim arises, it is important, even crucial, to retain an attorney early. Of course, the involvement of an attorney depends in part on the complexity and the dollar value of the case and on the contractor's knowledge of the processes in that state. However, if proper procedures are not followed within the proper time frame, recovery for either the owner or the contractor could be time-barred.

Awards of Damages

If damages are awarded on the state level, a legislative act could be required to settle those claims. Funds might have to be appropriated to pay the contractor. The governor of the state ultimately has to sign such bills, and political considerations may enter into his decisions. If a court has not directed the state to pay, but instead the directive comes from a panel or other administrative body, the legislature may act differently. Again, political matters may influence their decision to abide by the board's ruling, or whether or not they decide to pay, and if they do, how much. It is possible, in other words, that the legislature will be more attentive to the court's ruling.

State panels, boards, and legislatures, even those with competent individuals sitting, are generally not familiar with contract law or with the practices of the construction industry. They may be reluctant or not permitted to award great sums of public funds to contractors. By the same token, they may place onerous penalties on contractors out of misplaced zeal, lack of knowledge about the construction process and its problems, or even out of a wish to be punitive on behalf of the state. This can occur even when the individuals believe the claim to be reasonable or justified.

The American Bar Association has developed a Model Procurement Code, approved in 1979, which was intended as a guide for government bodies other than the federal government. One recommendation within this model code was that states and other governmental entities set up contract appeals boards similar to those of the federal government. Maryland has adopted a contract appeals board in accordance with those recommendations. As the model code is more widely promulgated, other states may follow suit. As we have mentioned, there are now other states with some sort of appeals board, or panel, but none except those following the ABA code recommendations can be expected to bear close resemblance to the federal procedures described in this book.

Cities and Other Local Government Bodies

Many of the statements we have made about states apply to local-level government: cities, towns, counties, or regional authorities. One major difference is this: sovereign immunity does not apply at the local level, so the government body, whichever it is, may be sued.

Some cities and other government bodies have review panels. A commissioner may have the authority to appoint such panels for cases involving his department. In some cases, the local commissioner of public works or another designated city official hears all disputes. These people may be lacking any experience in, or knowledge of, the construction field. The possibilities are tied to some extent to the size of the governmental unit in question.

There is no right to trial by jury in a claim against the state. Generally, however, this right does apply to claims against local governments. The right may be waived in the contract or by the parties in litigation, who may prefer to have the judge hear the case alone. Again this requires specialized knowledge of the laws and regulations of the local government.

Out-of-State Contractors

Generally, a contractor can only sue a local government body in a federal court if he is incorporated in a state other than the one in which the claim arose. States may not be sued in federal court on contract cases without explicit agreement and the existence of a federal question. Many contractors follow this procedure whenever they can because they have greater faith in the impartiality and the competency of the federal court system than they do in the local or state courts. These federal district courts provide for trial by jury unless the litigants agree to waive this right.

Clearly, the advice of an attorney should be sought before making such decisions.

Summary

Here, then, are some important points that bear on dealing with state and local governments in contract disputes.

1. The city or state may not have funds on hand to settle the construction claim. Taxes may have to be increased or a bond floated. Such considerations will certainly have an effect on the judgment of those individuals deliberating on the claim, especially if they sit on an administrative panel; or they even will have an effect on a local court if it is hearing the case. However, if sufficient funds are available and have been appropriated or earmarked for the project, claims are apt to be settled more swiftly.

    The matter of sufficient funds is increasingly complicated by the great number of public works projects that are funded on a sharing basis with a grantee. For example, costs may be shared on a 90–10 or a 75–25 basis. Providing for payment of damages with limited funds and another layer of restrictions can be a complicated and lengthy procedure.

2. Public bodies are generally not in a rush to settle a claim. The reason is that little or no interest is paid on such claims. In most cases, the public body saves money by waiting as long as possible to settle, while the contractor loses not only the use of his settlement cash, but of substantial interest as well. Cases against a state have been known to stretch on for ten years or more, with interest on the ultimate award lower than half the going rate. In addition, inflation eats away at the recovery figure. The federal government pays higher interest on claims than do most states or cities, but even the federal rate is below market rates.

3. If a claim does go to court, the level of defense raised by the state or local municipality may vary from the superb to the terrible. The same may apply to the contractor's presentation, of course, but he has some control over that level. He chooses his attorney; a local government may be restricted to using staff counsel.

   In cases involving large sums of money or matters that may establish precedents for future cases, it is not unusual for the state or city to retain, at considerable expense, the services of a law firm with expertise in this area. On the other hand, the government body in most cases will continue to exist regardless of the outcome of the case, and this may account for a less than spectacular defense. There are exceptions. Several major cities have been on the brink of default, and in those instances, a major settlement against those cities might have had serious repercussions. Again, the advice of one's attorney on the level of defense to be expected is useful.

   The contractor does have great incentive to invest in legal fees and to retain the top contract claims lawyer he can afford. His staying in business may well depend on the outcome of the case. Yet some contractors, operating under severe financial restrictions believe that they cannot afford the "best" counsel. A larger contractor often has the in-house staff to deal with claims, such as attorneys whose principal responsibility is the review of contracts, analysis of claims, and the prosecution of such claims through boards and courts. Even so, such a contractor may engage a local attorney because that individual knows local procedures, laws, and regulations, and in some cases, the reputations of the personalities involved.

4. Both parties should know the reputation of the other, especially in matters of claims and their resolution. If a contractor is con-

templating bidding on a major project for a government body, he should find out whether that body has been known to fight for years before paying on legitimate claims. The contractor, after all, has the choice of not bidding at all. Or he can bid high enough to cover the risk of a slow-pay or no-pay situation. The public body is obliged to accept the lowest bid, regardless whether the contractor has a litigious reputation or not, as long as he has the technical and financial capabilities to do the job. But the public owner should know about that reputation in order to deal effectively with the contractor on the project. The owner can be prepared; he can try to head off potential claims situations.

ATTORNEY-CLIENT RELATIONSHIPS

Throughout this book, the term "knowledgeable attorney" has been used. Naturally, both parties to a dispute want to retain an attorney likely to obtain a favorable result. The first step is to retain an attorney who has experience in the construction industry, with a particular emphasis on contracts and claims. A second step might be to assure that the attorney has argued before a federal contract appeals board, if that is the resolution forum. It's a good idea, though not essential, that the attorney have experience in the particular forum chosen.

Once the dispute has reached the litigation stage, however, parties will definitely want a lawyer with courtroom experience. This poses a catch-22 situation. Everyone wants an attorney with trial experience. How does a lawyer begin to get that experience? The set of such attorneys is rather small. More than 80 percent of construction cases are settled before they reach the courtroom, in pretrial phases.

Of course, an attorney needs experience to maneuver successfully here, too. In addition, firms specializing in contract law want to win. They are just as reluctant to assign an attorney without courtroom experience to a case as most clients are to retain such an individual. The number of attorneys who have actually argued a substantial construction case before a tribunal is not a large one.

Attorneys can gain experience by acting as assistants to trial attorneys by working on smaller cases. So there are suitable substitutes for an attorney with a lengthy courtroom record. The reputation of a firm carries some weight, too. A young attorney with a top-notch firm who has been an assistant to a trial attorney might be superior to someone else with a lengthy courtroom track record.

The scarcity of experienced trial attorneys in the contract field is reflected in fees: they can be high. Both owners and contractors face the problem of selecting the "right" attorney from that limited pool. Some public owners are restricted by law to using staff attorneys and are thus relieved of the task of seeking outside counsel. There are exceptions. When the dollars are considerable or there could be an important precedent set in the case, public bodies have retained outside counsel to supplement their in-house capability.

How does one find a suitable attorney? Mainly through word of mouth. Since the pool of firms that deal in construction law is not a large one, and the construction industry is relatively close-knit, reputations are not difficult to ascertain. Here are three points to bear in mind:

1. Actual trial experience should be weighed along with the lawyer's construction industry and contract law background. The advice a lawyer can offer based on his knowledge of attitudes, customs, and political considerations within a jurisdiction must also be taken into account. For example, the attorney might suggest moving the case to another jurisdiction (if that is possible) because of local political situations. Obviously, attorneys also know the temperament and reputations of judges in their district who might be assigned to the case.

2. The reputations and strengths of a law firm may change. A firm well thought of for years may be experiencing a stale period. A top attorney may have left the firm, or there may be internal conflict in the firm. A firm with a lesser reputation may have acquired an outstanding new attorney. Sometimes, it's a good idea to hire the attorney and the firm that comes with him (or her) than to hire the firm.

3. How can one find out what courtroom experience an attorney has? It is possible (but extraordinary) to obtain transcripts of previous trials and to read these. One can also seek the opinions of other attorneys, including house counsel if that applies. Finally, one can ask the attorney what cases he has handled and the names of persons that can be contacted about these cases—the judge, the parties, and other attorneys. All this research may seem excessive, but depending on the magnitude of the stakes, it can be worthwhile.

Once an attorney and that firm are retained, it's a good idea to monitor the proceedings. Here are some things to remember about law firms.

1. Ask if the attorney retained will also be the person in charge of billing your account. This establishes a clear line of responsibility in the firm. Sometimes a partner is assigned to bringing in new business; that "star" may turn the cases over to another attorney and not necessarily follow the progress of the litigation,

except to send the bills. This is an extreme example, but one that should be avoided. If it is made clear that the firm has been retained because of the reputation of attorney X, more of the responsibility of the case will reside with X. That means billing, too. This serves to centralize the firm's attention to the case.

2. Are large firms better than small ones? There are pros and cons to both. Larger firms may have greater resources, but smaller firms can guarantee that the case won't get lost in the shuffle. Regardless, make sure that the resources of any firm are adequate and that you are satisfied with the attorneys handling the case. There is probably nothing else that can be done in deciding the merits of a large or a small firm.

3. In general, it is considered more "professional" to retain an attorney on a time compensation basis rather than on a contingency basis. However, this provides no incentive for the attorney to move expeditiously. There is a trend toward modifying time compensation arrangements by reducing the hourly rate and adding a bonus, which is a percentage of recovery or the savings resulting from the final ruling. The idea is to provide incentive to prosecute the case in a timely manner.

4. Fees for attorneys are rarely recoverable. Such a clause may be put in the contract at the outset, as can almost any kind of clause if the parties agree. However, this is very unusual. If the case is extremely complicated and looks as though it will be costly to pursue—depositions requiring great distances for travel and so on—it could be worthwhile to examine the billing policy on these items. The firm's procedures on travel allowances and related expenses may be very liberal. Therefore, resentment over mounting costs may be avoided if these items are understood from the start. Of course, if the case is won, the memory of one's attorney "wining and dining" at the client's expense will dim quickly.

   It makes sense to ask that the lowest professional level feasible prepare appropriate portions of the case. It is unreasonable and extremely costly to insist that the senior partner or trial attorney perform research that an associate attorney can accomplish. The trial attorney's special skills and experience should be rationed out judiciously.

The Premise of Honesty

The best approach to settling any disputes is to be honest and straightforward. It should be understood by the attorney from the start that this is what the client wants. The client sets the tone in the attorney relationship. Most attorneys are eager to please the client and will conduct themselves in the manner they perceive to be what the client expects from them. This does not mean that an attorney will immediately fall in with the attitudes of an unscrupulous client. On the contrary, the relationship will be tortured and difficult. If the client makes it clear that he wants the case to be conducted in the most honest and credible manner possible, most attorneys will breathe a sigh of relief. The case will be easier to prosecute, and juries will more likely be convinced. If the client believes his case has merit, this shouldn't be difficult. As we walk through the next stage of litigation, it will be apparent why a straightforward approach must be maintained.

Preparation

Preparation for litigation is the most crucial stage of the process. The documents are the skeleton of the case, the witnesses are the flesh. The typical construction claims litigation is highly dependent on those documents. These include correspondence, meeting memos, invoices, time cards, daily reports—all the paper that accumulated in the course of a project. The attorney may have to read, organize, and absorb thousands of documents. There may be several sets in each category of documents; the contractor, the owner, and the engineer or architect may have maintained separate records, especially if each knew a dispute was brewing.

All documents must be read and understood. The attorney must know of every one. If a single piece of paper is introduced at trial that the attorney has not seen, something is wrong. This is part of the open and honest approach. The case may be lost on just one such omission, intentional or not. A client may hope the other side will not find a particular document and take his lawyer into his confidence. This is unacceptable and unfair to one's attorney, and of course, in the end the client may lose on the basis of that single unrevealed document. Still, there can be honest mistakes. A document can be lost in the mists of time, or it can surface because one side kept meticulous records and the other kept sloppy records. It doesn't matter how this happens, it should never happen.

No one can rely on anything remaining "secret." With today's discovery laws as to pretrial examination, almost every document will be accessible to both sides in the case. Clients must make certain that their attorneys see everything.

The same thorough preparation applies to witnesses. It is almost axiomatic that there are no bad witnesses, only bad attorneys. Witnesses must be well prepared and instructed to handle any contingency. Nothing should be left to chance. The courtroom is no place for off-the-cuff remarks. In fact, only part of that intense preparation will be needed. There is no way of knowing, however, just what part the opposition will concentrate on or what response will convince a jury. It's good sense to be prepared for anything.

During this preparation stage, witnesses should be instructed that jargon will only obscure the information they have to offer. Technical language is becoming a problem in jury trials. Newspapers have cited instances of lengthy antitrust or banking lawsuits where many highly educated persons could barely make sense out of the technical aspects, so it is unlikely that a jury selected more or less at random would be comfortable with technical language.

The danger in a construction claim case is that a jury may make judgments on grounds of sympathy or prejudice if testimony is unclear to them. Witnesses with fairly esoteric specialties, say in earthquake engineering or in a narrow area of soils or geology, may be world-renowned, but they won't do any good if the jury can't comprehend their points. On a less esoteric level, construction industry jargon can be confusing, too.

Finally, the judge and jury will want to hear the witnesses, not the attorney. Preparation should focus on an effective presentation by the witnesses, not on dramatic packaging by the attorney.

Pretrial Disclosure

Pretrial disclosure is not available in all the formal dispute resolution forums. But procedural rules in most courts grant this right, which allows each side to see documents in possession of the adversary and to question that adversary orally before a court stenographer.

In arbitration, for example, surprises can be sprung in the presence of the arbitrators, often to the detriment of one side's case. At best, an unexpected document or bit of information throws one's attorney off balance during the proceedings. The federal government boards of contract appeal permit pretrial disclosure; the various dispute forums at that state and local level may not. The attorney should know.

If some evidence that could jeopardize one side's position surfaces during the pretrial disclosure, the parties have the option of settling the case out of court. Of course, parties can always back out of other dispute resolution forums and decide to settle, but the earlier this information is disclosed, the easier that settlement may be. Once formal proceedings, say of arbitration, have begun, it becomes less easy to back out. The party whose case is strengthened by the disclosure may want to continue with the proceeding; his case may result in a larger award than the settlement otherwise would have been. In pretrial disclosure, the final formal proceedings of the courtroom have not yet begun, and in most cases, both parties would prefer to avoid the expense and time of a lawsuit if this is possible.

This pretrial process can take years. If no settlement is forthcoming, it could take years to set a trial date, and even more time to complete the trial. Under the rules of discovery, collecting and presenting all documents take time. Of course, in public contracts, the claimant contractor could ask for all documents under the freedom-of-information laws much earlier in the dispute since he is dealing with a public body. Often, however, the contractor does not do this while the parties are still at the negotiating table, out of fear of offending the other side by invoking these procedures. But once litigation is formally commenced, it's official. All and everything must be presented to both sides.

At this point, the attorney, depending on the information disclosed, will advise the client as to whether to proceed to trial or to try to settle. Recommendation to settle may be made on grounds other than that the other side possesses damning evidence. It's often cheaper to settle than to risk a more onerous court-mandated result. The time saved is saved for both parties.

THE COURTROOM

Once the decision has been made to move to trial, the advice given earlier about the straightforward, credible approach becomes even more important. Juries especially will be suspicious or offended by a ranting, raving, insinuating attorney. Clients don't have much to say once the attorney has behaved in such a fashion. But the client should previously have set the tone of the approach. If it has been made clear that an academy-award–winning performance is not the client's style, the attorney will probably avoid the histrionics. Juries often interpret attacks and delays in the courtroom as evidence of concealment or even of guilt. Similarly, dramatic courtroom maneuvers can be interpreted as purposeful distractions from the facts, an attempt to gloss over shaky facts. Reasonableness is less risky, and apparently more convincing. After all, the judge and jury are assessing the "reasonableness" of the damage figure and the degree of fault. While judges are accustomed to courtroom dramatics and are less likely to regard such tactics with the suspicion a jury might have, the judge is also more likely to see through a showy performance and to stick to the facts.

If the facts are on the side of the ranting attorney, the judge will see that, too, whether or not the performance is dramatic. It may sound naive, but the reasonable, honest approach works best in the courtroom.

Appendixes— Construction Claims: Prevention and Resolution

Appendix 1

STANDARD
GENERAL CONDITIONS
OF THE
CONSTRUCTION CONTRACT

Prepared by

ENGINEERS JOINT CONTRACT DOCUMENTS COMMITTEE

and

Issued and Published Jointly By

AMERICAN CONSULTING
ENGINEERS COUNCIL

**National Society of
Professional Engineers**
Professional Engineers in Private Practice

AMERICAN SOCIETY OF
CIVIL ENGINEERS

PROFESSIONAL ENGINEERS IN PRIVATE PRACTICE
a practice division of the
NATIONAL SOCIETY OF PROFESSIONAL ENGINEERS

AMERICAN CONSULTING ENGINEERS COUNCIL

AMERICAN SOCIETY OF CIVIL ENGINEERS

This document has been approved and endorsed by

The Associated General   Contractors of America

Construction Specifications Institute

Advancement
of Construction
Technology

These General Conditions have been prepared for use with the Owner-Contractor Agreements (No. 1910-8-A-1 or 1910-8-A-2) (1996 Editions). Their provisions are interrelated and a change in one may necessitate a change in the other. Comments concerning their usage are contained in the EJCDC User's Guide (No. 1910-50). For guidance in the preparation of Supplementary Conditions, see Guide to the Preparation of Supplementary Conditions (No. 1910-17) (1996 Edition).

EJCDC No. 1910-8 (1996 Edition)

TABLE OF CONTENTS

GENERAL CONDITIONS

ARTICLE 1 - DEFINITIONS AND TERMINOLOGY

1.01 *Defined Terms*

A. Wherever used in the Contract Documents and printed with initial or all capital letters, the terms listed below will have the meanings indicated which are applicable to both the singular and plural thereof.

1. *Addenda*--Written or graphic instruments issued prior to the opening of Bids which clarify, correct, or change the Bidding Requirements or the Contract Documents.

2. *Agreement*--The written instrument which is evidence of the agreement between OWNER and CONTRACTOR covering the Work.

3. *Application for Payment*--The form acceptable to ENGINEER which is to be used by CONTRACTOR during the course of the Work in requesting progress or final payments and which is to be accompanied by such supporting documentation as is required by the Contract Documents.

4. *Asbestos*--Any material that contains more than one percent asbestos and is friable or is releasing asbestos fibers into the air above current action levels established by the United States Occupational Safety and Health Administration.

5. *Bid*--The offer or proposal of a bidder submitted on the prescribed form setting forth the prices for the Work to be performed.

6. *Bidding Documents*--The Bidding Requirements and the proposed Contract Documents (including all Addenda issued prior to receipt of Bids).

7. *Bidding Requirements*--The Advertisement or Invitation to Bid, Instructions to Bidders, Bid security form, if any, and the Bid form with any supplements.

8. *Bonds*--Performance and payment bonds and other instruments of security.

9. *Change Order*--A document recommended by ENGINEER which is signed by CONTRACTOR and OWNER and authorizes an addition, deletion, or revision in the Work or an adjustment in the Contract Price or the Contract Times, issued on or after the Effective Date of the Agreement.

10. *Claim*--A demand or assertion by OWNER or CONTRACTOR seeking an adjustment of Contract Price or Contract Times, or both, or other relief with respect to the terms of the Contract. A demand for money or services by a third party is not a Claim.

11. *Contract*--The entire and integrated written agreement between the OWNER and CONTRACTOR concerning the Work. The Contract supersedes prior negotiations, representations, or agreements, whether written or oral.

12. *Contract Documents*--The Contract Documents establish the rights and obligations of the parties and include the Agreement, Addenda (which pertain to the Contract Documents), CONTRACTOR's Bid (including documentation accompanying the Bid and any post Bid documentation submitted prior to the Notice of Award) when attached as an exhibit to the Agreement, the Notice to Proceed, the Bonds, these General Conditions, the Supplementary Conditions, the Specifications and the Drawings as the same are more specifically identified in the Agreement, together with all Written Amendments, Change Orders, Work Change Directives, Field Orders, and ENGINEER's written interpretations and clarifications issued on or after the Effective Date of the Agreement. Approved Shop Drawings and the reports and drawings of subsurface and physical conditions are not Contract Documents. Only printed or hard copies of the items listed in this paragraph are Contract Documents. Files in electronic media format of text, data, graphics, and the like that may be furnished by OWNER to CONTRACTOR are not Contract Documents.

13. *Contract Price*--The moneys payable by OWNER to CONTRACTOR for completion of the Work in accordance with the Contract Documents as stated in the Agreement (subject to the provisions of paragraph 11.03 in the case of Unit Price Work).

14. *Contract Times*--The number of days or the dates stated in the Agreement to: (i) achieve Substantial Completion; and (ii) complete the Work so that it is ready for final payment as evidenced by ENGINEER's written recommendation of final payment.

15. *CONTRACTOR*--The individual or entity with whom OWNER has entered into the Agreement.

16. *Cost of the Work*--See paragraph 11.01.A for definition.

17. *Drawings*--That part of the Contract Documents prepared or approved by ENGINEER which graphically shows the scope, extent, and character of the Work to be performed by CONTRACTOR.   Shop Drawings and other CONTRACTOR submittals are not Drawings as so defined.

18. *Effective Date of the Agreement*--The date indicated in the Agreement on which it becomes effective, but if no such date is indicated, it means the date on which the Agreement is signed and delivered by the last of the two parties to sign and deliver.

19. *ENGINEER*--The individual or entity named as such in the Agreement.

20. *ENGINEER's Consultant*--An individual or entity having a contract with ENGINEER to furnish services as ENGINEER's independent professional associate or consultant with respect to the Project and who is identified as such in the Supplementary Conditions.

21. *Field Order*--A written order issued by ENGI-NEER which requires minor changes in the Work but which does not involve a change in the Contract Price or the Contract Times.

22. *General Requirements*--Sections of Division 1 of the Specifications. The General Requirements pertain to all sections of the Specifications.

23. *Hazardous Environmental Condition*--The presence at the Site of Asbestos, PCBs, Petroleum, Hazardous Waste, or Radioactive Material in such quantities or circumstances that may present a substantial danger to persons or property exposed thereto in connection with the Work.

24. *Hazardous Waste*--The term Hazardous Waste shall have the meaning provided in Section 1004 of the Solid Waste Disposal Act (42 USC Section 6903) as amended from time to time.

25. *Laws and Regulations; Laws or Regulations*--Any and all applicable laws, rules, regulations, ordinances, codes, and orders of any and all governmental bodies, agencies, authorities, and courts having jurisdiction.

26. *Liens*--Charges, security interests, or encumbrances upon Project funds, real property, or personal property.

27. *Milestone*--A principal event specified in the Contract Documents relating to an intermediate completion date or time prior to Substantial Completion of all the Work.

28. *Notice of Award*--The written notice by OWNER to the apparent successful bidder stating that upon timely compliance by the apparent successful bidder with the conditions precedent listed therein, OWNER will sign and deliver the Agreement.

29. *Notice to Proceed*--A written notice given by OWNER to CONTRACTOR fixing the date on which the Contract Times will commence to run and on which CONTRACTOR shall start to perform the Work under the Contract Documents.

30. *OWNER*--The individual, entity, public body, or authority with whom CONTRACTOR has entered into the Agreement and for whom the Work is to be performed.

31. *Partial Utilization*--Use by OWNER of a substantially completed part of the Work for the purpose for which it is intended (or a related purpose) prior to Substantial Completion of all the Work.

32. *PCBs*--Polychlorinated biphenyls.

33. *Petroleum*--Petroleum, including crude oil or any fraction thereof which is liquid at standard conditions of temperature and pressure (60 degrees Fahrenheit and 14.7 pounds per square inch absolute), such as oil, petroleum, fuel oil, oil sludge, oil refuse, gasoline, kerosene, and oil mixed with other non-Hazardous Waste and crude oils.

34. *Project*--The total construction of which the Work to be performed under the Contract Documents may be the whole, or a part as may be indicated elsewhere in the Contract Documents.

35. *Project Manual*--The bound documentary information prepared for bidding and constructing the Work. A listing of the contents of the Project Manual, which may be bound in one or more volumes, is contained in the table(s) of contents.

36. *Radioactive Material*--Source, special nuclear, or byproduct material as defined by the Atomic Energy Act of 1954 (42 USC Section 2011 et seq.) as amended from time to time.

37. *Resident Project Representative*--The authorized representative of ENGINEER who may be assigned to the Site or any part thereof.

38. *Samples*--Physical examples of materials, equipment, or workmanship that are representative of some portion of the Work and which establish the standards by which such portion of the Work will be judged.

39. *Shop Drawings*--All drawings, diagrams, illustrations, schedules, and other data or information which are specifically prepared or assembled by or for CONTRACTOR and submitted by CONTRACTOR to illustrate some portion of the Work.

40. *Site*--Lands or areas indicated in the Contract Documents as being furnished by OWNER upon which the Work is to be performed, including rights-of-way and easements for access thereto, and such other lands furnished by OWNER which are designated for the use of CONTRACTOR.

41. *Specifications*--That part of the Contract Documents consisting of written technical descriptions of materials, equipment, systems, standards, and workmanship as applied to the Work and certain administrative details applicable thereto.

42. *Subcontractor*--An individual or entity having a direct contract with CONTRACTOR or with any other Subcontractor for the performance of a part of the Work at the Site.

43. *Substantial Completion*--The time at which the Work (or a specified part thereof) has progressed to the point where, in the opinion of ENGINEER, the Work (or a specified part thereof) is sufficiently complete, in accordance with the Contract Documents, so that the Work (or a specified part thereof) can be utilized for the purposes for which it is intended. The terms "substantially complete" and "substantially completed" as applied to all or part of the Work refer to Substantial Completion thereof.

44. *Supplementary Conditions*--That part of the Contract Documents which amends or supplements these General Conditions.

45. *Supplier*--A manufacturer, fabricator, supplier, distributor, materialman, or vendor having a direct contract with CONTRACTOR or with any Subcontractor to furnish materials or equipment to be incorporated in the Work by CONTRACTOR or any Subcontractor.

46. *Underground Facilities*--All underground pipelines, conduits, ducts, cables, wires, manholes, vaults, tanks, tunnels, or other such facilities or attachments, and any encasements containing such facilities, including those that convey electricity, gases, steam, liquid petroleum products, telephone or other communications, cable television, water, wastewater, storm water, other liquids or chemicals, or traffic or other control systems.

47. *Unit Price Work*--Work to be paid for on the basis of unit prices.

48. *Work*--The entire completed construction or the various separately identifiable parts thereof required to be provided under the Contract Documents. Work includes and is the result of performing or providing all labor, services, and documentation necessary to produce such construction, and furnishing, installing, and incorporating all materials and equipment into such construction, all as required by the Contract Documents.

49. *Work Change Directive*--A written statement to CONTRACTOR issued on or after the Effective Date of the Agreement and signed by OWNER and recommended by ENGINEER ordering an addition, deletion, or revision in the Work, or responding to differing or unforeseen subsurface or physical conditions under which the Work is to be performed or to emergencies. A Work Change Directive will not change the Contract Price or the Contract Times but is evidence that the parties expect that the change ordered or documented by a Work Change Directive will be incorporated in a subsequently issued Change Order following negotiations by the parties as to its effect, if any, on the Contract Price or Contract Times.

50. *Written Amendment*--A written statement modifying the Contract Documents, signed by OWNER and CONTRACTOR on or after the Effective Date of the Agreement and normally dealing with the nonengineering or nontechnical rather than strictly construction-related aspects of the Contract Documents.

1.02   *Terminology*

A. *Intent of Certain Terms or Adjectives*

1.   Whenever in the Contract Documents the terms "as allowed," "as approved," or terms of like effect or import are used, or the adjectives "reasonable," "suitable," "acceptable," "proper," "satisfactory," or adjectives of like effect or import are used to describe an action or determination of ENGINEER as to the Work, it is intended that such action or determination will be solely to evaluate, in general, the completed Work for compliance with the requirements of and information in the Contract Documents and conformance with the design concept of the completed Project as a functioning whole as shown or indicated in the Contract Documents (unless there is a specific statement indicating otherwise). The

use of any such term or adjective shall not be effective to assign to ENGINEER any duty or authority to supervise or direct the performance of the Work or any duty or authority to undertake responsibility contrary to the provisions of paragraph 9.10 or any other provision of the Contract Documents.

B. *Day*

1. The word "day" shall constitute a calendar day of 24 hours measured from midnight to the next midnight.

C. *Defective*

1. The word "defective," when modifying the word "Work," refers to Work that is unsatisfactory, faulty, or deficient in that it does not conform to the Contract Documents or does not meet the requirements of any inspection, reference standard, test, or approval referred to in the Contract Documents, or has been damaged prior to ENGINEER's recommendation of final payment (unless responsibility for the protection thereof has been assumed by OWNER at Substantial Completion in accordance with paragraph 14.04 or 14.05).

D. *Furnish, Install, Perform, Provide*

1. The word "furnish," when used in connection with services, materials, or equipment, shall mean to supply and deliver said services, materials, or equipment to the Site (or some other specified location) ready for use or installation and in usable or operable condition.

2. The word "install," when used in connection with services, materials, or equipment, shall mean to put into use or place in final position said services, materials, or equipment complete and ready for intended use.

3. The words "perform" or "provide," when used in connection with services, materials, or equipment, shall mean to furnish and install said services, materials, or equipment complete and ready for intended use.

4. When "furnish," "install," "perform," or "provide" is not used in connection with services, materials, or equipment in a context clearly requiring an obligation of CONTRACTOR, "provide" is implied.

E. Unless stated otherwise in the Contract Documents, words or phrases which have a well-known technical or construction industry or trade meaning are used in the Contract Documents in accordance with such recognized meaning.

ARTICLE 2 - PRELIMINARY MATTERS

2.01   *Delivery of Bonds*

A. When CONTRACTOR delivers the executed Agreements to OWNER, CONTRACTOR shall also deliver to OWNER such Bonds as CONTRACTOR may be required to furnish.

2.02   *Copies of Documents*

A. OWNER shall furnish to CONTRACTOR up to ten copies of the Contract Documents. Additional copies will be furnished upon request at the cost of reproduction.

2.03   *Commencement of Contract Times; Notice to Proceed*

A. The Contract Times will commence to run on the thirtieth day after the Effective Date of the Agreement or, if a Notice to Proceed is given, on the day indicated in the Notice to Proceed. A Notice to Proceed may be given at any time within 30 days after the Effective Date of the Agreement. In no event will the Contract Times commence to run later than the sixtieth day after the day of Bid opening or the thirtieth day after the Effective Date of the Agreement, whichever date is earlier.

2.04   *Starting the Work*

A. CONTRACTOR shall start to perform the Work on the date when the Contract Times commence to run. No Work shall be done at the Site prior to the date on which the Contract Times commence to run.

2.05   *Before Starting Construction*

A. *CONTRACTOR's Review of Contract Documents:* Before undertaking each part of the Work, CONTRACTOR shall carefully study and compare the Contract Documents and check and verify pertinent figures therein and all applicable field measurements. CONTRACTOR shall promptly report in writing to ENGINEER any conflict, error, ambiguity, or discrepancy which CONTRACTOR may discover and shall obtain a written interpretation or clarification from ENGINEER before proceeding with any Work affected thereby; however, CONTRACTOR shall not be liable to OWNER or ENGINEER for failure to report any conflict, error, ambiguity, or discrepancy in the Contract Documents unless CONTRACTOR knew or reasonably should have known thereof.

B. *Preliminary Schedules:* Within ten days after the Effective Date of the Agreement (unless otherwise specified in the General Requirements), CONTRACTOR shall submit to ENGINEER for its timely review:

1. a preliminary progress schedule indicating the times (numbers of days or dates) for starting and completing the various stages of the Work, including any Milestones specified in the Contract Documents;

2. a preliminary schedule of Shop Drawing and Sample submittals which will list each required submittal and the times for submitting, reviewing, and processing such submittal; and

3. a preliminary schedule of values for all of the Work which includes quantities and prices of items which when added together equal the Contract Price and subdivides the Work into component parts in sufficient detail to serve as the basis for progress payments during performance of the Work. Such prices will include an appropriate amount of overhead and profit applicable to each item of Work.

C. *Evidence of Insurance:* Before any Work at the Site is started, CONTRACTOR and OWNER shall each deliver to the other, with copies to each additional insured identified in the Supplementary Conditions, certificates of insurance (and other evidence of insurance which either of them or any additional insured may reasonably request) which CONTRACTOR and OWNER respectively are required to purchase and maintain in accordance with Article 5.

2.06 *Preconstruction Conference*

A. Within 20 days after the Contract Times start to run, but before any Work at the Site is started, a conference attended by CONTRACTOR, ENGINEER, and others as appropriate will be held to establish a working understanding among the parties as to the Work and to discuss the schedules referred to in paragraph 2.05.B, procedures for handling Shop Drawings and other submittals, processing Applications for Payment, and maintaining required records.

2.07 *Initial Acceptance of Schedules*

A. Unless otherwise provided in the Contract Documents, at least ten days before submission of the first Application for Payment a conference attended by CONTRACTOR, ENGINEER, and others as appropriate will be held to review for acceptability to ENGINEER as provided below the schedules submitted in accordance with paragraph 2.05.B. CONTRACTOR shall have an additional ten days to make corrections and adjustments and to complete and resubmit the schedules. No progress payment shall be made to CONTRACTOR until acceptable schedules are submitted to ENGINEER.

1. The progress schedule will be acceptable to ENGINEER if it provides an orderly progression of the Work to completion within any specified Milestones and the Contract Times. Such acceptance will not impose on ENGINEER responsibility for the progress schedule, for sequencing, scheduling, or progress of the Work nor interfere with or relieve CONTRACTOR from CONTRACTOR's full responsibility therefor.

2. CONTRACTOR's schedule of Shop Drawing and Sample submittals will be acceptable to ENGINEER if it provides a workable arrangement for reviewing and processing the required submittals.

3. CONTRACTOR's schedule of values will be acceptable to ENGINEER as to form and substance if it provides a reasonable allocation of the Contract Price to component parts of the Work.

ARTICLE 3 - CONTRACT DOCUMENTS: INTENT, AMENDING, REUSE

3.01 *Intent*

A. The Contract Documents are complementary; what is called for by one is as binding as if called for by all.

B. It is the intent of the Contract Documents to describe a functionally complete Project (or part thereof) to be constructed in accordance with the Contract Documents. Any labor, documentation, services, materials, or equipment that may reasonably be inferred from the Contract Documents or from prevailing custom or trade usage as being required to produce the intended result will be provided whether or not specifically called for at no additional cost to OWNER.

C. Clarifications and interpretations of the Contract Documents shall be issued by ENGINEER as provided in Article 9.

3.02 *Reference Standards*

A. *Standards, Specifications, Codes, Laws, and Regulations*

1. Reference to standards, specifications, manuals, or codes of any technical society, organization, or association, or to Laws or Regulations, whether such reference be specific or by implication, shall mean the standard, specification, manual, code, or Laws or Regulations in effect at the time of opening of Bids (or on the Effective Date of the Agreement if there were no Bids),

except as may be otherwise specifically stated in the Contract Documents.

2. No provision of any such standard, specification, manual or code, or any instruction of a Supplier shall be effective to change the duties or responsibilities of OWNER, CONTRACTOR, or ENGINEER, or any of their subcontractors, consultants, agents, or employees from those set forth in the Contract Documents, nor shall any such provision or instruction be effective to assign to OWNER, ENGINEER, or any of ENGINEER's Consultants, agents, or employees any duty or authority to supervise or direct the performance of the Work or any duty or authority to undertake responsibility inconsistent with the provisions of the Contract Documents.

03   *Reporting and Resolving Discrepancies*

A. *Reporting Discrepancies*

1. If, during the performance of the Work, CONTRACTOR discovers any conflict, error, ambiguity, or discrepancy within the Contract Documents or between the Contract Documents and any provision of any Law or Regulation applicable to the performance of the Work or of any standard, specification, manual or code, or of any instruction of any Supplier, CONTRACTOR shall report it to ENGINEER in writing at once. CONTRACTOR shall not proceed with the Work affected thereby (except in an emergency as required by paragraph 6.16.A) until an amendment or supplement to the Contract Documents has been issued by one of the methods indicated in paragraph 3.04; provided, however, that CONTRACTOR shall not be liable to OWNER or ENGINEER for failure to report any such conflict, error, ambiguity, or discrepancy unless CONTRACTOR knew or reasonably should have known thereof.

B. *Resolving Discrepancies*

1. Except as may be otherwise specifically stated in the Contract Documents, the provisions of the Contract Documents shall take precedence in resolving any conflict, error, ambiguity, or discrepancy between the provisions of the Contract Documents and:

a. the provisions of any standard, specification, manual, code, or instruction (whether or not specifically incorporated by reference in the Contract Documents); or

b. the provisions of any Laws or Regulations applicable to the performance of the Work (unless such an interpretation of the provisions of the Contract Documents would result in violation of such Law or Regulation).

3.04   *Amending and Supplementing Contract Documents*

A. The Contract Documents may be amended to provide for additions, deletions, and revisions in the Work or to modify the terms and conditions thereof in one or more of the following ways: (i) a Written Amendment; (ii) a Change Order; or (iii) a Work Change Directive.

B. The requirements of the Contract Documents may be supplemented, and minor variations and deviations in the Work may be authorized, by one or more of the following ways: (i) a Field Order; (ii) ENGINEER's approval of a Shop Drawing or Sample; or (iii) ENGINEER's written interpretation or clarification.

3.05   *Reuse of Documents*

A. CONTRACTOR and any Subcontractor or Supplier or other individual or entity performing or furnishing any of the Work under a direct or indirect contract with OWNER: (i) shall not have or acquire any title to or ownership rights in any of the Drawings, Specifications, or other documents (or copies of any thereof) prepared by or bearing the seal of ENGINEER or ENGINEER's Consultant, including electronic media editions; and (ii) shall not reuse any of such Drawings, Specifications, other documents, or copies thereof on extensions of the Project or any other project without written consent of OWNER and ENGINEER and specific written verification or adaption by ENGINEER. This prohibition will survive final payment, completion, and acceptance of the Work, or termination or completion of the Contract. Nothing herein shall preclude CONTRACTOR from retaining copies of the Contract Documents for record purposes.

ARTICLE 4 - AVAILABILITY OF LANDS; SUBSURFACE AND PHYSICAL CONDITIONS; REFERENCE POINTS

4.01   *Availability of Lands*

A. OWNER shall furnish the Site. OWNER shall notify CONTRACTOR of any encumbrances or restrictions not of general application but specifically related to use of the Site with which CONTRACTOR must comply in performing the Work. OWNER will obtain in a timely manner and pay for easements for permanent structures or permanent changes in existing facilities. If CONTRACTOR and OWNER are unable to agree on entitlement to or on the amount or extent, if any, of any adjustment in the Contract Price or Contract Times, or both, as a result of any delay in OWNER's furnishing the Site, CONTRACTOR may make a Claim therefor as provided in paragraph 10.05.

B. Upon reasonable written request, OWNER shall furnish CONTRACTOR with a current statement of record legal title and legal description of the lands upon which the Work is to be performed and OWNER's interest therein as necessary for giving notice of or filing a mechanic's or construction lien against such lands in accordance with applicable Laws and Regulations.

C. CONTRACTOR shall provide for all additional lands and access thereto that may be required for temporary construction facilities or storage of materials and equipment.

4.02   *Subsurface and Physical Conditions*

A. *Reports and Drawings:*   The Supplementary Conditions identify:

1. those reports of explorations and tests of subsurface conditions at or contiguous to the Site that ENGINEER has used in preparing the Contract Documents; and

2. those drawings of physical conditions in or relating to existing surface or subsurface structures at or contiguous to the Site (except Underground Facilities) that ENGINEER has used in preparing the Contract Documents.

B. *Limited Reliance by CONTRACTOR on Technical Data Authorized:*   CONTRACTOR may rely upon the general accuracy of the "technical data" contained in such reports and drawings, but such reports and drawings are not Contract Documents. Such "technical data" is identified in the Supplementary Conditions. Except for such reliance on such "technical data," CONTRACTOR may not rely upon or make any Claim against OWNER, ENGINEER, or any of ENGINEER's Consultants with respect to:

1. the completeness of such reports and drawings for CONTRACTOR's purposes, including, but not limited to, any aspects of the means, methods, techniques, sequences, and procedures of construction to be employed by CONTRACTOR, and safety precautions and programs incident thereto; or

2. other data, interpretations, opinions, and information contained in such reports or shown or indicated in such drawings; or

3. any CONTRACTOR interpretation of or conclusion drawn from any "technical data" or any such other data, interpretations, opinions, or information.

4.03   *Differing Subsurface or Physical Conditions*

A. *Notice:* If CONTRACTOR believes that any subsurface or physical condition at or contiguous to the Site that is uncovered or revealed either:

1. is of such a nature as to establish that any "technical data" on which CONTRACTOR is entitled to rely as provided in paragraph 4.02 is materially inaccurate; or

2. is of such a nature as to require a change in the Contract Documents; or

3. differs materially from that shown or indicated in the Contract Documents; or

4. is of an unusual nature, and differs materially from conditions ordinarily encountered and generally recognized as inherent in work of the character provided for in the Contract Documents;

then CONTRACTOR shall, promptly after becoming aware thereof and before further disturbing the subsurface or physical conditions or performing any Work in connection therewith (except in an emergency as required by paragraph 6.16.A), notify OWNER and ENGINEER in writing about such condition. CONTRACTOR shall not further disturb such condition or perform any Work in connection therewith (except as aforesaid) until receipt of written order to do so.

B. *ENGINEER's Review:* After receipt of written notice as required by paragraph 4.03.A, ENGINEER will promptly review the pertinent condition, determine the necessity of OWNER's obtaining additional exploration or tests with respect thereto, and advise OWNER in writing (with a copy to CONTRACTOR) of ENGINEER's findings and conclusions.

C. *Possible Price and Times Adjustments*

1. The Contract Price or the Contract Times, or both, will be equitably adjusted to the extent that the existence of such differing subsurface or physical condition causes an increase or decrease in CONTRACTOR's cost of, or time required for, performance of the Work; subject, however, to the following:

a. such condition must meet any one or more of the categories described in paragraph 4.03.A; and

b. with respect to Work that is paid for on a Unit Price Basis, any adjustment in Contract Price will be subject to the provisions of paragraphs 9.08 and 11.03.

2. CONTRACTOR shall not be entitled to any adjustment in the Contract Price or Contract Times if:

a. CONTRACTOR knew of the existence of such conditions at the time CONTRACTOR made a final commitment to OWNER in respect of Contract Price and Contract Times by the submission of a Bid or becoming bound under a negotiated contract; or

b. the existence of such condition could reasonably have been discovered or revealed as a result of any examination, investigation, exploration, test, or study of the Site and contiguous areas required by the Bidding Requirements or Contract Documents to be conducted by or for CONTRACTOR prior to CONTRACTOR's making such final commitment; or

c. CONTRACTOR failed to give the written notice within the time and as required by paragraph 4.03.A.

3. If OWNER and CONTRACTOR are unable to agree on entitlement to or on the amount or extent, if any, of any adjustment in the Contract Price or Contract Times, or both, a Claim may be made therefor as provided in paragraph 10.05. However, OWNER, ENGINEER, and ENGINEER's Consultants shall not be liable to CONTRACTOR for any claims, costs, losses, or damages (including but not limited to all fees and charges of engineers, architects, attorneys, and other professionals and all court or arbitration or other dispute resolution costs) sustained by CONTRACTOR on or in connection with any other project or anticipated project.

4.04   *Underground Facilities*

A. *Shown or Indicated:* The information and data shown or indicated in the Contract Documents with respect to existing Underground Facilities at or contiguous to the Site is based on information and data furnished to OWNER or ENGINEER by the owners of such Underground Facilities, including OWNER, or by others. Unless it is otherwise expressly provided in the Supplementary Conditions:

1. OWNER and ENGINEER shall not be responsible for the accuracy or completeness of any such information or data; and

2. the cost of all of the following will be included in the Contract Price, and CONTRACTOR shall have full responsibility for:

a. reviewing and checking all such information and data,

b. locating all Underground Facilities shown or indicated in the Contract Documents,

c. coordination of the Work with the owners of such Underground Facilities, including OWNER, during construction, and

d. the safety and protection of all such Underground Facilities and repairing any damage thereto resulting from the Work.

B. *Not Shown or Indicated*

1. If an Underground Facility is uncovered or revealed at or contiguous to the Site which was not shown or indicated, or not shown or indicated with reasonable accuracy in the Contract Documents, CONTRACTOR shall, promptly after becoming aware thereof and before further disturbing conditions affected thereby or performing any Work in connection therewith (except in an emergency as required by paragraph 6.16.A), identify the owner of such Underground Facility and give written notice to that owner and to OWNER and ENGINEER. ENGINEER will promptly review the Underground Facility and determine the extent, if any, to which a change is required in the Contract Documents to reflect and document the consequences of the existence or location of the Underground Facility. During such time, CONTRACTOR shall be responsible for the safety and protection of such Underground Facility.

2. If ENGINEER concludes that a change in the Contract Documents is required, a Work Change Directive or a Change Order will be issued to reflect and document such consequences. An equitable adjustment shall be made in the Contract Price of Contract Times, or both, to the extent that they are attributable to the existence or location of any Underground Facility that was not shown or indicated or not shown or indicated with reasonable accuracy in the Contract Documents and that CONTRACTOR did not know of and could not reasonably have been expected to be aware of or to have anticipated. If OWNER and CONTRACTOR are unable to agree on entitlement to or on the amount or extent, if any, of any such adjustment in Contract Price or Contract Times, OWNER or CONTRACTOR may make a Claim therefor as provided in paragraph 10.05.

4.05   *Reference Points*

A. OWNER shall provide engineering surveys to establish reference points for construction which in ENGINEER's judgment are necessary to enable CONTRACTOR to proceed with the Work. CONTRACTOR shall be responsible for laying out the Work, shall protect and preserve the established reference points and property monuments, and shall make no changes or relocations without the prior written approval of OWNER. CONTRACTOR shall report to ENGINEER whenever any reference point or property monument is lost or destroyed or requires relocation because of necessary changes in grades or locations, and shall be responsible for the accurate replacement or relocation of such reference points or property monuments by professionally qualified personnel.

4.06   *Hazardous Environmental Condition at Site*

A. *Reports and Drawings:* Reference is made to the Supplementary Conditions for the identification of those reports and drawings relating to a Hazardous Environmental Condition identified at the Site, if any, that have been utilized by the ENGINEER in the preparation of the Contract Documents.

B. *Limited Reliance by CONTRACTOR on Technical Data Authorized:* CONTRACTOR may rely upon the general accuracy of the "technical data" contained in such reports and drawings, but such reports and drawings are not Contract Documents. Such "technical data" is identified in the Supplementary Conditions. Except for such reliance on such "technical data," CONTRACTOR may not rely upon or make any Claim against OWNER, ENGINEER or any of ENGINEER's Consultants with respect to:

1. the completeness of such reports and drawings for CONTRACTOR's purposes, including, but not limited to, any aspects of the means, methods, techniques, sequences and procedures of construction to be employed by CONTRACTOR and safety precautions and programs incident thereto; or

2. other data, interpretations, opinions and information contained in such reports or shown or indicated in such drawings; or

3. any CONTRACTOR interpretation of or conclusion drawn from any "technical data" or any such other data, interpretations, opinions or information.

C. CONTRACTOR shall not be responsible for any Hazardous Environmental Condition uncovered or revealed at the Site which was not shown or indicated in Drawings or Specifications or identified in the Contract Documents to be within the scope of the Work. CONTRACTOR shall be responsible for a Hazardous Environmental Condition created with any materials brought to the Site by CONTRACTOR, Subcontractors, Suppliers, or anyone else for whom CON-TRACTOR is responsible.

D. If CONTRACTOR encounters a Hazardous Environmental Condition or if CONTRACTOR or anyone for whom CONTRACTOR is responsible creates a Hazardous

Environmental Condition, CONTRACTOR shall immediately: (i) secure or otherwise isolate such condition; (ii) stop all Work in connection with such condition and in any area affected thereby (except in an emergency as required by paragraph 6.16); and (iii) notify OWNER and ENGINEER (and promptly thereafter confirm such notice in writing). OWNER shall promptly consult with ENGINEER concerning the necessity for OWNER to retain a qualified expert to evaluate such condition or take corrective action, if any.

E. CONTRACTOR shall not be required to resume Work in connection with such condition or in any affected area until after OWNER has obtained any required permits related thereto and delivered to CONTRACTOR written notice: (i) specifying that such condition and any affected area is or has been rendered safe for the resumption of Work; or (ii) specifying any special conditions under which Work may be resumed safely.   If OWNER and CONTRACTOR cannot agree as to entitlement to or on the amount or extent, if any, of any adjustment in Contract Price or Contract Times, or both, as a result of such Work stoppage or such special conditions under which Work is agreed to be resumed by CONTRACTOR, either party may make a Claim therefor as provided in paragraph 10.05.

F. If after receipt of such written notice CONTRACTOR does not agree to resume such Work based on a reasonable belief it is unsafe, or does not agree to resume such Work under such special conditions, then OWNER may order the portion of the Work that is in the area affected by such condition to be deleted from the Work. If OWNER and CONTRACTOR cannot agree as to entitlement to or on the amount or extent, if any, of an adjustment in Contract Price or Contract Times as a result of deleting such portion of the Work, then either party may make a Claim therefor as provided in paragraph 10.05. OWNER may have such deleted portion of the Work performed by OWNER's own forces or others in accordance with Article 7.

G. To the fullest extent permitted by Laws and Regulations, OWNER shall indemnify and hold harmless CONTRACTOR, Subcontractors, ENGINEER, ENGINEER's Consultants and the officers, directors, partners, employees, agents, other consultants, and subcontractors of each and any of them from and against all claims, costs, losses, and damages (including but not limited to all fees and charges of engineers, architects, attorneys, and other professionals and all court or arbitration or other dispute resolution costs) arising out of or relating to a Hazardous Environmental Condition, provided that such Hazardous Environmental Condition: (i) was not shown or indicated in the Drawings or Specifications or identified in the Contract Documents to be included within the scope of the Work, and (ii) was not created by CONTRACTOR or by anyone for whom CONTRACTOR is responsible. Nothing in this paragraph 4.06.E shall obligate OWNER to indemnify any individual or entity from and against the consequences of that individual's or entity's own negligence.

H. To the fullest extent permitted by Laws and Regulations, CONTRACTOR shall indemnify and hold harmless OWNER, ENGINEER, ENGINEER's Consultants, and the officers, directors, partners, employees, agents, other consultants, and subcontractors of each and any of them from and against all claims, costs, losses, and damages (including but not limited to all fees and charges of engineers, architects, attorneys, and other professionals and all court or arbitration or other dispute resolution costs) arising out of or relating to a Hazardous Environmental Condition created by CONTRACTOR or by anyone for whom CONTRACTOR is responsible. Nothing in this paragraph 4.06.F shall obligate CONTRACTOR to indemnify any individual or entity from and against the consequences of that individual's or entity's own negligence.

I. The provisions of paragraphs 4.02, 4.03, and 4.04 are not intended to apply to a Hazardous Environmental Condition uncovered or revealed at the Site.

ARTICLE 5 - BONDS AND INSURANCE

5.01 *Performance, Payment, and Other Bonds*

A. CONTRACTOR shall furnish performance and payment Bonds, each in an amount at least equal to the Contract Price as security for the faithful performance and payment of all CONTRACTOR's obligations under the Contract Documents. These Bonds shall remain in effect at least until one year after the date when final payment becomes due, except as provided otherwise by Laws or Regulations or by the Contract Documents. CONTRACTOR shall also furnish such other Bonds as are required by the Contract Documents.

B. All Bonds shall be in the form prescribed by the Contract Documents except as provided otherwise by Laws or Regulations, and shall be executed by such sureties as are named in the current list of "Companies Holding Certificates of Authority as Acceptable Sureties on Federal Bonds and as Acceptable Reinsuring Companies" as published in Circular 570 (amended) by the Financial Management Service, Surety Bond Branch, U.S. Department of the Treasury. All Bonds signed by an agent must be accompanied by a certified copy of such agent's authority to act.

C. If the surety on any Bond furnished by CON-TRACTOR is declared bankrupt or becomes insolvent or its right to do business is terminated in any state where any part of the Project is located or it ceases to meet the requirements of paragraph 5.01.B, CONTRACTOR shall within 20 days thereafter substitute another Bond and surety, both of which shall comply with the requirements of paragraphs 5.01.B and 5.02.

5.02 *Licensed Sureties and Insurers*

A. All Bonds and insurance required by the Contract Documents to be purchased and maintained by OWNER or CONTRACTOR shall be obtained from surety or insurance companies that are duly licensed or authorized in the jurisdiction in which the Project is located to issue Bonds or insurance policies for the limits and coverages so required. Such surety and insurance companies shall also meet such additional requirements and qualifications as may be provided in the Supplementary Conditions.

5.03 *Certificates of Insurance*

A. CONTRACTOR shall deliver to OWNER, with copies to each additional insured identified in the Supplementary Conditions, certificates of insurance (and other evidence of insurance requested by OWNER or any other additional insured) which CONTRACTOR is required to purchase and maintain. OWNER shall deliver to CONTRACTOR, with copies to each additional insured identified in the Supplementary Conditions, certificates of insurance (and other evidence of insurance requested by CONTRACTOR or any other additional insured) which OWNER is required to purchase and maintain.

5.04 *CONTRACTOR's Liability Insurance*

A. CONTRACTOR shall purchase and maintain such liability and other insurance as is appropriate for the Work being performed and as will provide protection from claims set forth below which may arise out of or result from CONTRACTOR's performance of the Work and CONTRACTOR's other obligations under the Contract Documents, whether it is to be performed by CONTRACTOR, any Subcontractor or Supplier, or by anyone directly or indirectly employed by any of them to perform any of the Work, or by anyone for whose acts any of them may be liable:

1. claims under workers' compensation, disability benefits, and other similar employee benefit acts;

2. claims for damages because of bodily injury, occupational sickness or disease, or death of CONTRACTOR's employees;

3. claims for damages because of bodily injury, sickness or disease, or death of any person other than CONTRACTOR's employees;

4. claims for damages insured by reasonably available personal injury liability coverage which are sustained: (i) by any person as a result of an offense directly or indirectly related to the employment of such person by CONTRACTOR, or (ii) by any other person for any other reason;

5. claims for damages, other than to the Work itself, because of injury to or destruction of tangible property wherever located, including loss of use resulting therefrom; and

6. claims for damages because of bodily injury or death of any person or property damage arising out of the ownership, maintenance or use of any motor vehicle.

B. The policies of insurance so required by this paragraph 5.04 to be purchased and maintained shall:

1. with respect to insurance required by paragraphs 5.04.A.3 through 5.04.A.6 inclusive, include as additional insureds (subject to any customary exclusion in respect of professional liability) OWNER, ENGINEER, ENGINEER's Consultants, and any other individuals or entities identified in the Supplementary Conditions, all of whom shall be listed as additional insureds, and include coverage for the respective officers, directors, partners, employees, agents, and other consultants and subcontractors of each and any of all such additional insureds, and the insurance afforded to these additional insureds shall provide primary coverage for all claims covered thereby;

2. include at least the specific coverages and be written for not less than the limits of liability provided in the Supplementary Conditions or required by Laws or Regulations, whichever is greater;

3. include completed operations insurance;

4. include contractual liability insurance covering CONTRACTOR's indemnity obligations under paragraphs 6.07, 6.11, and 6.20;

5. contain a provision or endorsement that the coverage afforded will not be canceled, materially changed or renewal refused until at least thirty days prior written notice has been given to OWNER and CONTRACTOR and to each other additional insured identified in the Supplementary Conditions to whom a certificate of insurance has been issued (and the certificates of insurance furnished by the CONTRACTOR pursuant to paragraph 5.03 will so provide);

6. remain in effect at least until final payment and at all times thereafter when CONTRACTOR may be

correcting, removing, or replacing defective Work in accordance with paragraph 13.07; and

7. with respect to completed operations insurance, and any insurance coverage written on a claims-made basis, remain in effect for at least two years after final payment (and CONTRACTOR shall furnish OWNER and each other additional insured identified in the Supplementary Conditions, to whom a certificate of insurance has been issued, evidence satisfactory to OWNER and any such additional insured of continuation of such insurance at final payment and one year thereafter).

5.05 *OWNER's Liability Insurance*

A. In addition to the insurance required to be provided by CONTRACTOR under paragraph 5.04, OWNER, at OWNER's option, may purchase and maintain at OWNER's expense OWNER's own liability insurance as will protect OWNER against claims which may arise from operations under the Contract Documents.

5.06 *Property Insurance*

A. Unless otherwise provided in the Supplementary Conditions, OWNER shall purchase and maintain property insurance upon the Work at the Site in the amount of the full replacement cost thereof (subject to such deductible amounts as may be provided in the Supplementary Conditions or required by Laws and Regulations). This insurance shall:

1. include the interests of OWNER, CONTRACTOR, Subcontractors, ENGINEER, ENGINEER's Consultants, and any other individuals or entities identified in the Supplementary Conditions, and the officers, directors, partners, employees, agents, and other consultants and subcontractors of each and any of them, each of whom is deemed to have an insurable interest and shall be listed as an additional insured;

2. be written on a Builder's Risk "all-risk" or open peril or special causes of loss policy form that shall at least include insurance for physical loss or damage to the Work, temporary buildings, false work, and materials and equipment in transit, and shall insure against at least the following perils or causes of loss: fire, lightning, extended coverage, theft, vandalism and malicious mischief, earthquake, collapse, debris removal, demolition occasioned by enforcement of Laws and Regulations, water damage, and such other perils or causes of loss as may be specifically required by the Supplementary Conditions;

3. include expenses incurred in the repair or replacement of any insured property (including but not limited to fees and charges of engineers and architects);

4. cover materials and equipment stored at the Site or at another location that was agreed to in writing by OWNER prior to being incorporated in the Work, provided that such materials and equipment have been included in an Application for Payment recommended by ENGINEER;

5. allow for partial utilization of the Work by OWNER;

6. include testing and startup; and

7. be maintained in effect until final payment is made unless otherwise agreed to in writing by OWNER, CONTRACTOR, and ENGINEER with 30 days written notice to each other additional insured to whom a certificate of insurance has been issued.

B. OWNER shall purchase and maintain such boiler and machinery insurance or additional property insurance as may be required by the Supplementary Conditions or Laws and Regulations which will include the interests of OWNER, CONTRACTOR, Subcontractors, ENGINEER, ENGINEER's Consultants, and any other individuals or entities identified in the Supplementary Conditions, each of whom is deemed to have an insurable interest and shall be listed as an insured or additional insured.

C. All the policies of insurance (and the certificates or other evidence thereof) required to be purchased and maintained in accordance with paragraph 5.06 will contain a provision or endorsement that the coverage afforded will not be canceled or materially changed or renewal refused until at least 30 days prior written notice has been given to OWNER and CONTRACTOR and to each other additional insured to whom a certificate of insurance has been issued and will contain waiver provisions in accordance with paragraph 5.07.

D. OWNER shall not be responsible for purchasing and maintaining any property insurance specified in this paragraph 5.06 to protect the interests of CONTRACTOR, Subcontractors, or others in the Work to the extent of any deductible amounts that are identified in the Supplementary Conditions. The risk of loss within such identified deductible amount will be borne by CONTRACTOR, Subcontractors, or others suffering any such loss, and if any of them wishes property insurance coverage within the limits of such amounts, each may purchase and maintain it at the purchaser's own expense.

E. If CONTRACTOR requests in writing that other special insurance be included in the property insurance policies provided under paragraph 5.06, OWNER shall, if possible, include such insurance, and the cost thereof will be charged to CONTRACTOR by appropriate Change Order or Written Amendment. Prior to commencement of the Work

at the Site, OWNER shall in writing advise CONTRACTOR whether or not such other insurance has been procured by OWNER.

5.07 *Waiver of Rights*

A. OWNER and CONTRACTOR intend that all policies purchased in accordance with paragraph 5.06 will protect OWNER, CONTRACTOR, Subcontractors, ENGINEER, ENGINEER's Consultants, and all other individuals or entities identified in the Supplementary Conditions to be listed as insureds or additional insureds (and the officers, directors, partners, employees, agents, and other consultants and subcontractors of each and any of them) in such policies and will provide primary coverage for all losses and damages caused by the perils or causes of loss covered thereby. All such policies shall contain provisions to the effect that in the event of payment of any loss or damage the insurers will have no rights of recovery against any of the insureds or additional insureds thereunder. OWNER and CONTRACTOR waive all rights against each other and their respective officers, directors, partners, employees, agents, and other consultants and subcontractors of each and any of them for all losses and damages caused by, arising out of or resulting from any of the perils or causes of loss covered by such policies and any other property insurance applicable to the Work; and, in addition, waive all such rights against Subcontractors, ENGINEER, ENGINEER's Consultants, and all other individuals or entities identified in the Supplementary Conditions to be listed as insureds or additional insureds (and the officers, directors, partners, employees, agents, and other consultants and subcontractors of each and any of them) under such policies for losses and damages so caused. None of the above waivers shall extend to the rights that any party making such waiver may have to the proceeds of insurance held by OWNER as trustee or otherwise payable under any policy so issued.

B. OWNER waives all rights against CONTRACTOR, Subcontractors, ENGINEER, ENGINEER's Consultants, and the officers, directors, partners, employees, agents, and other consultants and subcontractors of each and any of them for:

1. loss due to business interruption, loss of use, or other consequential loss extending beyond direct physical loss or damage to OWNER's property or the Work caused by, arising out of, or resulting from fire or other peril whether or not insured by OWNER; and

2. loss or damage to the completed Project or part thereof caused by, arising out of, or resulting from fire or other insured peril or cause of loss covered by any property insurance maintained on the completed Project or part thereof by OWNER during partial utilization pursuant to paragraph 14.05, after Substantial Completion

pursuant to paragraph 14.04, or after final payment pursuant to paragraph 14.07.

C. Any insurance policy maintained by OWNER covering any loss, damage or consequential loss referred to in paragraph 5.07.B shall contain provisions to the effect that in the event of payment of any such loss, damage, or consequential loss, the insurers will have no rights of recovery against CONTRACTOR, Subcontractors, ENGINEER, or ENGINEER's Consultants and the officers, directors, partners, employees, agents, and other consultants and subcontractors of each and any of them.

5.08   Receipt and Application of Insurance Proceeds

A. Any insured loss under the policies of insurance required by paragraph 5.06 will be adjusted with OWNER and made payable to OWNER as fiduciary for the insureds, as their interests may appear, subject to the requirements of any applicable mortgage clause and of paragraph 5.08.B. OWNER shall deposit in a separate account any money so received and shall distribute it in accordance with such agreement as the parties in interest may reach. If no other special agreement is reached, the damaged Work shall be repaired or replaced, the moneys so received applied on account thereof, and the Work and the cost thereof covered by an appropriate Change Order or Written Amendment.

B. OWNER as fiduciary shall have power to adjust and settle any loss with the insurers unless one of the parties in interest shall object in writing within 15 days after the occurrence of loss to OWNER's exercise of this power. If such objection be made, OWNER as fiduciary shall make settlement with the insurers in accordance with such agreement as the parties in interest may reach. If no such agreement among the parties in interest is reached, OWNER as fiduciary shall adjust and settle the loss with the insurers and, if required in writing by any party in interest, OWNER as fiduciary shall give bond for the proper performance of such duties.

5.09   Acceptance of Bonds and Insurance; Option to Replace

A. If either OWNER or CONTRACTOR has any objection to the coverage afforded by or other provisions of the Bonds or insurance required to be purchased and maintained by the other party in accordance with Article 5 on the basis of non-conformance with the Contract Documents, the objecting party shall so notify the other party in writing within 10 days after receipt of the certificates (or other evidence requested) required by paragraph 2.05.C. OWNER and CONTRACTOR shall each provide to the other such additional information in respect of insurance provided as the other may reasonably request. If either party does not purchase or maintain all of the Bonds and insurance required

of such party by the Contract Documents, such party shall notify the other party in writing of such failure to purchase prior to the start of the Work, or of such failure to maintain prior to any change in the required coverage. Without prejudice to any other right or remedy, the other party may elect to obtain equivalent Bonds or insurance to protect such other party's interests at the expense of the party who was required to provide such coverage, and a Change Order shall be issued to adjust the Contract Price accordingly.

5.10   Partial Utilization, Acknowledgment of Property Insurer

A. If OWNER finds it necessary to occupy or use a portion or portions of the Work prior to Substantial Completion of all the Work as provided in paragraph 14.05, no such use or occupancy shall commence before the insurers providing the property insurance pursuant to paragraph 5.06 have acknowledged notice thereof and in writing effected any changes in coverage necessitated thereby. The insurers providing the property insurance shall consent by endorsement on the policy or policies, but the property insurance shall not be canceled or permitted to lapse on account of any such partial use or occupancy.

ARTICLE 6 - CONTRACTOR'S RESPONSIBILITIES

6.01   Supervision and Superintendence

A. CONTRACTOR shall supervise, inspect, and direct the Work competently and efficiently, devoting such attention thereto and applying such skills and expertise as may be necessary to perform the Work in accordance with the Contract Documents. CONTRACTOR shall be solely responsible for the means, methods, techniques, sequences, and procedures of construction, but CONTRACTOR shall not be responsible for the negligence of OWNER or ENGINEER in the design or specification of a specific means, method, technique, sequence, or procedure of construction which is shown or indicated in and expressly required by the Contract Documents. CONTRACTOR shall be responsible to see that the completed Work complies accurately with the Contract Documents.

B. At all times during the progress of the Work, CONTRACTOR shall assign a competent resident superintendent thereto who shall not be replaced without written notice to OWNER and ENGINEER except under extraordinary circumstances. The superintendent will be CONTRACTOR's representative at the Site and shall have authority to act on behalf of CONTRACTOR. All communications given to or received from the superintendent shall be binding on CONTRACTOR.

6.02 Labor; Working Hours

A. CONTRACTOR shall provide competent, suitably qualified personnel to survey, lay out, and construct the Work as required by the Contract Documents. CONTRACTOR shall at all times maintain good discipline and order at the Site.

B. Except as otherwise required for the safety or protection of persons or the Work or property at the Site or adjacent thereto, and except as otherwise stated in the Contract Documents, all Work at the Site shall be performed during regular working hours, and CONTRACTOR will not permit overtime work or the performance of Work on Saturday, Sunday, or any legal holiday without OWNER's written consent (which will not be unreasonably withheld) given after prior written notice to ENGINEER.

6.03 Services, Materials, and Equipment

A. Unless otherwise specified in the General Requirements, CONTRACTOR shall provide and assume full responsibility for all services, materials, equipment, labor, transportation, construction equipment and machinery, tools, appliances, fuel, power, light, heat, telephone, water, sanitary facilities, temporary facilities, and all other facilities and incidentals necessary for the performance, testing, start-up, and completion of the Work.

B. All materials and equipment incorporated into the Work shall be as specified or, if not specified, shall be of good quality and new, except as otherwise provided in the Contract Documents. All warranties and guarantees specifically called for by the Specifications shall expressly run to the benefit of OWNER. If required by ENGINEER, CONTRACTOR shall furnish satisfactory evidence (including reports of required tests) as to the source, kind, and quality of materials and equipment. All materials and equipment shall be stored, applied, installed, connected, erected, protected, used, cleaned, and conditioned in accordance with instructions of the applicable Supplier, except as otherwise may be provided in the Contract Documents.

6.04 Progress Schedule

A. CONTRACTOR shall adhere to the progress schedule established in accordance with paragraph 2.07 as it may be adjusted from time to time as provided below.

1. CONTRACTOR shall submit to ENGINEER for acceptance (to the extent indicated in paragraph 2.07) proposed adjustments in the progress schedule that will not result in changing the Contract Times (or Milestones). Such adjustments will conform generally to the progress schedule then in effect and additionally will comply with

any provisions of the General Requirements applicable thereto.

2. Proposed adjustments in the progress schedule that will change the Contract Times (or Milestones) shall be submitted in accordance with the requirements of Article 12. Such adjustments may only be made by a Change Order or Written Amendment in accordance with Article 12.

6.05 Substitutes and "Or-Equals"

A. Whenever an item of material or equipment is specified or described in the Contract Documents by using the name of a proprietary item or the name of a particular Supplier, the specification or description is intended to establish the type, function, appearance, and quality required. Unless the specification or description contains or is followed by words reading that no like, equivalent, or "or-equal" item or no substitution is permitted, other items of material or equipment or material or equipment of other Suppliers may be submitted to ENGINEER for review under the circumstances described below.

1. "Or-Equal" Items: If in ENGINEER's sole discretion an item of material or equipment proposed by CONTRACTOR is functionally equal to that named and sufficiently similar so that no change in related Work will be required, it may be considered by ENGINEER as an "or-equal" item, in which case review and approval of the proposed item may, in ENGINEER's sole discretion, be accomplished without compliance with some or all of the requirements for approval of proposed substitute items. For the purposes of this paragraph 6.05.A.1, a proposed item of material or equipment will be considered functionally equal to an item so named if:

a. in the exercise of reasonable judgment ENGINEER determines that: (i) it is at least equal in quality, durability, appearance, strength, and design characteristics; (ii) it will reliably perform at least equally well the function imposed by the design concept of the completed Project as a functioning whole, and;

b. CONTRACTOR certifies that: (i) there is no increase in cost to the OWNER; and (ii) it will conform substantially, even with deviations, to the detailed requirements of the item named in the Contract Documents.

2. Substitute Items

a. If in ENGINEER's sole discretion an item of material or equipment proposed by CONTRACTOR does not qualify as an "or-equal" item under paragraph 6.05.A.1, it will be considered a proposed substitute item.

b. CONTRACTOR shall submit sufficient information as provided below to allow ENGINEER to determine that the item of material or equipment proposed is essentially equivalent to that named and an acceptable substitute therefor. Requests for review of proposed substitute items of material or equipment will not be accepted by ENGINEER from anyone other than CONTRACTOR.

c. The procedure for review by ENGINEER will be as set forth in paragraph 6.05.A.2.d, as supplemented in the General Requirements and as ENGINEER may decide is appropriate under the circumstances.

d. CONTRACTOR shall first make written application to ENGINEER for review of a proposed substitute item of material or equipment that CONTRACTOR seeks to furnish or use. The application shall certify that the proposed substitute item will perform adequately the functions and achieve the results called for by the general design, be similar in substance to that specified, and be suited to the same use as that specified. The application will state the extent, if any, to which the use of the proposed substitute item will prejudice CONTRACTOR's achievement of Substantial Completion on time, whether or not use of the proposed substitute item in the Work will require a change in any of the Contract Documents (or in the provisions of any other direct contract with OWNER for work on the Project) to adapt the design to the proposed substitute item and whether or not incorporation or use of the proposed substitute item in connection with the Work is subject to payment of any license fee or royalty. All variations of the proposed substitute item from that specified will be identified in the application, and available engineering, sales, maintenance, repair, and replacement services will be indicated. The application will also contain an itemized estimate of all costs or credits that will result directly or indirectly from use of such substitute item, including costs of redesign and claims of other contractors affected by any resulting change, all of which will be considered by ENGINEER in evaluating the proposed substitute item. ENGINEER may require CONTRACTOR to furnish additional data about the proposed substitute item.

B. *Substitute Construction Methods or Procedures:* If a specific means, method, technique, sequence, or procedure of construction is shown or indicated in and expressly required by the Contract Documents, CONTRACTOR may furnish or utilize a substitute means, method, technique, sequence, or procedure of construction approved by ENGINEER. CONTRACTOR shall submit sufficient information to allow ENGINEER, in ENGINEER's sole discretion, to determine that the substitute proposed is equivalent to that expressly called for by the Contract Documents. The procedure for review by ENGINEER will be similar to that provided in subparagraph 6.05.A.2.

C. *Engineer's Evaluation:* ENGINEER will be allowed a reasonable time within which to evaluate each proposal or submittal made pursuant to paragraphs 6.05.A and 6.05.B. ENGINEER will be the sole judge of acceptability. No "or-equal" or substitute will be ordered, installed or utilized until ENGINEER's review is complete, which will be evidenced by either a Change Order for a substitute or an approved Shop Drawing for an "or equal." ENGINEER will advise CONTRACTOR in writing of any negative determination.

D. *Special Guarantee:* OWNER may require CONTRACTOR to furnish at CONTRACTOR's expense a special performance guarantee or other surety with respect to any substitute.

E. *ENGINEER's Cost Reimbursement:* ENGINEER will record time required by ENGINEER and ENGINEER's Consultants in evaluating substitute proposed or submitted by CONTRACTOR pursuant to paragraphs 6.05.A.2 and 6.05.B and in making changes in the Contract Documents (or in the provisions of any other direct contract with OWNER for work on the Project) occasioned thereby. Whether or not ENGINEER approves a substitute item so proposed or submitted by CONTRACTOR, CONTRACTOR shall reimburse OWNER for the charges of ENGINEER and ENGINEER's Consultants for evaluating each such proposed substitute.

F. *CONTRACTOR's Expense:* CONTRACTOR shall provide all data in support of any proposed substitute or "or-equal" at CONTRACTOR's expense.

6.06   *Concerning Subcontractors, Suppliers, and Others*

A. CONTRACTOR shall not employ any Subcontractor, Supplier, or other individual or entity (including those acceptable to OWNER as indicated in paragraph 6.06.B), whether initially or as a replacement, against whom OWNER may have reasonable objection. CONTRACTOR shall not be required to employ any Subcontractor, Supplier, or other individual or entity to furnish or perform any of the Work against whom CONTRACTOR has reasonable objection.

B. If the Supplementary Conditions require the identity of certain Subcontractors, Suppliers, or other individuals or entities to be submitted to OWNER in advance for acceptance by OWNER by a specified date prior to the Effective Date of the Agreement, and if CONTRACTOR has submitted a list thereof in accordance with the Supplementary Conditions, OWNER's acceptance (either in writing or by failing to make written objection thereto by the date indicated for acceptance or objection in the Bidding Documents or the Contract Documents) of any such Subcontractor, Supplier, or other individual or entity so identified may be revoked on the basis of reasonable objection after due investigation.   CON-TRACTOR shall submit an acceptable replacement for the rejected Subcontractor, Supplier, or other individual or entity, and the Contract Price will be adjusted by the difference in the cost occasioned by such replacement, and an appropriate Change Order will be issued or Written Amendment signed. No acceptance by OWNER of any such Subcontractor, Supplier, or other individual or entity, whether initially or as a replacement, shall constitute a waiver of any right of OWNER or ENGINEER to reject defective Work.

C. CONTRACTOR shall be fully responsible to OWNER and ENGINEER for all acts and omissions of the Subcontractors, Suppliers, and other individuals or entities performing or furnishing any of the Work just as CONTRACTOR is responsible for CONTRACTOR's own acts and omissions. Nothing in the Contract Documents shall create for the benefit of any such Subcontractor, Supplier, or other individual or entity any contractual relationship between OWNER or ENGINEER and any such Subcontractor, Supplier or other individual or entity, nor shall it create any obligation on the part of OWNER or ENGINEER to pay or to see to the payment of any moneys due any such Subcontractor, Supplier, or other individual or entity except as may otherwise be required by Laws and Regulations.

D. CONTRACTOR shall be solely responsible for scheduling and coordinating the Work of Subcontractors, Suppliers, and other individuals or entities performing or furnishing any of the Work under a direct or indirect contract with CONTRACTOR.

E. CONTRACTOR shall require all Subcontractors, Suppliers, and such other individuals or entities performing or furnishing any of the Work to communicate with ENGI-NEER through CONTRACTOR.

F. The divisions and sections of the Specifications and the identifications of any Drawings shall not control CONTRACTOR in dividing the Work among Subcontractors or Suppliers or delineating the Work to be performed by any specific trade.

G. All Work performed for CONTRACTOR by a Subcontractor or Supplier will be pursuant to an appropriate agreement between CONTRACTOR and the Subcontractor

or Supplier which specifically binds the Subcontractor or Supplier to the applicable terms and conditions of the Contract Documents for the benefit of OWNER and ENGINEER.   Whenever any such agreement is with a Subcontractor or Supplier who is listed as an additional insured on the property insurance provided in paragraph 5.06, the agreement between the CONTRACTOR and the Subcontractor or Supplier will contain provisions whereby the Subcontractor or Supplier waives all rights against OWNER, CONTRACTOR, ENGINEER, ENGINEER's Consultants, and all other individuals or entities identified in the Supplementary Conditions to be listed as insureds or additional insureds (and the officers, directors, partners, employees, agents, and other consultants and subcontractors of each and any of them) for all losses and damages caused by, arising out of, relating to, or resulting from any of the perils or causes of loss covered by such policies and any other property insurance applicable to the Work. If the insurers on any such policies require separate waiver forms to be signed by any Subcontractor or Supplier, CONTRAC-TOR will obtain the same.

6.07   *Patent Fees and Royalties*

A. CONTRACTOR shall pay all license fees and royalties and assume all costs incident to the use in the performance of the Work or the incorporation in the Work of any invention, design, process, product, or device which is the subject of patent rights or copyrights held by others. If a particular invention, design, process, product, or device is specified in the Contract Documents for use in the performance of the Work and if to the actual knowledge of OWNER or ENGINEER its use is subject to patent rights or copyrights calling for the payment of any license fee or royalty to others, the existence of such rights shall be disclosed by OWNER in the Contract Documents. To the fullest extent permitted by Laws and Regulations, CONTRACTOR shall indemnify and hold harmless OWNER, ENGINEER, ENGINEER's Consultants, and the officers, directors, partners, employees or agents, and other consultants of each and any of them from and against all claims, costs, losses, and damages (including but not limited to all fees and charges of engineers, architects, attorneys, and other professionals and all court or arbitration or other dispute resolution costs) arising out of or relating to any infringement of patent rights or copyrights incident to the use in the performance of the Work or resulting from the incorporation in the Work of any invention, design, process, product, or device not specified in the Contract Documents.

6.08   *Permits*

A. Unless otherwise provided in the Supplementary Conditions, CONTRACTOR shall obtain and pay for all construction permits and licenses. OWNER shall assist CONTRACTOR, when necessary, in obtaining such permits and licenses. CONTRACTOR shall pay all governmental charges and inspection fees necessary for the prosecution of the Work which are applicable at the time of opening of Bids, or, if there are no Bids, on the Effective Date of the Agreement. CONTRACTOR shall pay all charges of utility owners for connections to the Work, and OWNER shall pay all charges of such utility owners for capital costs related thereto, such as plant investment fees.

6.09 Laws and Regulations

A. CONTRACTOR shall give all notices and comply with all Laws and Regulations applicable to the performance of the Work. Except where otherwise expressly required by applicable Laws and Regulations, neither OWNER nor ENGINEER shall be responsible for monitoring CONTRACTOR's compliance with any Laws or Regulations.

B. If CONTRACTOR performs any Work knowing or having reason to know that it is contrary to Laws or Regulations, CONTRACTOR shall bear all claims, costs, losses, and damages (including but not limited to all fees and charges of engineers, architects, attorneys, and other professionals and all court or arbitration or other dispute resolution costs) arising out of or relating to such Work; however, it shall not be CONTRACTOR's primary responsibility to make certain that the Specifications and Drawings are in accordance with Laws and Regulations, but this shall not relieve CONTRACTOR of CONTRACTOR's obligations under paragraph 3.03.

C. Changes in Laws or Regulations not known at the time of opening of Bids (or, on the Effective Date of the Agreement if there were no Bids) having an effect on the cost or time of performance of the Work may be the subject of an adjustment in Contract Price or Contract Times. If OWNER and CONTRACTOR are unable to agree on entitlement to or on the amount or extent, if any, of any such adjustment, a Claim may be made therefor as provided in paragraph 10.05.

6.10 Taxes

A. CONTRACTOR shall pay all sales, consumer, use, and other similar taxes required to be paid by CONTRACTOR in accordance with the Laws and Regulations of the place of the Project which are applicable during the performance of the Work.

6.11 Use of Site and Other Areas

A. Limitation on Use of Site and Other Areas

1. CONTRACTOR shall confine construction equipment, the storage of materials and equipment, and the operations of workers to the Site and other areas permitted by Laws and Regulations, and shall not unreasonably encumber the Site and other areas with construction equipment or other materials or equipment. CONTRACTOR shall assume full responsibility for any damage to any such land or area, or to the owner or occupant thereof, or of any adjacent land or areas resulting from the performance of the Work.

2. Should any claim be made by any such owner or occupant because of the performance of the Work, CONTRACTOR shall promptly settle with such other party by negotiation or otherwise resolve the claim by arbitration or other dispute resolution proceeding or at law.

3. To the fullest extent permitted by Laws and Regulations, CONTRACTOR shall indemnify and hold harmless OWNER, ENGINEER, ENGINEER's Consultant, and the officers, directors, partners, employees, agents, and other consultants of each and any of them from and against all claims, costs, losses, and damages (including but not limited to all fees and charges of engineers, architects, attorneys, and other professionals and all court or arbitration or other dispute resolution costs) arising out of or relating to any claim or action, legal or equitable, brought by any such owner or occupant against OWNER, ENGINEER, or any other party indemnified hereunder to the extent caused by or based upon CONTRACTOR's performance of the Work.

B. Removal of Debris During Performance of the Work: During the progress of the Work CONTRACTOR shall keep the Site and other areas free from accumulations of waste materials, rubbish, and other debris. Removal and disposal of such waste materials, rubbish, and other debris shall conform to applicable Laws and Regulations.

C. Cleaning: Prior to Substantial Completion of the Work CONTRACTOR shall clean the Site and make it ready for utilization by OWNER. At the completion of the Work CONTRACTOR shall remove from the Site all tools, appliances, construction equipment and machinery, and surplus materials and shall restore to original condition all property not designated for alteration by the Contract Documents.

D. Loading Structures: CONTRACTOR shall not load nor permit any part of any structure to be loaded in any manner that will endanger the structure, nor shall CONTRACTOR subject any part of the Work or adjacent property to stresses or pressures that will endanger it.

6.12 Record Documents

A. CONTRACTOR shall maintain in a safe place at the Site one record copy of all Drawings, Specifications, Addenda, Written Amendments, Change Orders, Work

Change Directives, Field Orders, and written interpretations and clarifications in good order and annotated to show changes made during construction. These record documents together with all approved Samples and a counterpart of all approved Shop Drawings will be available to ENGINEER for reference. Upon completion of the Work, these record documents, Samples, and Shop Drawings will be delivered to ENGINEER for OWNER.

6.13   Safety and Protection

A. CONTRACTOR shall be solely responsible for initiating, maintaining and supervising all safety precautions and programs in connection with the Work. CONTRACTOR shall take all necessary precautions for the safety of, and shall provide the necessary protection to prevent damage, injury or loss to:

1. all persons on the Site or who may be affected by the Work;

2. all the Work and materials and equipment to be incorporated therein, whether in storage on or off the Site; and

3. other property at the Site or adjacent thereto, including trees, shrubs, lawns, walks, pavements, roadways, structures, utilities, and Underground Facilities not designated for removal, relocation, or replacement in the course of construction.

B. CONTRACTOR shall comply with all applicable Laws and Regulations relating to the safety of persons or property, or to the protection of persons or property from damage, injury, or loss; and shall erect and maintain all necessary safeguards for such safety and protection. CONTRACTOR shall notify owners of adjacent property and of Underground Facilities and other utility owners when prosecution of the Work may affect them, and shall cooperate with them in the protection, removal, relocation, and replacement of their property. All damage, injury, or loss to any property referred to in paragraph 6.13.A.2 or 6.13.A.3 caused, directly or indirectly, in whole or in part, by CONTRACTOR, any Subcontractor, Supplier, or any other individual or entity directly or indirectly employed by any of them to perform any of the Work, or anyone for whose acts any of them may be liable, shall be remedied by CONTRACTOR (except damage or loss attributable to the fault of Drawings or Specifications or to the acts or omissions of OWNER or ENGINEER or ENGINEER's Consultant, or anyone employed by any of them, or anyone for whose acts any of them may be liable, and not attributable, directly or indirectly, in whole or in part, to the fault of negligence of CONTRACTOR or any Subcontractor, Supplier, or other individual or entity directly or indirectly employed by any of them). CONTRACTOR's duties and responsibilities for safety and for protection of the Work shall continue until such time as all the Work is completed and ENGINEER has issued a notice to OWNER and CONTRACTOR in accordance with paragraph 14.07.B that the Work is acceptable (except as otherwise expressly provided in connection with Substantial Completion).

6.14   Safety Representative

A. CONTRACTOR shall designate a qualified and experienced safety representative at the Site whose duties and responsibilities shall be the prevention of accidents and the maintaining and supervising of safety precautions and programs.

6.15   Hazard Communication Programs

A. CONTRACTOR shall be responsible for coordinating any exchange of material safety data sheets or other hazard communication information required to be made available to or exchanged between or among employers at the Site in accordance with Laws or Regulations.

6.16   Emergencies

A. In emergencies affecting the safety or protection of persons or the Work or property at the Site or adjacent thereto, CONTRACTOR is obligated to act to prevent threatened damage, injury, or loss. CONTRACTOR shall give ENGINEER prompt written notice if CONTRACTOR believes that any significant changes in the Work or variations from the Contract Documents have been caused thereby or are required as a result thereof. If ENGINEER determines that a change in the Contract Documents is required because of the action taken by CONTRACTOR in response to such an emergency, a Work Change Directive or Change Order will be issued.

6.17   Shop Drawings and Samples

A. CONTRACTOR shall submit Shop Drawings to ENGINEER for review and approval in accordance with the acceptable schedule of Shop Drawings and Sample submittals. All submittals will be identified as ENGINEER may require and in the number of copies specified in the General Requirements. The data shown on the Shop Drawings will be complete with respect to quantities, dimensions, specified performance and design criteria, materials, and similar data to show ENGINEER the services, materials, and equipment CONTRACTOR proposes to provide and to enable ENGINEER to review the information for the limited purposes required by paragraph 6.17.E.

B. CONTRACTOR shall also submit Samples to ENGINEER for review and approval in accordance with the acceptable schedule of Shop Drawings and Sample submittals. Each Sample will be identified clearly as to material, Supplier, pertinent data such as catalog numbers, and the use for which intended and otherwise as ENGINEER may require to enable ENGINEER to review the submittal for the limited purposes required by paragraph 6.17.E. The numbers of each Sample to be submitted will be as specified in the Specifications.

C. Where a Shop Drawing or Sample is required by the Contract Documents or the schedule of Shop Drawings and Sample submittals acceptable to ENGINEER as required by paragraph 2.07, any related Work performed prior to ENGINEER's review and approval of the pertinent submittal will be at the sole expense and responsibility of CONTRACTOR.

D. *Submittal Procedures*

1. Before submitting each Shop Drawing or Sample, CONTRACTOR shall have determined and verified:

a. all field measurements, quantities, dimensions, specified performance criteria, installation requirements, materials, catalog numbers, and similar information with respect thereto;

b. all materials with respect to intended use, fabrication, shipping, handling, storage, assembly, and installation pertaining to the performance of the Work;

c. all information relative to means, methods, techniques, sequences, and procedures of construction and safety precautions and programs incident thereto; and

d. CONTRACTOR shall also have reviewed and coordinated each Shop Drawing or Sample with other Shop Drawings and Samples and with the requirements of the Work and the Contract Documents.

2. Each submittal shall bear a stamp or specific written indication that CONTRACTOR has satisfied CONTRACTOR's obligations under the Contract Documents with respect to CONTRACTOR's review and approval of that submittal.

3. At the time of each submittal, CONTRACTOR shall give ENGINEER specific written notice of such variations, if any, that the Shop Drawing or Sample submitted may have from the requirements of the Contract Documents, such notice to be in a written communication separate from the submittal; and, in addition, shall cause a specific notation to be made on each Shop

Drawing and Sample submitted to ENGINEER for review and approval of each such variation.

E. *ENGINEER's Review*

1. ENGINEER will timely review and approve Shop Drawings and Samples in accordance with the schedule of Shop Drawings and Sample submittals acceptable to ENGINEER. ENGINEER's review and approval will be only to determine if the items covered by the submittals will, after installation or incorporation in the Work, conform to the information given in the Contract Documents and be compatible with the design concept of the completed Project as a functioning whole as indicated by the Contract Documents.

2. ENGINEER's review and approval will not extend to means, methods, techniques, sequences, or procedures of construction (except where a particular means, method, technique, sequence, or procedure of construction is specifically and expressly called for by the Contract Documents) or to safety precautions or programs incident thereto. The review and approval of a separate item as such will not indicate approval of the assembly in which the item functions.

3. ENGINEER's review and approval of Shop Drawings or Samples shall not relieve CONTRACTOR from responsibility for any variation from the requirements of the Contract Documents unless CONTRACTOR has in writing called ENGINEER's attention to each such variation at the time of each submittal as required by paragraph 6.17.D.3 and ENGINEER has given written approval of each such variation by specific written notation thereof incorporated in or accompanying the Shop Drawing or Sample approval; nor will any approval by ENGINEER relieve CONTRACTOR from responsibility for complying with the requirements of paragraph 6.17.D.1.

F. *Resubmittal Procedures*

1. CONTRACTOR shall make corrections required by ENGINEER and shall return the required number of corrected copies of Shop Drawings and submit as required new Samples for review and approval. CONTRACTOR shall direct specific attention in writing to revisions other than the corrections called for by ENGINEER on previous submittals.

6.18   *Continuing the Work*

A. CONTRACTOR shall carry on the Work and adhere to the progress schedule during all disputes or disagreements with OWNER. No Work shall be delayed or postponed pending resolution of any disputes or disagreements, except as permitted by paragraph 15.04 or as OWNER and CONTRACTOR may otherwise agree in writing.

6.19   CONTRACTOR's General Warranty and Guarantee

A.   CONTRACTOR warrants and guarantees to OWNER, ENGINEER, and ENGINEER's Consultants that all Work will be in accordance with the Contract Documents and will not be defective. CONTRACTOR's warranty and guarantee hereunder excludes defects or damage caused by:

1.   abuse, modification, or improper maintenance or operation by persons other than CONTRACTOR, Subcontractors, Suppliers, or any other individual or entity for whom CONTRACTOR is responsible; or

2.   normal wear and tear under normal usage.

B.   CONTRACTOR's obligation to perform and complete the Work in accordance with the Contract Documents shall be absolute. None of the following will constitute an acceptance of Work that is not in accordance with the Contract Documents or a release of CONTRACTOR's obligation to perform the Work in accordance with the Contract Documents:

1.   observations by ENGINEER;

2.   recommendation by ENGINEER or payment by OWNER of any progress or final payment;

3.   the issuance of a certificate of Substantial Completion by ENGINEER or any payment related thereto by OWNER;

4.   use or occupancy of the Work or any part thereof by OWNER;

5.   any acceptance by OWNER or any failure to do so;

6.   any review and approval of a Shop Drawing or Sample submittal or the issuance of a notice of acceptability by ENGINEER;

7.   any inspection, test, or approval by others; or

8.   any correction of defective Work by OWNER.

6.20   Indemnification

A.   To the fullest extent permitted by Laws and Regulations, CONTRACTOR shall indemnify and hold harmless OWNER, ENGINEER, ENGINEER's Consultants, and the officers, directors, partners, employees, agents, and other consultants and subcontractors of each and any of them from and against all claims, costs, losses, and damages (including but not limited to all fees and charges of engineers, architects, attorneys, and other professionals and all court or arbitration or other dispute resolution costs) arising out of or relating to the performance of the Work, provided that any such claim, cost, loss, or damage:

1.   is attributable to bodily injury, sickness, disease, or death, or to injury to or destruction of tangible property (other than the Work itself), including the loss of use resulting therefrom; and

2.   is caused in whole or in part by any negligent act or omission of CONTRACTOR, any Subcontractor, any Supplier, or any individual or entity directly or indirectly employed by any of them to perform any of the Work or anyone for whose acts any of them may be liable, regardless of whether or not caused in part by any negligence or omission of an individual or entity indemnified hereunder or whether liability is imposed upon such indemnified party by Laws and Regulations regardless of the negligence of any such individual or entity.

B.   In any and all claims against OWNER or ENGINEER or any of their respective consultants, agents, officers, directors, partners, or employees by any employee (or the survivor or personal representative of such employee) of CONTRACTOR, any Subcontractor, any Supplier, or any individual or entity directly or indirectly employed by any of them to perform any of the Work, or anyone for whose acts any of them may be liable, the indemnification obligation under paragraph 6.20.A shall not be limited in any way by any limitation on the amount or type of damages, compensation, or benefits payable by or for CONTRACTOR or any such Subcontractor, Supplier, or other individual or entity under workers' compensation acts, disability benefit acts, or other employee benefit acts.

C.   The indemnification obligations of CONTRACTOR under paragraph 6.20.A shall not extend to the liability of ENGINEER and ENGINEER's Consultants or to the officers, directors, partners, employees, agents, and other consultants and subcontractors of each and any of them arising out of:

1.   the preparation or approval of, or the failure to prepare or approve, maps, Drawings, opinions, reports, surveys, Change Orders, designs, or Specifications; or

2.   giving directions or instructions, or failing to give them, if that is the primary cause of the injury or damage.

ARTICLE 7 - OTHER WORK

7.01 *Related Work at Site*

A. OWNER may perform other work related to the Project at the Site by OWNER's employees, or let other direct contracts therefor, or have other work performed by utility owners. If such other work is not noted in the Contract Documents, then:

1. written notice thereof will be given to CONTRACTOR prior to starting any such other work; and

2. if OWNER and CONTRACTOR are unable to agree on entitlement to or on the amount or extent, if any, of any adjustment in the Contract Price or Contract Times that should be allowed as a result of such other work, a Claim may be made therefor as provided in paragraph 10.05.

B. CONTRACTOR shall afford each other contractor who is a party to such a direct contract and each utility owner (and OWNER, if OWNER is performing the other work with OWNER's employees) proper and safe access to the Site and a reasonable opportunity for the introduction and storage of materials and equipment and the execution of such other work and shall properly coordinate the Work with theirs. Unless otherwise provided in the Contract Documents, CONTRACTOR shall do all cutting, fitting, and patching of the Work that may be required to properly connect or otherwise make its several parts come together and properly integrate with such other work. CONTRACTOR shall not endanger any work of others by cutting, excavating, or otherwise altering their work and will only cut or alter their work with the written consent of ENGINEER and the others whose work will be affected. The duties and responsibilities of CONTRACTOR under this paragraph are for the benefit of such utility owners and other contractors to the extent that there are comparable provisions for the benefit of CONTRACTOR in said direct contracts between OWNER and such utility owners and other contractors.

C. If the proper execution or results of any part of CONTRACTOR's Work depends upon work performed by others under this Article 7, CONTRACTOR shall inspect such other work and promptly report to ENGINEER in writing any delays, defects, or deficiencies in such other work that render it unavailable or unsuitable for the proper execution and results of CONTRACTOR's Work. CONTRACTOR's failure to so report will constitute an acceptance of such other work as fit and proper for integration with CONTRACTOR's Work except for latent defects and deficiencies in such other work.

7.02 *Coordination*

A. If OWNER intends to contract with others for the performance of other work on the Project at the Site, the following will be set forth in Supplementary Conditions:

1. the individual or entity who will have authority and responsibility for coordination of the activities among the various contractors will be identified;

2. the specific matters to be covered by such authority and responsibility will be itemized; and

3. the extent of such authority and responsibilities will be provided.

B. Unless otherwise provided in the Supplementary Conditions, OWNER shall have sole authority and responsibility for such coordination.

ARTICLE 8 - OWNER'S RESPONSIBILITIES

8.01 *Communications to Contractor*

A. Except as otherwise provided in these General Conditions, OWNER shall issue all communications to CONTRACTOR through ENGINEER.

8.02 *Replacement of ENGINEER*

A. In case of termination of the employment of ENGINEER, OWNER shall appoint an engineer to whom CONTRACTOR makes no reasonable objection, whose status under the Contract Documents shall be that of the former ENGINEER.

8.03 *Furnish Data*

A. OWNER shall promptly furnish the data required of OWNER under the Contract Documents.

8.04 *Pay Promptly When Due*

A. OWNER shall make payments to CONTRACTOR promptly when they are due as provided in paragraphs 14.02.C and 14.07.C.

8.05 *Lands and Easements; Reports and Tests*

A. OWNER's duties in respect of providing lands and easements and providing engineering surveys to establish reference points are set forth in paragraphs 4.01 and 4.05. Paragraph 4.02 refers to OWNER's identifying and making available to CONTRACTOR copies of reports of explorations and tests of subsurface conditions and drawings of physical conditions in or relating to existing surface or subsurface structures at or contiguous to the Site that have been utilized by ENGINEER in preparing the Contract Documents.

8.06 *Insurance*

A. OWNER's responsibilities, if any, in respect to purchasing and maintaining liability and property insurance are set forth in Article 5.

8.07 *Change Orders*

A. OWNER is obligated to execute Change Orders as indicated in paragraph 10.03.

8.08 *Inspections, Tests, and Approvals*

A. OWNER's responsibility in respect to certain inspections, tests, and approvals is set forth in paragraph 13.03.B.

8.09 *Limitations on OWNER's Responsibilities*

A. The OWNER shall not supervise, direct, or have control or authority over, nor be responsible for, CONTRACTOR's means, methods, techniques, sequences, or procedures of construction, or the safety precautions and programs incident thereto, or for any failure of CONTRACTOR to comply with Laws and Regulations applicable to the performance of the Work. OWNER will not be responsible for CONTRACTOR's failure to perform the Work in accordance with the Contract Documents.

8.10 *Undisclosed Hazardous Environmental Condition*

A. OWNER's responsibility in respect to an undisclosed Hazardous Environmental Condition is set forth in paragraph 4.06.

8.11 *Evidence of Financial Arrangements*

A. If and to the extent OWNER has agreed to furnish CONTRACTOR reasonable evidence that financial arrangements have been made to satisfy OWNER's obligations under the Contract Documents, OWNER's responsibility in respect thereof will be as set forth in the Supplementary Conditions.

ARTICLE 9 - ENGINEER'S STATUS DURING CONSTRUCTION

9.01 *OWNER'S Representative*

A. ENGINEER will be OWNER's representative during the construction period. The duties and responsibilities and the limitations of authority of ENGINEER as OWNER's representative during construction are set forth in the Contract Documents and will not be changed without written consent of OWNER and ENGINEER.

9.02 *Visits to Site*

A. ENGINEER will make visits to the Site at intervals appropriate to the various stages of construction as ENGINEER deems necessary in order to observe as an experienced and qualified design professional the progress that has been made and the quality of the various aspects of CONTRACTOR's executed Work. Based on information obtained during such visits and observations, ENGINEER, for the benefit of OWNER, will determine, in general, if the Work is proceeding in accordance with the Contract Documents. ENGINEER will not be required to make exhaustive or continuous inspections on the Site to check the quality or quantity of the Work. ENGINEER's efforts will be directed toward providing for OWNER a greater degree of confidence that the completed Work will conform generally to the Contract Documents. On the basis of such visits and observations, ENGINEER will keep OWNER informed of the progress of the Work and will endeavor to guard OWNER against defective Work.

B. ENGINEER's visits and observations are subject to all the limitations on ENGINEER's authority and responsibility set forth in paragraph 9.10, and particularly, but without limitation, during or as a result of ENGINEER's visits or observations of CONTRACTOR's Work ENGINEER will not supervise, direct, control, or have authority over or be responsible for CONTRACTOR's means, methods, techniques, sequences, or procedures of construction, or the safety precautions and programs incident thereto, or for any failure of CONTRACTOR to comply with Laws and Regulations applicable to the performance of the Work.

9.03 *Project Representative*

A. If OWNER and ENGINEER agree, ENGINEER will furnish a Resident Project Representative to assist ENGINEER in providing more extensive observation of the Work. The responsibilities and authority and limitations thereon of any such Resident Project Representative and assistants will be as provided in paragraph 9.10 and in the Supplementary Conditions. If OWNER designates another representative or agent to represent OWNER at the Site who is not ENGINEER's Consultant, agent or employee, the responsibilities and authority and limitations thereon of such other individual or entity will be as provided in the Supplementary Conditions.

9.04 Clarifications and Interpretations

A. ENGINEER will issue with reasonable promptness such written clarifications or interpretations of the requirements of the Contract Documents as ENGINEER may determine necessary, which shall be consistent with the intent of and reasonably inferable from the Contract Documents. Such written clarifications and interpretations will be binding on OWNER and CONTRACTOR. If OWNER and CONTRACTOR are unable to agree on entitlement to or on the amount or extent, if any, of any adjustment in the Contract Price or Contract Times, or both, that should be allowed as a result of a written clarification or interpretation, a Claim may be made therefor as provided in paragraph 10.05.

9.05 Authorized Variations in Work

A. ENGINEER may authorize minor variations in the Work from the requirements of the Contract Documents which do not involve an adjustment in the Contract Price or the Contract Times and are compatible with the design concept of the completed Project as a functioning whole as indicated by the Contract Documents. These may be accomplished by a Field Order and will be binding on OWNER and also on CONTRACTOR, who shall perform the Work involved promptly. If OWNER and CONTRACTOR are unable to agree on entitlement to or on the amount or extent, if any, of any adjustment in the Contract Price or Contract Times, or both, as a result of a Field Order, a Claim may be made therefor as provided in paragraph 10.05.

9.06 Rejecting Defective Work

A. ENGINEER will have authority to disapprove or reject Work which ENGINEER believes to be defective, or that ENGINEER believes will not produce a completed Project that conforms to the Contract Documents or that will prejudice the integrity of the design concept of the completed Project as a functioning whole as indicated by the Contract Documents. ENGINEER will also have authority to require special inspection or testing of the Work as provided in paragraph 13.04, whether or not the Work is fabricated, installed, or completed.

9.07 Shop Drawings, Change Orders and Payments

A. In connection with ENGINEER's authority as to Shop Drawings and Samples, see paragraph 6.17.

B. In connection with ENGINEER's authority as to Change Orders, see Articles 10, 11, and 12.

C. In connection with ENGINEER's authority as to Applications for Payment, see Article 14.

9.08 Determinations for Unit Price Work

A. ENGINEER will determine the actual quantities and classifications of Unit Price Work performed by CONTRACTOR. ENGINEER will review with CONTRACTOR the ENGINEER's preliminary determinations on such matters before rendering a written decision thereon (by recommendation of an Application for Payment or otherwise). ENGINEER's written decision thereon will be final and binding (except as modified by ENGINEER to reflect changed factual conditions or more accurate data) upon OWNER and CONTRACTOR, subject to the provisions of paragraph 10.05.

9.09 Decisions on Requirements of Contract Documents and Acceptability of Work

A. ENGINEER will be the initial interpreter of the requirements of the Contract Documents and judge of the acceptability of the Work thereunder. Claims, disputes and other matters relating to the acceptability of the Work, the quantities and classifications of Unit Price Work, the interpretation of the requirements of the Contract Documents pertaining to the performance of the Work, and Claims seeking changes in the Contract Price or Contract Times will be referred initially to ENGINEER in writing, in accordance with the provisions of paragraph 10.05, with a request for a formal decision.

B. When functioning as interpreter and judge under this paragraph 9.09, ENGINEER will not show partiality to OWNER or CONTRACTOR and will not be liable in connection with any interpretation or decision rendered in good faith in such capacity. The rendering of a decision by ENGINEER pursuant to this paragraph 9.09 with respect to any such Claim, dispute, or other matter (except any which have been waived by the making or acceptance of final payment as provided in paragraph 14.07) will be a condition precedent to any exercise by OWNER or CONTRACTOR of such rights or remedies as either may otherwise have under the Contract Documents or by Laws or Regulations in respect of any such Claim, dispute, or other matter.

9.10 Limitations on ENGINEER's Authority and Responsibilities

A. Neither ENGINEER's authority or responsibility under this Article 9 or under any other provision of the Contract Documents nor any decision made by ENGINEER in good faith either to exercise or not exercise such authority or responsibility or the undertaking, exercise, or performance of any authority or responsibility by ENGINEER shall create, impose, or give rise to any duty in contract, tort, or otherwise owed by ENGINEER to CONTRACTOR, any Subcontractor, any Supplier, any other individual or entity, or to any surety for or employee or agent of any of them.

B. ENGINEER will not supervise, direct, control, or have authority over or be responsible for CONTRACTOR's means, methods, techniques, sequences, or procedures of construction, or the safety precautions and programs incident thereto, or for any failure of CONTRACTOR to comply with Laws and Regulations applicable to the performance of the Work.   ENGINEER will not be responsible for CONTRACTOR's failure to perform the Work in accordance with the Contract Documents.

C. ENGINEER will not be responsible for the acts or omissions of CONTRACTOR or of any Subcontractor, any Supplier, or of any other individual or entity performing any of the Work.

D. ENGINEER's review of the final Application for Payment and accompanying documentation and all maintenance and operating instructions, schedules, guarantees, Bonds, certificates of inspection, tests and approvals, and other documentation required to be delivered by paragraph 14.07.A will only be to determine generally that their content complies with the requirements of, and in the case of certificates of inspections, tests, and approvals that the results certified indicate compliance with, the Contract Documents.

E. The limitations upon authority and responsibility set forth in this paragraph 9.10 shall also apply to ENGINEER's Consultants, Resident Project Representative, and assistants.

ARTICLE 10 - CHANGES IN THE WORK; CLAIMS

10.01   Authorized Changes in the Work

A. Without invalidating the Agreement and without notice to any surety, OWNER may, at any time or from time to time, order additions, deletions, or revisions in the Work by a Written Amendment, a Change Order, or a Work Change Directive. Upon receipt of any such document, CONTRACTOR shall promptly proceed with the Work involved which will be performed under the applicable conditions of the Contract Documents (except as otherwise specifically provided).

B. If OWNER and CONTRACTOR are unable to agree on entitlement to, or on the amount or extent, if any, of an adjustment in the Contract Price or Contract Times, or both, that should be allowed as a result of a Work Change

Directive, a Claim may be made therefor as provided in paragraph 10.05.

10.02   Unauthorized Changes in the Work

A. CONTRACTOR shall not be entitled to an increase in the Contract Price or an extension of the Contract Times with respect to any work performed that is not required by the Contract Documents as amended, modified, or supplemented as provided in paragraph 3.04, except in the case of an emergency as provided in paragraph 6.16 or in the case of uncovering Work as provided in paragraph 13.04.B.

10.03   Execution of Change Orders

A. OWNER and CONTRACTOR shall execute appropriate Change Orders recommended by ENGINEER (or Written Amendments) covering:

1. changes in the Work which are: (i) ordered by OWNER pursuant to paragraph 10.01.A, (ii) required because of acceptance of defective Work under paragraph 13.08.A or OWNER's correction of defective Work under paragraph 13.09, or (iii) agreed to by the parties;

2. changes in the Contract Price or Contract Times which are agreed to by the parties, including any undisputed sum or amount of time for Work actually performed in accordance with a Work Change Directive; and

3. changes in the Contract Price or Contract Times which embody the substance of any written decision rendered by ENGINEER pursuant to paragraph 10.05; provided that, in lieu of executing any such Change Order, an appeal may be taken from any such decision in accordance with the provisions of the Contract Documents and applicable Laws and Regulations, but during any such appeal, CONTRACTOR shall carry on the Work and adhere to the progress schedule as provided in paragraph 6.18.A.

10.04   Notification to Surety

A. If notice of any change affecting the general scope of the Work or the provisions of the Contract Documents (including, but not limited to, Contract Price or Contract Times) is required by the provisions of any Bond to be given to a surety, the giving of any such notice will be CONTRACTOR's responsibility. The amount of each applicable Bond will be adjusted to reflect the effect of any such change.

10.05   *Claims and Disputes*

A. *Notice:* Written notice stating the general nature of each Claim, dispute, or other matter shall be delivered by the claimant to ENGINEER and the other party to the Contract promptly (but in no event later than 30 days) after the start of the event giving rise thereto. Notice of the amount or extent of the Claim, dispute, or other matter with supporting data shall be delivered to the ENGINEER and the other party to the Contract within 60 days after the start of such event (unless ENGINEER allows additional time for claimant to submit additional or more accurate data in support of such Claim, dispute, or other matter). A Claim for an adjustment in Contract Price shall be prepared in accordance with the provisions of paragraph 12.01.B. A Claim for an adjustment in Contract Time shall be prepared in accordance with the provisions of paragraph 12.02.B. Each Claim shall be accompanied by claimant's written statement that the adjustment claimed is the entire adjustment to which the claimant believes it is entitled as a result of said event. The opposing party shall submit any response to ENGINEER and the claimant within 30 days after receipt of the claimant's last submittal (unless ENGINEER allows additional time).

B. *ENGINEER's Decision:* ENGINEER will render a formal decision in writing within 30 days after receipt of the last submittal of the claimant or the last submittal of the opposing party, if any. ENGINEER's written decision on such Claim, dispute, or other matter will be final and binding upon OWNER and CONTRACTOR unless:

1. an appeal from ENGINEER's decision is taken within the time limits and in accordance with the dispute resolution procedures set forth in Article 16; or

2. if no such dispute resolution procedures have been set forth in Article 16, a written notice of intention to appeal from ENGINEER's written decision is delivered by OWNER or CONTRACTOR to the other and to ENGINEER within 30 days after the date of such decision, and a formal proceeding is instituted by the appealing party in a forum of competent jurisdiction within 60 days after the date of such decision or within 60 days after Substantial Completion, whichever is later (unless otherwise agreed in writing by OWNER and CONTRACTOR), to exercise such rights or remedies as the appealing party may have with respect to such Claim, dispute, or other matter in accordance with applicable Laws and Regulations.

C. If ENGINEER does not render a formal decision in writing within the time stated in paragraph 10.05.B, a decision denying the Claim in its entirety shall be deemed to have been issued 31 days after receipt of the last submittal of the claimant or the last submittal of the opposing party, if any.

D. No Claim for an adjustment in Contract Price or Contract Times (or Milestones) will be valid if not submitted in accordance with this paragraph 10.05.

ARTICLE 11 - COST OF THE WORK; CASH ALLOWANCES; UNIT PRICE WORK

11.01   *Cost of the Work*

A. *Costs Included:* The term Cost of the Work means the sum of all costs necessarily incurred and paid by CONTRACTOR in the proper performance of the Work. When the value of any Work covered by a Change Order or when a Claim for an adjustment in Contract Price is determined on the basis of Cost of the Work, the costs to be reimbursed to CONTRACTOR will be only those additional or incremental costs required because of the change in the Work or because of the event giving rise to the Claim. Except as otherwise may be agreed to in writing by OWNER, such costs shall be in amounts no higher than those prevailing in the locality of the Project, shall include only the following items, and shall not include any of the costs itemized in paragraph 11.01.B.

1. Payroll costs for employees in the direct employ of CONTRACTOR in the performance of the Work under schedules of job classifications agreed upon by OWNER and CONTRACTOR. Such employees shall include without limitation superintendents, foremen, and other personnel employed full time at the Site. Payroll costs for employees not employed full time on the Work shall be apportioned on the basis of their time spent on the Work. Payroll costs shall include, but not be limited to, salaries and wages plus the cost of fringe benefits, which shall include social security contributions, unemployment, excise, and payroll taxes, workers' compensation, health and retirement benefits, bonuses, sick leave, vacation and holiday pay applicable thereto. The expenses of performing Work outside of regular working hours, on Saturday, Sunday, or legal holidays, shall be included in the above to the extent authorized by OWNER.

2. Cost of all materials and equipment furnished and incorporated in the Work, including costs of transportation and storage thereof, and Suppliers' field services required in connection therewith. All cash discounts shall accrue to CONTRACTOR unless OWNER deposits funds with CONTRACTOR with which to make payments, in which case the cash discounts shall accrue to OWNER. All trade discounts, rebates and refunds and returns from sale of surplus materials and equipment shall accrue to OWNER, and CONTRACTOR shall make provisions so that they may be obtained.

3. Payments made by CONTRACTOR to Subcontractors for Work performed by Subcontractors. If required by OWNER, CONTRACTOR shall obtain competitive bids from subcontractors acceptable to OWNER and CONTRACTOR and shall deliver such bids to OWNER, who will then determine, with the advice of ENGINEER, which bids, if any, will be acceptable. If any subcontract provides that the Subcontractor is to be paid on the basis of Cost of the Work plus a fee, the Subcontractor's Cost of the Work and fee shall be determined in the same manner as CONTRACTOR's Cost of the Work and fee as provided in this paragraph 11.01.

4. Costs of special consultants (including but not limited to engineers, architects, testing laboratories, surveyors, attorneys, and accountants) employed for services specifically related to the Work.

5. Supplemental costs including the following:

a. The proportion of necessary transportation, travel, and subsistence expenses of CONTRACTOR's employees incurred in discharge of duties connected with the Work.

b. Cost, including transportation and maintenance, of all materials, supplies, equipment, machinery, appliances, office, and temporary facilities at the Site, and hand tools not owned by the workers, which are consumed in the performance of the Work, and cost, less market value, of such items used but not consumed which remain the property of CONTRACTOR.

c. Rentals of all construction equipment and machinery, and the parts thereof whether rented from CONTRACTOR or others in accordance with rental agreements approved by OWNER with the advice of ENGINEER, and the costs of transportation, loading, unloading, assembly, dismantling, and removal thereof. All such costs shall be in accordance with the terms of said rental agreements. The rental of any such equipment, machinery, or parts shall cease when the use thereof is no longer necessary for the Work.

d. Sales, consumer, use, and other similar taxes related to the Work, and for which CONTRACTOR is liable, imposed by Laws and Regulations.

e. Deposits lost for causes other than negligence of CONTRACTOR, any Subcontractor, or anyone directly or indirectly employed by any of them or for whose acts any of them may be liable,

and royalty payments and fees for permits and licenses.

f. Losses and damages (and related expenses) caused by damage to the Work, not compensated by insurance or otherwise, sustained by CONTRACTOR in connection with the performance of the Work (except losses and damages within the deductible amounts of property insurance established in accordance with paragraph 5.06.D), provided such losses and damages have resulted from causes other than the negligence of CONTRACTOR, any Subcontractor, or anyone directly or indirectly employed by any of them or for whose acts any of them may be liable. Such losses shall include settlements made with the written consent and approval of OWNER. No such losses, damages, and expenses shall be included in the Cost of the Work for the purpose of determining CONTRACTOR's fee.

g. The cost of utilities, fuel, and sanitary facilities at the Site.

h. Minor expenses such as telegrams, long distance telephone calls, telephone service at the Site, expressage, and similar petty cash items in connection with the Work.

i. When the Cost of the Work is used to determine the value of a Change Order or of a Claim, the cost of premiums for additional Bonds and insurance required because of the changes in the Work or caused by the event giving rise to the Claim.

j. When all the Work is performed on the basis of cost-plus, the costs of premiums for all Bonds and insurance CONTRACTOR is required by the Contract Documents to purchase and maintain.

B. *Costs Excluded:* The term Cost of the Work shall not include any of the following items:

1. Payroll costs and other compensation of CONTRACTOR's officers, executives, principals (of partnerships and sole proprietorships), general managers, engineers, architects, estimators, attorneys, auditors, accountants, purchasing and contracting agents, expediters, timekeepers, clerks, and other personnel employed by CONTRACTOR, whether at the Site or in CONTRACTOR's principal or branch office for general administration of the Work and not specifically included in the agreed upon schedule of job classifications referred to in paragraph 11.01.A.1 or specifically covered by paragraph 11.01.A.4, all of which are to be

considered administrative costs covered by the CONTRACTOR's fee.

2. Expenses of CONTRACTOR's principal and branch offices other than CONTRACTOR's office at the Site.

3. Any part of CONTRACTOR's capital expenses, including interest on CONTRACTOR's capital employed for the Work and charges against CONTRACTOR for delinquent payments.

4. Costs due to the negligence of CONTRACTOR, any Subcontractor, or anyone directly or indirectly employed by any of them or for whose acts any of them may be liable, including but not limited to, the correction of defective Work, disposal of materials or equipment wrongly supplied, and making good any damage to property.

5. Other overhead or general expense costs of any kind and the costs of any item not specifically and expressly included in paragraphs 11.01.A and 11.01.B.

C. *CONTRACTOR's Fee:* When all the Work is performed on the basis of cost-plus, CONTRACTOR's fee shall be determined as set forth in the Agreement. When the value of any Work covered by a Change Order or when a Claim for an adjustment in Contract Price is determined on the basis of Cost of the Work, CONTRACTOR's fee shall be determined as set forth in paragraph 12.01.C.

D. *Documentation:* Whenever the Cost of the Work for any purpose is to be determined pursuant to paragraphs 11.01.A and 11.01.B, CONTRACTOR will establish and maintain records thereof in accordance with generally accepted accounting practices and submit in a form acceptable to ENGINEER an itemized cost breakdown together with supporting data.

11.02   *Cash Allowances*

A. It is understood that CONTRACTOR has included in the Contract Price all allowances so named in the Contract Documents and shall cause the Work so covered to be performed for such sums as may be acceptable to OWNER and ENGINEER. CONTRACTOR agrees that:

1. the allowances include the cost to CONTRAC-TOR (less any applicable trade discounts) of materials and equipment required by the allowances to be delivered at the Site, and all applicable taxes; and

2. CONTRACTOR's costs for unloading and handling on the Site, labor, installation costs, overhead, profit, and other expenses contemplated for the allow-

ances have been included in the Contract Price and not in the allowances, and no demand for additional payment on account of any of the foregoing will be valid.

B. Prior to final payment, an appropriate Change Order will be issued as recommended by ENGINEER to reflect actual amounts due CONTRACTOR on account of Work covered by allowances, and the Contract Price shall be correspondingly adjusted.

11.03   *Unit Price Work*

A. Where the Contract Documents provide that all or part of the Work is to be Unit Price Work, initially the Contract Price will be deemed to include for all Unit Price Work an amount equal to the sum of the unit price for each separately identified item of Unit Price Work times the estimated quantity of each item as indicated in the Agreement. The estimated quantities of items of Unit Price Work are not guaranteed and are solely for the purpose of comparison of Bids and determining an initial Contract Price. Determinations of the actual quantities and classifications of Unit Price Work performed by CONTRACTOR will be made by ENGINEER subject to the provisions of paragraph 9.08.

B. Each unit price will be deemed to include an amount considered by CONTRACTOR to be adequate to cover CONTRACTOR's overhead and profit for each separately identified item.

C. OWNER or CONTRACTOR may make a Claim for an adjustment in the Contract Price in accordance with paragraph 10.05 if:

1. the quantity of any item of Unit Price Work performed by CONTRACTOR differs materially and significantly from the estimated quantity of such item indicated in the Agreement; and

2. there is no corresponding adjustment with respect any other item of Work; and

3. if CONTRACTOR believes that CONTRACTOR is entitled to an increase in Contract Price as a result of having incurred additional expense or OWNER believes that OWNER is entitled to a decrease in Contract Price and the parties are unable to agree as to the amount of any such increase or decrease.

ARTICLE 12 - CHANGE OF CONTRACT PRICE; CHANGE OF CONTRACT TIMES

12.01   *Change of Contract Price*

A. The Contract Price may only be changed by a Change Order or by a Written Amendment. Any Claim for an adjustment in the Contract Price shall be based on written notice submitted by the party making the Claim to the ENGINEER and the other party to the Contract in accordance with the provisions of paragraph 10.05.

B. The value of any Work covered by a Change Order or of any Claim for an adjustment in the Contract Price will be determined as follows:

1. where the Work involved is covered by unit prices contained in the Contract Documents, by application of such unit prices to the quantities of the items involved (subject to the provisions of paragraph 11.03); or

2. where the Work involved is not covered by unit prices contained in the Contract Documents, by a mutually agreed lump sum (which may include an allowance for overhead and profit not necessarily in accordance with paragraph 12.01.C.2); or

3. where the Work involved is not covered by unit prices contained in the Contract Documents and agreement to a lump sum is not reached under paragraph 12.01.B.2, on the basis of the Cost of the Work (determined as provided in paragraph 11.01) plus a CONTRACTOR's fee for overhead and profit (determined as provided in paragraph 12.01.C).

C. *CONTRACTOR's Fee:* The CONTRACTOR's fee for overhead and profit shall be determined as follows:

1. a mutually acceptable fixed fee; or

2. if a fixed fee is not agreed upon, then a fee based on the following percentages of the various portions of the Cost of the Work:

a. for costs incurred under paragraphs 11.01.A.1 and 11.01.A.2, the CONTRACTOR's fee shall be 15 percent;

b. for costs incurred under paragraph 11.01.A.3, the CONTRACTOR's fee shall be five percent;

c. where one or more tiers of subcontracts are on the basis of Cost of the Work plus a fee and no

fixed fee is agreed upon, the intent of paragraph 12.01.C.2.a is that the Subcontractor who actually performs the Work, at whatever tier, will be paid a fee of 15 percent of the costs incurred by such Subcontractor under paragraphs 11.01.A.1 and 11.01.A.2 and that any higher tier Subcontractor and CONTRACTOR will each be paid a fee of five percent of the amount paid to the next lower tier Subcontractor;

d. no fee shall be payable on the basis of costs itemized under paragraphs 11.01.A.4, 11.01.A.5, and 11.01.B;

e. the amount of credit to be allowed by CONTRACTOR to OWNER for any change which results in a net decrease in cost will be the amount of the actual net decrease in cost plus a deduction in CONTRACTOR's fee by an amount equal to five percent of such net decrease; and

f. when both additions and credits are involved in any one change, the adjustment in CONTRACTOR's fee shall be computed on the basis of the net change in accordance with paragraphs 12.01.C.2.a through 12.01.C.2.e, inclusive.

12.02   *Change of Contract Times*

A. The Contract Times (or Milestones) may only be changed by a Change Order or by a Written Amendment. Any Claim for an adjustment in the Contract Times (or Milestones) shall be based on written notice submitted by the party making the claim to the ENGINEER and the other party to the Contract in accordance with the provisions of paragraph 10.05.

B. Any adjustment of the Contract Times (or Milestones) covered by a Change Order or of any Claim for an adjustment in the Contract Times (or Milestones) will be determined in accordance with the provisions of this Article 12.

12.03   *Delays Beyond CONTRACTOR's Control*

A. Where CONTRACTOR is prevented from completing any part of the Work within the Contract Times (or Milestones) due to delay beyond the control of CONTRACTOR, the Contract Times (or Milestones) will be extended in an amount equal to the time lost due to such delay if a Claim is made therefor as provided in paragraph 12.02.A. Delays beyond the control of CONTRACTOR shall include, but not be limited to, acts or neglect by OWNER, acts or neglect of utility owners or other contractors performing other work as contemplated by

Article 7, fires, floods, epidemics, abnormal weather conditions, or acts of God.

12.04 Delays Within CONTRACTOR's Control

A. The Contract Times (or Milestones) will not be extended due to delays within the control of CONTRACTOR. Delays attributable to and within the control of a Subcontractor or Supplier shall be deemed to be delays within the control of CONTRACTOR.

12.05 Delays Beyond OWNER's and CONTRACTOR's Control

A. Where CONTRACTOR is prevented from completing any part of the Work within the Contract Times (or Milestones) due to delay beyond the control of both OWNER and CONTRACTOR, an extension of the Contract Times (or Milestones) in an amount equal to the time lost due to such delay shall be CONTRACTOR's sole and exclusive remedy for such delay.

12.06 Delay Damages

A. In no event shall OWNER or ENGINEER be liable to CONTRACTOR, any Subcontractor, any Supplier, or any other person or organization, or to any surety for or employee or agent of any of them, for damages arising out of or resulting from:

1. delays caused by or within the control of CONTRACTOR; or

2. delays beyond the control of both OWNER and CONTRACTOR including but not limited to fires, floods, epidemics, abnormal weather conditions, acts of God, or acts or neglect by utility owners or other contractors performing other work as contemplated by Article 7.

B. Nothing in this paragraph 12.06 bars a change in Contract Price pursuant to this Article 12 to compensate CONTRACTOR due to delay, interference, or disruption directly attributable to actions or inactions of OWNER or anyone for whom OWNER is responsible.

ARTICLE 13 - TESTS AND INSPECTIONS; CORRECTION, REMOVAL OR ACCEPTANCE OF DEFECTIVE WORK

13.01 Notice of Defects

A. Prompt notice of all defective Work of which OWNER or ENGINEER has actual knowledge will be given to CONTRACTOR. All defective Work may be rejected, corrected, or accepted as provided in this Article 13.

13.02 Access to Work

A. OWNER, ENGINEER, ENGINEER's Consultants, other representatives and personnel of OWNER, independent testing laboratories, and governmental agencies with jurisdictional interests will have access to the Site and the Work at reasonable times for their observation, inspecting, and testing. CONTRACTOR shall provide them proper and safe conditions for such access and advise them of CONTRACTOR's Site safety procedures and programs so that they may comply therewith as applicable.

13.03 Tests and Inspections

A. CONTRACTOR shall give ENGINEER timely notice of readiness of the Work for all required inspections, tests, or approvals and shall cooperate with inspection and testing personnel to facilitate required inspections or tests.

B. OWNER shall employ and pay for the services of an independent testing laboratory to perform all inspections, tests, or approvals required by the Contract Documents except:

1. for inspections, tests, or approvals covered by paragraphs 13.03.C and 13.03.D below;

2. that costs incurred in connection with tests or inspections conducted pursuant to paragraph 13.04.B shall be paid as provided in said paragraph 13.04.B; and

3. as otherwise specifically provided in the Contract Documents.

C. If Laws or Regulations of any public body having jurisdiction require any Work (or part thereof) specifically to be inspected, tested, or approved by an employee or other representative of such public body, CONTRACTOR shall assume full responsibility for arranging and obtaining such inspections, tests, or approvals, pay all costs in connection therewith, and furnish ENGINEER the required certificates of inspection or approval.

D. CONTRACTOR shall be responsible for arranging and obtaining and shall pay all costs in connection with any inspections, tests, or approvals required for OWNER's and ENGINEER's acceptance of materials or equipment to be incorporated in the Work; or acceptance of materials, mix designs, or equipment submitted for approval prior to CONTRACTOR's purchase thereof for incorporation in the Work. Such inspections, tests, or approvals shall be performed by organizations acceptable to OWNER and ENGINEER.

E. If any Work (or the work of others) that is to be inspected, tested, or approved is covered by CONTRACTOR without written concurrence of ENGINEER, it must, if requested by ENGINEER, be uncovered for observation.

F. Uncovering Work as provided in paragraph 13.03.E shall be at CONTRACTOR's expense unless CONTRACTOR has given ENGINEER timely notice of CONTRACTOR's intention to cover the same and ENGINEER has not acted with reasonable promptness in response to such notice.

13.04   Uncovering Work

A. If any Work is covered contrary to the written request of ENGINEER, it must, if requested by ENGINEER, be uncovered for ENGINEER's observation and replaced at CONTRACTOR's expense.

B. If ENGINEER considers it necessary or advisable that covered Work be observed by ENGINEER or inspected or tested by others, CONTRACTOR, at ENGINEER's request, shall uncover, expose, or otherwise make available for observation, inspection, or testing as ENGINEER may require, that portion of the Work in question, furnishing all necessary labor, material, and equipment. If it is found that such Work is defective, CONTRACTOR shall pay all Claims, costs, losses, and damages (including but not limited to all fees and charges of engineers, architects, attorneys, and other professionals and all court or arbitration or other dispute resolution costs) arising out of or relating to such uncovering, exposure, observation, inspection, and testing, and of satisfactory replacement or reconstruction (including but not limited to all costs of repair or replacement of work of others); and OWNER shall be entitled to an appropriate decrease in the Contract Price. If the parties are unable to agree as to the amount thereof, OWNER may make a Claim therefor as provided in paragraph 10.05. If, however, such Work is not found to be defective, CONTRACTOR shall be allowed an increase in the Contract Price or an extension of the Contract Times (or Milestones), or both, directly attributable to such uncovering, exposure, observation, inspection, testing, replacement, and reconstruction. If the parties are unable to agree as to the amount or extent thereof, CONTRACTOR may make a Claim therefor as provided in paragraph 10.05.

13.05   OWNER May Stop the Work

A. If the Work is defective, or CONTRACTOR fails to supply sufficient skilled workers or suitable materials or equipment, or fails to perform the Work in such a way that the completed Work will conform to the Contract Documents, OWNER may order CONTRACTOR to stop the Work, or any portion thereof, until the cause for such order has been eliminated; however, this right of OWNER to stop

the Work shall not give rise to any duty on the part of OWNER to exercise this right for the benefit of CONTRACTOR, any Subcontractor, any Supplier, any other individual or entity, or any surety for, or employee or agent of any of them.

13.06   Correction or Removal of Defective Work

A. CONTRACTOR shall correct all defective Work, whether or not fabricated, installed, or completed, or, if the Work has been rejected by ENGINEER, remove it from the Project and replace it with Work that is not defective. CONTRACTOR shall pay all Claims, costs, losses, and damages (including but not limited to all fees and charges of engineers, architects, attorneys, and other professionals and all court or arbitration or other dispute resolution costs) arising out of or relating to such correction or removal (including but not limited to all costs of repair or replacement of work of others).

13.07   Correction Period

A. If within one year after the date of Substantial Completion or such longer period of time as may be prescribed by Laws or Regulations or by the terms of any applicable special guarantee required by the Contract Documents or by any specific provision of the Contract Documents, any Work is found to be defective, or if the repair of any damages to the land or areas made available for CONTRACTOR's use by OWNER or permitted by Laws and Regulations as contemplated in paragraph 6.11.A is found to be defective, CONTRACTOR shall promptly, without cost to OWNER and in accordance with OWNER's written instructions: (i) repair such defective land or areas, or (ii) correct such defective Work or, if the defective Work has been rejected by OWNER, remove it from the Project and replace it with Work that is not defective, and (iii) satisfactorily correct or repair or remove and replace any damage to other Work, to the work of others or other land or areas resulting therefrom. If CONTRACTOR does not promptly comply with the terms of such instructions, or in an emergency where delay would cause serious risk of loss or damage, OWNER may have the defective Work corrected or repaired or may have the rejected Work removed and replaced, and all Claims, costs, losses, and damages (including but not limited to all fees and charges of engineers, architects, attorneys, and other professionals and all court or arbitration or other dispute resolution costs) arising out of or relating to such correction or repair or such removal and replacement (including but not limited to all costs of repair or replacement of work of others) will be paid by CONTRACTOR.

B. In special circumstances where a particular item of equipment is placed in continuous service before Substantial Completion of all the Work, the correction period for that item may start to run from an earlier date if so provided in the Specifications or by Written Amendment.

C. Where defective Work (and damage to other Work resulting therefrom) has been corrected or removed and replaced under this paragraph 13.07, the correction period hereunder with respect to such Work will be extended for an additional period of one year after such correction or removal and replacement has been satisfactorily completed.

D. CONTRACTOR's obligations under this paragraph 13.07 are in addition to any other obligation or warranty. The provisions of this paragraph 13.07 shall not be construed as a substitute for or a waiver of the provisions of any applicable statute of limitation or repose.

13.08   *Acceptance of Defective Work*

A. If, instead of requiring correction or removal and replacement of defective Work, OWNER (and, prior to ENGINEER's recommendation of final payment, ENGINEER) prefers to accept it, OWNER may do so. CONTRACTOR shall pay all Claims, costs, losses, and damages (including but not limited to all fees and charges of engineers, architects, attorneys, and other professionals and all court or arbitration or other dispute resolution costs) attributable to OWNER's evaluation of and determination to accept such defective Work (such costs to be approved by ENGINEER as to reasonableness) and the diminished value of the Work to the extent not otherwise paid by CONTRACTOR pursuant to this sentence. If any such acceptance occurs prior to ENGINEER's recommendation of final payment, a Change Order will be issued incorporating the necessary revisions in the Contract Documents with respect to the Work, and OWNER shall be entitled to an appropriate decrease in the Contract Price, reflecting the diminished value of Work so accepted. If the parties are unable to agree as to the amount thereof, OWNER may make a Claim therefor as provided in paragraph 10.05. If the acceptance occurs after such recommendation, an appropriate amount will be paid by CONTRACTOR to OWNER.

13.09   *OWNER May Correct Defective Work*

A. If CONTRACTOR fails within a reasonable time after written notice from ENGINEER to correct defective Work or to remove and replace rejected Work as required by ENGINEER in accordance with paragraph 13.06.A, or if CONTRACTOR fails to perform the Work in accordance with the Contract Documents, or if CONTRACTOR fails to comply with any other provision of the Contract Documents, OWNER may, after seven days written notice to CONTRACTOR, correct and remedy any such deficiency.

B. In exercising the rights and remedies under this paragraph, OWNER shall proceed expeditiously. In connection with such corrective and remedial action, OWNER may exclude CONTRACTOR from all or part of the Site, take possession of all or part of the Work and suspend CONTRACTOR's services related thereto, take possession of CONTRACTOR's tools, appliances, construction equipment and machinery at the Site, and incorporate in the Work all materials and equipment stored at the Site or for which OWNER has paid CONTRACTOR but which are stored elsewhere. CONTRACTOR shall allow OWNER, OWNER's representatives, agents and employees, OWNER's other contractors, and ENGINEER and ENGINEER's Consultants access to the Site to enable OWNER to exercise the rights and remedies under this paragraph.

C. All Claims, costs, losses, and damages (including but not limited to all fees and charges of engineers, architects, attorneys, and other professionals and all court or arbitration or other dispute resolution costs) incurred or sustained by OWNER in exercising the rights and remedies under this paragraph 13.09 will be charged against CONTRACTOR, and a Change Order will be issued incorporating the necessary revisions in the Contract Documents with respect to the Work; and OWNER shall be entitled to an appropriate decrease in the Contract Price. If the parties are unable to agree as to the amount of the adjustment, OWNER may make a Claim therefor as provided in paragraph 10.05. Such claims, costs, losses and damages will include but not be limited to all costs of repair, or replacement of work of others destroyed or damaged by correction, removal, or replacement of CONTRACTOR's defective Work.

D. CONTRACTOR shall not be allowed an extension of the Contract Times (or Milestones) because of any delay in the performance of the Work attributable to the exercise by OWNER of OWNER's rights and remedies under this paragraph 13.09.

ARTICLE 14 - PAYMENTS TO CONTRACTOR AND COMPLETION

14.01   *Schedule of Values*

A. The schedule of values established as provided in paragraph 2.07.A will serve as the basis for progress payments and will be incorporated into a form of Application for Payment acceptable to ENGINEER. Progress payments on account of Unit Price Work will be based on the number of units completed.

14.02   Progress Payments

A. Applications for Payments

1.   At least 20 days before the date established for each progress payment (but not more often than once a month), CONTRACTOR shall submit to ENGINEER for review an Application for Payment filled out and signed by CONTRACTOR covering the Work completed as of the date of the Application and accompanied by such supporting documentation as is required by the Contract Documents. If payment is requested on the basis of materials and equipment not incorporated in the Work but delivered and suitably stored at the Site or at another location agreed to in writing, the Application for Payment shall also be accompanied by a bill of sale, invoice, or other documentation warranting that OWNER has received the materials and equipment free and clear of all Liens and evidence that the materials and equipment are covered by appropriate property insurance or other arrangements to protect OWNER's interest therein, all of which must be satisfactory to OWNER.

2.   Beginning with the second Application for Payment, each Application shall include an affidavit of CONTRACTOR stating that all previous progress payments received on account of the Work have been applied on account to discharge CONTRACTOR's legitimate obligations associated with prior Applications for Payment.

3.   The amount of retainage with respect to progress payments will be as stipulated in the Agreement.

B. Review of Applications

1.   ENGINEER will, within 10 days after receipt of each Application for Payment, either indicate in writing a recommendation of payment and present the Application to OWNER or return the Application to CONTRACTOR indicating in writing ENGINEER's reasons for refusing to recommend payment. In the latter case, CONTRACTOR may make the necessary corrections and resubmit the Application.

2.   ENGINEER's recommendation of any payment requested in an Application for Payment will constitute a representation by ENGINEER to OWNER, based on ENGINEER's observations on the Site of the executed Work as an experienced and qualified design professional and on ENGINEER's review of the Application for Payment and the accompanying data and schedules, that to the best of ENGINEER's knowledge, information and belief:

a. the Work has progressed to the point indicated;

b. the quality of the Work is generally in accordance with the Contract Documents (subject to an evaluation of the Work as a functioning whole prior to or upon Substantial Completion, to the results of any subsequent tests called for in the Contract Documents, to a final determination of quantities and classifications for Unit Price Work under paragraph 9.08, and to any other qualifications stated in the recommendation); and

c. the   conditions   precedent   to CONTRACTOR's being entitled to such payment appear to have been fulfilled in so far as it is ENGINEER's responsibility to observe the Work.

3.   By recommending any such payment ENGINEER will not thereby be deemed to have represented that: (i) inspections made to check the quality or the quantity of the Work as it has been performed have been exhaustive, extended to every aspect of the Work in progress, or involved detailed inspections of the Work beyond the responsibilities specifically assigned to ENGINEER in the Contract Documents; or (ii) that there may not be other matters or issues between the parties that might entitle CONTRACTOR to be paid additionally by OWNER or entitle OWNER to withhold payment to CONTRACTOR.

4.   Neither   ENGINEER's   review   of CONTRACTOR's Work for the purposes of recommending payments nor ENGINEER's recommendation of any payment, including final payment, will impose responsibility on ENGINEER to supervise, direct, or control the Work or for the means, methods, techniques, sequences, or procedures of construction, or the safety precautions and programs incident thereto, or for CONTRACTOR's failure to comply with Laws and Regulations applicable to CONTRACTOR's performance of the Work. Additionally, said review or recommendation will not impose responsibility on ENGINEER to make any examination to ascertain how or for what purposes CONTRACTOR has used the moneys paid on account of the Contract Price, or to determine that title to any of the Work, materials, or equipment has passed to OWNER free and clear of any Liens.

5.   ENGINEER may refuse to recommend the whole or any part of any payment if, in ENGINEER's opinion, it would be incorrect to make the representations to OWNER referred to in paragraph 14.02.B.2. ENGINEER may also refuse to recommend any such payment or, because of subsequently discovered evidence or the results of subsequent inspections or tests,

revise or revoke any such payment recommendation previously made, to such extent as may be necessary in ENGINEER's opinion to protect OWNER from loss because:

    a. the Work is defective, or completed Work has been damaged, requiring correction or replacement;

    b. the Contract Price has been reduced by Written Amendment or Change Orders;

    c. OWNER has been required to correct defective Work or complete Work in accordance with paragraph 13.09; or

    d. ENGINEER has actual knowledge of the occurrence of any of the events enumerated in paragraph 15.02.A.

C. *Payment Becomes Due*

1. Ten days after presentation of the Application for Payment to OWNER with ENGINEER's recommendation, the amount recommended will (subject to the provisions of paragraph 14.02.D) become due, and when due will be paid by OWNER to CONTRACTOR.

D. *Reduction in Payment*

1. OWNER may refuse to make payment of the full amount recommended by ENGINEER because:

    a. claims have been made against OWNER on account of CONTRACTOR's performance or furnishing of the Work;

    b. Liens have been filed in connection with the Work, except where CONTRACTOR has delivered a specific Bond satisfactory to OWNER to secure the satisfaction and discharge of such Liens;

    c. there are other items entitling OWNER to a set-off against the amount recommended; or

    d. OWNER has actual knowledge of the occurrence of any of the events enumerated in paragraphs 14.02.B.5.a through 14.02.B.5.c or paragraph 15.02.A.

2. If OWNER refuses to make payment of the full amount recommended by ENGINEER, OWNER must give CONTRACTOR immediate written notice (with a copy to ENGINEER) stating the reasons for such action and promptly pay CONTRACTOR any amount remaining after deduction of the amount so withheld.

OWNER shall promptly pay CONTRACTOR the amount so withheld, or any adjustment thereto agreed to by OWNER and CONTRACTOR, when CONTRACTOR corrects to OWNER's satisfaction the reasons for such action.

3. If it is subsequently determined that OWNER's refusal of payment was not justified, the amount wrongfully withheld shall be treated as an amount due as determined by paragraph 14.02.C.1.

14.03 *CONTRACTOR's Warranty of Title*

A. CONTRACTOR warrants and guarantees that title to all Work, materials, and equipment covered by any Application for Payment, whether incorporated in the Project or not, will pass to OWNER no later than the time of payment free and clear of all Liens.

14.04 *Substantial Completion*

A. When CONTRACTOR considers the entire Work ready for its intended use CONTRACTOR shall notify OWNER and ENGINEER in writing that the entire Work is substantially complete (except for items specifically listed by CONTRACTOR as incomplete) and request that ENGINEER issue a certificate of Substantial Completion. Promptly thereafter, OWNER, CONTRACTOR, and ENGINEER shall make an inspection of the Work to determine the status of completion. If ENGINEER does not consider the Work substantially complete, ENGINEER will notify CONTRACTOR in writing giving the reasons therefor. If ENGINEER considers the Work substantially complete, ENGINEER will prepare and deliver to OWNER a tentative certificate of Substantial Completion which shall fix the date of Substantial Completion. There shall be attached to the certificate a tentative list of items to be completed or corrected before final payment. OWNER shall have seven days after receipt of the tentative certificate during which to make written objection to ENGINEER as to any provisions of the certificate or attached list. If, after considering such objections, ENGINEER concludes that the Work is not substantially complete, ENGINEER will within 14 days after submission of the tentative certificate to OWNER notify CONTRACTOR in writing, stating the reasons therefor. If, after consideration of OWNER's objections, ENGINEER considers the Work substantially complete, ENGINEER will within said 14 days execute and deliver to OWNER and CONTRACTOR a definitive certificate of Substantial Completion (with a revised tentative list of items to be completed or corrected) reflecting such changes from the tentative certificate as ENGINEER believes justified after consideration of any objections from OWNER. At the time of delivery of the tentative certificate of Substantial Completion ENGINEER will deliver to OWNER and CONTRACTOR a written recommendation as to division of responsibilities pending final payment between OWNER and CONTRACTOR with respect to security, operation, safety, and protection of the Work, maintenance, heat, utilities, insurance, and warranties and guarantees. Unless OWNER and CONTRACTOR agree otherwise in writing and so inform ENGINEER in writing prior to ENGINEER's issuing the definitive certificate of Substantial Completion, ENGINEER's aforesaid recommendation will be binding on OWNER and CONTRACTOR until final payment.

B. OWNER shall have the right to exclude CONTRACTOR from the Site after the date of Substantial Completion, but OWNER shall allow CONTRACTOR reasonable access to complete or correct items on the tentative list.

14.05 *Partial Utilization*

A. Use by OWNER at OWNER's option of any substantially completed part of the Work which has specifically been identified in the Contract Documents, or which OWNER, ENGINEER, and CONTRACTOR agree constitutes a separately functioning and usable part of the Work that can be used by OWNER for its intended purpose without significant interference with CONTRACTOR's performance of the remainder of the Work, may be accomplished prior to Substantial Completion of all the Work subject to the following conditions.

1. OWNER at any time may request CON-TRACTOR in writing to permit OWNER to use any such part of the Work which OWNER believes to be ready for its intended use and substantially complete. If CONTRACTOR agrees that such part of the Work is substantially complete, CONTRACTOR will certify to OWNER and ENGINEER that such part of the Work is substantially complete and request ENGINEER to issue a certificate of Substantial Completion for that part of the Work. CONTRACTOR at any time may notify OWNER and ENGINEER in writing that CONTRACTOR considers any such part of the Work ready for its intended use and substantially complete and request ENGINEER to issue a certificate of Substantial Completion for that part of the Work. Within a reasonable time after either such request, OWNER, CONTRACTOR, and ENGINEER shall make an inspection of that part of the Work to determine its status of completion. If ENGINEER does not consider that part of the Work to be substantially complete, ENGINEER will notify OWNER and CONTRACTOR in writing giving the reasons therefor. If ENGINEER considers that part of the Work to be substantially complete, the provisions of paragraph 14.04 will apply with respect to certification of Substantial Completion of that part of the Work and the division of responsibility in respect thereof and access thereto.

2. No occupancy or separate operation of part of the Work may occur prior to compliance with the requirements of paragraph 5.10 regarding property insurance.

14.06 *Final Inspection*

A. Upon written notice from CONTRACTOR that the entire Work or an agreed portion thereof is complete, ENGINEER will promptly make a final inspection with OWNER and CONTRACTOR and will notify CON-TRACTOR in writing of all particulars in which this inspection reveals that the Work is incomplete or defective. CONTRACTOR shall immediately take such measures as are necessary to complete such Work or remedy such deficiencies.

14.07 *Final Payment*

A. *Application for Payment*

1. After CONTRACTOR has, in the opinion of ENGINEER, satisfactorily completed all corrections identified during the final inspection and has delivered, in accordance with the Contract Documents, all maintenance and operating instructions, schedules, guarantees, Bonds, certificates or other evidence of insurance certificates of inspection, marked-up record documents (as provided in paragraph 6.12), and other documents, CONTRACTOR may make application for final payment following the procedure for progress payments.

2. The final Application for Payment shall be accompanied (except as previously delivered) by: (i) all documentation called for in the Contract Documents, including but not limited to the evidence of insurance required by subparagraph 5.04.B.7; (ii) consent of the surety, if any, to final payment; and (iii) complete and legally effective releases or waivers (satisfactory to OWNER) of all Lien rights arising out of or Liens filed in connection with the Work.

3. In lieu of the releases or waivers of Liens specified in paragraph 14.07.A.2 and as approved by OWNER, CONTRACTOR may furnish receipts or releases in full and an affidavit of CONTRACTOR that: (i) the releases and receipts include all labor, services, material, and equipment for which a Lien could be filed; and (ii) all payrolls, material and equipment bills, and other indebtedness connected with the Work for which OWNER or OWNER's property might in any way be responsible have been paid or otherwise satisfied. If any Subcontractor or Supplier fails to furnish such a release or receipt in full, CONTRACTOR may furnish a Bond or other collateral satisfactory to OWNER to indemnify OWNER against any Lien.

B. *Review of Application and Acceptance*

1. If, on the basis of ENGINEER's observation of the Work during construction and final inspection, and ENGINEER's review of the final Application for Payment and accompanying documentation as required by the Contract Documents, ENGINEER is satisfied that the Work has been completed and CONTRACTOR's other obligations under the Contract Documents have been fulfilled, ENGINEER will, within ten days after receipt of the final Application for Payment, indicate in writing ENGINEER's recommendation of payment and present the Application for Payment to OWNER for payment. At the same time ENGINEER will also give written notice to OWNER and CONTRACTOR that the Work is acceptable subject to the provisions of paragraph 14.09. Otherwise, ENGINEER will return the Application for Payment to CONTRACTOR, indicating in writing the reasons for refusing to recommend final payment, in which case CONTRACTOR shall make the necessary corrections and resubmit the Application for Payment.

C. *Payment Becomes Due*

1. Thirty days after the presentation to OWNER of the Application for Payment and accompanying documentation, the amount recommended by ENGINEER will become due and, when due, will be paid by OWNER to CONTRACTOR.

14.08   *Final Completion Delayed*

A. If, through no fault of CONTRACTOR, final completion of the Work is significantly delayed, and if ENGINEER so confirms, OWNER shall, upon receipt of CONTRACTOR's final Application for Payment and recommendation of ENGINEER, and without terminating the Agreement, make payment of the balance due for that portion of the Work fully completed and accepted. If the remaining balance to be held by OWNER for Work not fully completed or corrected is less than the retainage stipulated in the Agreement, and if Bonds have been furnished as required in paragraph 5.01, the written consent of the surety to the payment of the balance due for that portion of the Work fully completed and accepted shall be submitted by CONTRACTOR to ENGINEER with the Application for such payment. Such payment shall be made under the terms and conditions governing final payment, except that it shall not constitute a waiver of Claims.

14.09   *Waiver of Claims*

A. The making and acceptance of final payment will constitute:

1. a waiver of all Claims by OWNER against CONTRACTOR, except Claims arising from unsettled Liens, from defective Work appearing after final inspection pursuant to paragraph 14.06, from failure to comply with the Contract Documents or the terms of any special guarantees specified therein, or from CONTRACTOR's continuing obligations under the Contract Documents; and

2. a waiver of all Claims by CONTRACTOR against OWNER other than those previously made in writing which are still unsettled.

ARTICLE 15 - SUSPENSION OF WORK AND TERMINATION

15.01   *OWNER May Suspend Work*

A. At any time and without cause, OWNER may suspend the Work or any portion thereof for a period of not more than 90 consecutive days by notice in writing to CONTRACTOR and ENGINEER which will fix the date on which Work will be resumed. CONTRACTOR shall resume the Work on the date so fixed. CONTRACTOR shall be allowed an adjustment in the Contract Price or an extension of the Contract Times, or both, directly attributable to any such suspension if CONTRACTOR makes a Claim therefor as provided in paragraph 10.05.

15.02   *OWNER May Terminate for Cause*

A. The occurrence of any one or more of the following events will justify termination for cause:

1. CONTRACTOR's persistent failure to perform the Work in accordance with the Contract Documents (including, but not limited to, failure to supply sufficient skilled workers or suitable materials or equipment or failure to adhere to the progress schedule established under paragraph 2.07 as adjusted from time to time pursuant to paragraph 6.04);

2. CONTRACTOR's disregard of Laws or Regulations of any public body having jurisdiction;

3. CONTRACTOR's disregard of the authority of ENGINEER; or

4. CONTRACTOR's violation in any substantial way of any provisions of the Contract Documents.

B. If one or more of the events identified in paragraph 15.02.A occur, OWNER may, after giving CONTRACTOR (and the surety, if any) seven days written notice, terminate the services of CONTRACTOR, exclude CONTRACTOR from the Site, and take possession of the Work and of all CONTRACTOR's tools, appliances, construction equipment, and machinery at the Site, and use the same to the full extent they could be used by CONTRACTOR (without liability to CONTRACTOR for trespass or conversion), incorporate in the Work all materials and equipment stored at the Site or for which OWNER has paid CONTRACTOR but which are stored elsewhere, and finish the Work as OWNER may deem expedient. In such case, CONTRACTOR shall not be entitled to receive any further payment until the Work is finished. If the unpaid balance of the Contract Price exceeds all claims, costs, losses, and damages (including but not limited to all fees and charges of engineers, architects, attorneys, and other professionals and all court or arbitration or other dispute resolution costs) sustained by OWNER arising out of or relating to completing the Work, such excess will be paid to CONTRACTOR. If such claims, costs, losses, and damages exceed such unpaid balance, CONTRACTOR shall pay the difference to OWNER. Such claims, costs, losses, and damages incurred by OWNER will be reviewed by ENGINEER as to their reasonableness and, when so approved by ENGINEER, incorporated in a Change Order. When exercising any rights or remedies under this paragraph OWNER shall not be required to obtain the lowest price for the Work performed.

C. Where CONTRACTOR's services have been so terminated by OWNER, the termination will not affect any rights or remedies of OWNER against CONTRACTOR then existing or which may thereafter accrue. Any retention or payment of moneys due CONTRACTOR by OWNER will not release CONTRACTOR from liability.

15.03 *OWNER May Terminate For Convenience*

A. Upon seven days written notice to CONTRACTOR and ENGINEER, OWNER may, without cause and without prejudice to any other right or remedy of OWNER, elect to terminate the Contract. In such case, CONTRACTOR shall be paid (without duplication of any items):

   1. for completed and acceptable Work executed in accordance with the Contract Documents prior to the effective date of termination, including fair and reasonable sums for overhead and profit on such Work;

   2. for expenses sustained prior to the effective date of termination in performing services and furnishing labor, materials, or equipment as required by the Contract Documents in connection with uncompleted Work, plus fair and reasonable sums for overhead and profit on such expenses;

   3. for all claims, costs, losses, and damages (including but not limited to all fees and charges of engineers, architects, attorneys, and other professionals and all court or arbitration or other dispute resolution costs) incurred in settlement of terminated contracts with Subcontractors, Suppliers, and others; and

   4. for reasonable expenses directly attributable to termination.

B. CONTRACTOR shall not be paid on account of loss of anticipated profits or revenue or other economic loss arising out of or resulting from such termination.

15.04 *CONTRACTOR May Stop Work or Terminate*

A. If, through no act or fault of CONTRACTOR, the Work is suspended for more than 90 consecutive days by OWNER or under an order of court or other public authority, or ENGINEER fails to act on any Application for Payment within 30 days after it is submitted, or OWNER fails for 30 days to pay CONTRACTOR any sum finally determined to be due, then CONTRACTOR may, upon seven days written notice to OWNER and ENGINEER, and provided OWNER or ENGINEER do not remedy such suspension or failure within that time, terminate the Contract and recover from OWNER payment on the same terms as provided in paragraph 15.03. In lieu of terminating the Contract and without prejudice to any other right or remedy, if ENGI-NEER has failed to act on an Application for Payment within 30 days after it is submitted, or OWNER has failed for 30 days to pay CONTRACTOR any sum finally determined to be due, CONTRACTOR may, seven days after written notice to OWNER and ENGINEER, stop the Work until payment is made of all such amounts due CONTRACTOR, including interest thereon. The provisions of this paragraph 15.04 are not intended to preclude CONTRACTOR from making a Claim under paragraph 10.05 for an adjustment in Contract Price or Contract Times or otherwise for expenses or damage directly attributable to CONTRACTOR's stopping the Work as permitted by this paragraph.

ARTICLE 16 - DISPUTE RESOLUTION

16.01 *Methods and Procedures*

A. Dispute resolution methods and procedures, if any, shall be as set forth in the Supplementary Conditions. If no method and procedure has been set forth, and subject to the provisions of paragraphs 9.09 and 10.05, OWNER and CONTRACTOR may exercise such rights or remedies as either may otherwise have under the Contract Documents or by Laws or Regulations in respect of any dispute.

ARTICLE 17 - MISCELLANEOUS

17.01   *Giving Notice*

A. Whenever any provision of the Contract Documents requires the giving of written notice, it will be deemed to have been validly given if delivered in person to the individual or to a member of the firm or to an officer of the corporation for whom it is intended, or if delivered at or sent by registered or certified mail, postage prepaid, to the last business address known to the giver of the notice.

17.02   *Computation of Times*

A. When any period of time is referred to in the Contract Documents by days, it will be computed to exclude the first and include the last day of such period. If the last day of any such period falls on a Saturday or Sunday or on a day made a legal holiday by the law of the applicable jurisdiction, such day will be omitted from the computation.

17.03   *Cumulative Remedies*

A. The duties and obligations imposed by these General Conditions and the rights and remedies available hereunder to the parties hereto are in addition to, and are not to be construed in any way as a limitation of, any rights and remedies available to any or all of them which are otherwise imposed or available by Laws or Regulations, by special warranty or guarantee, or by other provisions of the Contract Documents, and the provisions of this paragraph will be as effective as if repeated specifically in the Contract Documents in connection with each particular duty, obligation, right, and remedy to which they apply.

17.04   *Survival of Obligations*

A. All representations, indemnifications, warranties, and guarantees made in, required by, or given in accordance with the Contract Documents, as well as all continuing obligations indicated in the Contract Documents, will survive final payment, completion, and acceptance of the Work or termination or completion of the Agreement.

17.05   *Controlling Law*

A. This Contract is to be governed by the law of the state in which the Project is located.

Complete copies of this document may be obtained from the National Society of Professional Engineers, 1420 King Street, Alexandria, VA 22314; the American Consulting Engineers Council, 1015 15th Street, N.W., Washington, D.C. 20005; the American Society of Civil Engineers, 1801 Alexander Bell Drive, Reston, VA 20191; and the Construction Specifications Institute, 601 Madison St., Alexandria, VA 22314.

Appendix 2

ABA Forum on the Construction Industry— 1985 Survey on Construction Arbitration

REPORT ON THE AAA ARBITRATION SURVEY TO THE FORUM COMMITTEE ON THE CONSTRUCTION INDUSTRY*

Because of the multi-party nature of a construction project, along with the fact that substantial projects take a number of years to build, the construction industry is one that often engenders contract disputes involving complex factual circumstances and large dollar consequences. Time is indeed money, and the delay in opening the "hotel" on time is as significant to the owner as is the cost to the contractor whose manpower and equipment is tied up while awaiting a change in the plans and specifications.

In recent years there appears to have been a growing trend to seek third-party dispute resolution rather than resolution among the parties themselves. The disputes often involve numerous complex issues, a large number of parties, and large sums of money. As might be expected, the marshalling of facts and the linking of damages to these "causative" events has resulted in the use of sophisticated technological analyses such as critical path scheduling which, of course, requires the use of computerization. In short, a construction dispute is often as difficult to unravel, after the fact, as it was to manage and build the actual project.

The traditional method of third-party dispute resolution, in courts of general jurisdiction, is subject to growing criticism in the belief that it is ill-suited to the specialized problems of construction. Interest continues to mount for alternative means of dispute resolution. Even prior to the congestion of court dockets, arbitration was a means frequently employed to resolve construction disputes. Arbitration was often thought of as a fast and economical method by which the disputing parties could have their issues heard by people with expertise in the construction industry. This assumption is one of the reasons the American Institute of Architects' and the Engineers Joint Contract Documents Committee's standard forms of contract require that all disputes be resolved by means of arbitration administered pursuant to the Construction Industry Arbitration Rules of the American Arbitration Association. The American Arbitration Association is a national organization which, for a fee, administers the arbitration of construction disputes by a national panel of construction arbitrators.

Arbitration, like the court system, has its severe critics and staunch defenders. Is arbitration "better" than a jury trial or a bench trial? Is arbitration faster or cheaper than having one's day in court? Are the

* The report on the survey of arbitration and the analysis of survey responses first appeared in the *Construction Law* Volume 7, No. 3, August 1987, which is the publication of the Forum Committee on the construction industry of the American Bar Association. We are most grateful to the American Bar Association Section of Litigation and its Construction Litigation Committee, the authors of the report and the survey for their kind permission to reproduce the results. We also appreciate the assistance given by Mr. James J. Myers, a member of our Editorial Advisory Board, facilitating the publication of this material. © American Bar Association Section of Litigation (1987). No reproduction of any part of this material is permissible without the prior permission of the copyright holders.

This material appeared in *The Construction Lawyer*, a publication of the Forum Committee on the Construction Industry. Reproduced with permission.

arbitrators who hear disputes more or less competent than a judge and jury? Should American Arbitration Association administration be recommended? Can the arbitration process be improved? The questions are legion.

In 1985 the Dispute Avoidance and Resolution Sub-committee of the Forum Committee on the Construction Industry and the Construction Committee of the Litigation Section of the American Bar Association jointly conducted a survey of their membership, in the form of the attached questionnaire. Representatives of the American Arbitration Association also participated in its drafting.

The data collected in this study reflects the opinions of approximately 600 attorneys who have actually been involved in the arbitration process in the field of construction. To our knowledge, no similar arbitration survey has ever before been conducted.

The questionnaire responses are tabulated below on the actual form that was used. An italic typeface differentiates the tabulated results from the original questionnaire. The questionnaire had numerous spaces for comments, as well as "yes"/"no" answers. When compiled, the comments consume over 200 single spaced, typewritten pages. No attempt has been made to characterize the comments.

We believe that the readers of this survey will, like ourselves, find some surprising results. We hope that the results will provide the basis for further discussions among attorneys who are interested in improving the process by which construction disputes can be resolved and also form the basis for discussions with representatives of the American Arbitration Association with a view toward modifying its rules and administration where needed.

Both committees wish to extend their thanks to Michael F. Hoellering, Esq., General Counsel of the American Arbitration Association, for his efforts and support.

Respectfully submitted,

Kenneth M. Cushman
Pepper, Hamilton & Scheetz
Philadelphia, PA

Robert A. Rubin
Postner & Rubin
New York, NY

Responses to Survey

1. How many times have you participated as an attorney in a construction arbitration?

   a. Number where amount in dispute exceeded $250,000.

   <u>1,842</u>   *The 576 attorneys who*
   responded have participated in 1,847
   arbitrations exceeding $250,000.

b. Of these, number that were administered by the American Arbitration Association (AAA).
<u>1,603</u>   *87%*

c. Number where amount in dispute was less than $250,000.
<u>2,726</u>

d. Of these, number that were administered by the AAA.
<u>2,278</u>   *83%*

2. How many times have you served as an arbitrator on a construction case?

a. Number where amount in dispute exceeded $250,000.
<u>448</u>

b. Of these, number that were administered by the AAA.
<u>387</u>

c. Number where amount in dispute was less than $250,000.
<u>857</u>

d. Of these, number that were administered by the AAA.
<u>668</u>

3. Do the cases you are involved with generally have party-appointed arbitrators?
Yes <u>191</u>   No <u>300</u>   *61% do not have party appointed arbitrators*

R = Total Response

4. Have you ever participated in a pre-hearing conference before an American Arbitration Association representative?
Yes <u>247</u>   No <u>264</u>   *R–511*   *48% have*
If Yes, was the pre-hearing conference helpful?
Yes <u>197</u>   No <u>52</u>
Comment   *79% found it helpful*

5. Have you ever participated in a preliminary hearing before the panel of arbitrators?
Yes <u>203</u>   No <u>309</u>   *R–512*   *39.6% have*
If Yes, was the preliminary hearing helpful?
Yes <u>171</u>   No <u>29</u>
Comment   *85.5% found it helpful*

6. Is the method by which the American Arbitration Association describes the qualifications of potential arbitrators satisfactory?
Yes <u>241</u>   No <u>59</u>   *R–300*   *80.3% found method of description satisfactory.*
If No, how can the method be improved?

7. Should the arbitration rules provide for a mandatory counterclaim?
Yes <u>304</u>   No <u>184</u>   *R–488 62% want rules to require mandatory c.c.*
Comment

8. Should specific time schedules or rules be imposed on arbitration procedures and the parties so as to further promote prompt resolutions of disputes?
   Yes <u>253</u>   No <u>206</u>   *R–459*   *55% responded yes*
   If Yes, how do you propose to establish such time periods?

9. Should arbitration proceedings be continuous even if an arbitrary time limitation on each party is required to complete the proceeding within the available scheduled time?
   Yes <u>217</u>   No <u>257</u>   *R–474*   *54% responded no.*
   Comment

10. Should the arbitration rules provide that the decision of the arbitrator(s) is final, binding and conclusive except in cases of fraud, arbitrariness and capriciousness of the arbitrators?
    Yes <u>353</u>   No <u>141</u>   *R–494*   *71.4% would have* rules *make arbitrators' decision final and binding*

    Comment

11. Do you believe that the arbitration rules should provide for a broader scope of appeal?
    Yes <u>144</u>   No <u>346</u>   *R–490*   *70.6% would not enlarge appeal rights.*
    If Yes, what do you suggest as the appropriate appellate review standard?

12. Has the fee structure of the American Arbitration Association influenced your decision as to whether you recommend that your clients agree to arbitrate pursuant to the Construction Industry Rules of the American Arbitration Association?
    Yes <u>204</u>   No <u>297</u>   *R–501*   *59% responded fee structure of AAA has not influenced decision to arbitrate.*
    If Yes, describe how.

13. Has it been your experience that arbitrators unjustifiably render a compromise decision?
    Yes <u>216</u>   No <u>257</u>   *R–473*   *54% responded no unjustifiable compromise of verdict.*
    Comment

14. Should the arbitration rules provide that arbitrators have the power to award punitive damages?
    Yes <u>153</u>   No <u>353</u>   *R–506*   *69.7% responded rules should not permit punitive damages.*
    Comment

15. Should the arbitration rules provide that arbitrators are permitted to award attorneys' fees?
Yes <u>345</u>   No <u>160</u>   *R–505*   *68% responded rules should allow for award of attorneys' fees.*

If Yes, what guidelines do you suggest for the awarding of the attorneys' fees?

Disputes in Excess of $250,000

Please check the box which best characterizes your overall experience with the arbitration process.

	Total Responses	% Excellent	% Good	% Fair	% Poor	% Very Poor
1. Speed						
a. Time in which the case was first scheduled for hearing	430	15.3	49.3	23.2	7.9	4.1
b. Speed of actual hearing	391	14.8	27.8	32.7	16.6	7.9
c. Speed of arbitrators' decision	400	24.5	40.7	24.2	6.5	4.0
2. As an economic means of resolving disputes	424	14.8	30.1	35.8	11.0	8.0
3. Qualification of arbitrators	426	15.4	45.5	30.7	7.5	0.7
4. Fairness of decision	414	13.7	47.5	28.7	8.2	1.6
5. Predictability of result	401	3.9	44.3	34.9	12.7	3.9

<u>COMMENTS</u>   *Of 876 attorneys who responded that there was an abnormal delay, fault was apportioned as follows:*

1. If there was an abnormal delay in the arbitration proceedings, was it primarily due to the:
   a. arbitrators   Yes <u>136</u>   No <u>201</u> *15%*
   b. attorneys   Yes <u>245</u>   No <u>160</u> *28%*
   c. the claimant or respondent
   Yes <u>117</u>   No <u>209</u> *13%*
   d. collateral litigation (i.e., suits to enjoin arbitration proceedings)
   Yes <u>116</u>   No <u>199</u> *13%*
   e. type of problem being arbitrated
   Yes <u>194</u>   No <u>228</u> *22%*
   f. administrative problems
   Yes <u>68</u>   No <u>268</u> *18%*

Disputes Under $250,000

Please check the box which best characterizes your overall experience with the arbitration process.

	Total Responses	% Excellent	% Good	% Fair	% Poor	% Very Poor
1. Speed						
a. Time in which the case was first scheduled for hearing	395	25.0	48.8	16.2	7.8	2.0
b. Speed of actual hearing	377	29.9	47.2	15.9	5.8	1.0
c. Speed of arbitrators' decision	386	25.9	48.7	22.5	2.8	0.0
2. As an economic means of resolving disputes	382	24.0	42.4	23.5	7.3	2.6
3. Qualification of arbitrators	373	21.4	46.6	24.9	6.9	0.0
4. Fairness of decision	488	11.0	40.7	21.5	26.0	0.6
5. Predictability of result	473	4.8	38.0	23.4	30.0	3.5

COMMENTS   *Of 421 responses who thought there was an abnormal delay, fault was apportioned as follows:*

1. If there was an abnormal delay in the arbitration proceedings, was it primarily due to the:

   a. arbitrators   Yes 62   No 141 .... *14.7%*
   b. attorneys   Yes 152   No 90 .... *36%*
   c. the claimant or respondent
      Yes 80   No 147 .... *19%*
   d. collateral litigation (i.e., suits to enjoin arbitration proceedings)
      Yes 42   No 185 .... *9.9%*
   e. type of problem being arbitrated
      Yes 45   No 194 .... *9.5%*
   f. administrative problems
      Yes 45   No 194 .... *10.6%*

2. Should the arbitration rules give the arbitrators the specific power to require each party to:

   a. produce all relevant documents prior to the arbitration proceeding?
      Yes 288   No 93   R–381   *75.5% yes*

b. produce all documents which the attorney intends to introduce as evidence at the arbitration proceeding?
Yes <u>323</u>   No <u>51</u>   R–374   *86% yes*

c. not require production of any documents prior to the arbitration proceeding?
Yes <u>46</u>   No <u>306</u>   R–352   *13% would not require*
the production of any
documents prior to arbitration.

3. Should the arbitration rules give the arbitrators the specific power to permit discovery in arbitration proceedings?
Yes <u>277</u>   No <u>99</u>   R–375   *73.8% yes*
If Yes, how, if at all, would you limit discovery rights?

4. Should the arbitration rules give the arbitrators the specific power to require that a party must file a detailed "pretrial" or statement of claim prior to the arbitration proceeding?
Yes <u>270</u>   No <u>108</u>   R–378   *71.4% yes*
If Yes, state the extent of detail required.

5. Should the arbitration rules give the arbitrators the specific power to impose sanctions for failure to comply with the arbitration rules with respect to arbitration matters?
Yes <u>308</u>   No <u>64</u>   R–372   *82.7% yes*

6. Should the arbitration rules give the arbitrators the specific power to permit the joinder of parties and consolidate disputes?
Yes <u>240</u>   No <u>129</u>   R–369   *65% yes*

7. Should arbitrators be required to write findings of fact?
Yes <u>175</u>   No <u>196</u>   R–371   *47% yes*
Comment

8. Should arbitrators be required to write conclusions of law?
Yes <u>116</u>   No <u>256</u>   R–372   *31% yes*
Comment

9. General Comments

Arbitration as Compared to Jury Trial

Please check the box which best characterizes your overall opinion as to your experience.

In order to compare arbitration with a jury trial, please state whether arbitration was better, the same, or worse with respect to the following categories:

	Total Responses	% Better	% Same	% Worse
1. Speed				
a. Time in which the case was first scheduled for proceedings	460	87.1	9.5	3.2
b. Speed of actual hearing	477	59.7	18.2	22.0
c. Speed of ultimate resolution of entire dispute from when "complaint/demand" is filed	458	73.7	15.2	10.9
2. Costs and Expenses (including discovery costs)	453	56.5	29.3	14.1
3. Quality of trier of fact:				
a. Fairness of decision	450	39.1	42.4	18.4
b. Predictability of result	448	37.7	40.1	22.0

Arbitration as Compared to Bench Trial

Please check the box which best characterizes your overall opinion as to your experience.

In order to compare arbitration with a bench trial, please state whether arbitration was better, the same, or worse with respect to the following categories:

	Total Responses	% Better	% Same	% Worse
1. Speed				
a. Time in which the case was first scheduled for proceedings	454	74.6	19.3	5.9
b. Speed of actual hearing	456	42.3	31.5	26.0
c. Speed of ultimate resolution of entire dispute from when "complaint/demand" is filed	464	64.4	23.9	11.6
2. Costs and Expenses (including discovery costs)	446	52.4	30.4	17.0
3. Quality of trier of fact:				
a. Fairness of decision	455	23.5	51.6	24.8
b. Predictability of result	454	22.6	47.3	29.9

Client Reaction—Disputes in Excess of $250,000

Please check the box which best characterizes your overall experience with the arbitration process.

How have your <u>clients</u> reacted to arbitration after its conclusion with respect to the following categories:

	Total Responses	% Excellent	% Good	% Fair	% Poor	% Very Poor
1. Speed a. Time in which the case was first scheduled for hearing	373	12.0	51.4	22.7	9.1	4.5
b. Speed of actual hearing	365	7.9	37.5	29.0	15.6	9.8
c. Speed of arbitrat decision	360	10.2	47.7	29.1	9.4	3.3
2. As an economic means of resolving disputes	361	8.8	28.2	39.6	14.1	9.1
3. Qualification of arbitrators	365	8.7	41.3	30.4	15.6	3.8
4. Fairness of decision	361	8.3	36.8	37.1	13.8	3.8

<u>COMMENTS</u>:

1. After the arbitrators' decision, did your clients believe that the proceedings would have been "better resolved" had there been required pre-hearing discovery?
   Yes <u>114</u>   No <u>190</u>   *R–304*   *37.5% yes*

2. Would your client select arbitration as opposed to litigation the next time around?
   Yes <u>160</u>   No <u>130</u>   *R–290*   *55% yes*

3. General Comments

Client Reaction—Disputes Under $250,000

Please check the box which best characterizes your overall experience with the arbitration process.

How have your <u>clients</u> reacted to arbitration after its conclusion with respect to the following categories:

	Total Responses	% Excellent	% Good	% Fair	% Poor	% Very Poor
1. Speed						
a. Time in which the case was first scheduled for hearing	360	21.1	53.6	19.7	3.0	2.5
b. Speed of actual hearing	310	21.9	45.8	21.2	7.0	3.8
c. Speed of arbitrators' decision	355	18.8	52.9	22.8	3.6	1.6
2. As an economic means of resolving disputes	358	17.8	38.8	30.1	9.2	3.6
3. Qualification of arbitrators	352	12.7	44.0	30.9	9.6	2.5
4. Fairness of decision	352	11.0	41.7	32.6	10.5	3.9

<u>COMMENTS</u>:

1. After the arbitrators' decision, did your clients believe that the proceedings would have been "better resolved" had there been required pre-hearing discovery?
   Yes <u>96</u>   No <u>229</u>   R–325   29.5% yes

2. Would your client select arbitration as opposed to litigation the next time around?
   Yes <u>212</u>   No <u>91</u>   R–303   69.9% yes

3. General Comments

Experience with the American Arbitration Association

Please check the box which best characterizes your overall experience with the arbitration process.

If you have participated in an arbitration proceeding administered by the American Arbitration Association, please answer the following:

List the regional office where arbitration was administered and, as to each, evaluate the following:

Regional Office
OVERALL
(city)

	Grade Point	% Excellent	% Good	% Fair	% Poor	% Very Poor
1. Quality of arbitrators	3.7	101	311	155	33	13
2. Training of arbitrators	3.3	44	201	221	70	25
3. Information given about arbitrators	3.0	29	151	282	113	24
4. Performance of tribunal administrator	3.7	124	293	112	48	31

Number of responses shown in boxes.
Grade Point Average shown on scale 1 to 5.

COMMENTS:

Respondents were asked to rate their experiences with individual American Arbitration Association regional offices. In all, there were 613 responses covering 29 American Arbitration Association offices. Because the number of responses on several American Arbitration Association offices was inadequate to yield statistically significant data, only the combined, overall data has been set forth.

Appendix 3

A Preconstruction Conference, Federal Government Project

A preconstruction conference was held in the Conference Room of the Southern Nevada Construction Office at 9:00 AM, November 16, 1998 with (Smith/Jones JV) contractor for construction of (XXXXXX) and No. 7B Pumping Plants, (XXXXXXX) Switchyard a Modifications to Pumping Plants No. 1, 4 and 5, and additions to Switchyards No. 1, 1A and 2 under specifications No. DC-7372, Southern Nevada Water Project, Second Stage. The meeting was held in Englewood, Nevada. In attendance were:

Water and Power Resources Service:
Project Construction Engineer
Office Engineer
Field Engineer
Safety Manager
Chief, Contract Administration
Chief, Mechanical Inspection
Chief, Electrical Inspection
Resident Engineer
Chief, Construction Surveys

(Smith/Jones JV):
Five individuals from this joint-venture were present, no titles listed.

Other agencies:
Other individuals are listed from the Division of Colorado River Resources, the Southern Nevada Water System, and another firm.

(Within the minutes, we have substituted XX for names of participants.)

Preconstruction Conference
November 16, 1998
Specifications No. DC-7372

Jones/Smith Inc.

AGENDA

A. Introductions and Organization

    1. SNCO
    2. Contractor
    3. Contracting Officer and authorized representative

B. General Provisions

    1. Specifications and drawings—differences
    2. Changes
    3. Differing site conditions
    4. Disputes
    5. Material and workmanship
    6. Inspection and acceptance
    7. Superintendence by contractor (G.P. 11 and 1.2.11)
    8. Permits
    9. Other contracts (G.P. 14 and 1.2.9)
    10. Equal opportunity
    11. Taxes
    12. Supplements to General Provisions

C. Contract Administration

    1. Labor and wage provisions
    2. Subcontracts
    3. Commencement, prosecution and completion of work (1.2.3)
    4. Liquidated damages (1.2.4)
    5. Construction program (1.2.6)
    6. Small Business and Minority Subcontracting (1.2.12 and 1.2.13)
    7. Payments and adjustments
      a. Funds available for earning (1.3.1)
      b. Progress payments (1.3.2)
      c. Posting securities in lieu of retained percentages (1.3.3)
    8. Cost reduction incentive (1.3.14)
    9. Correspondence
   10. Submittals

D. Safety and Health

    1. General requirement (1.1.4)
    2. Safety of the Public (1.5.10)
    3. Rollover protective structures (1.5.11)
    4. Aerial lifts (1.5.12)
    5. Protection of existing installations (1.5.15)
    6. Security and access to work sites (1.5.9)
    7. Electrical power for construction purposes (1.5.18)
    8. Water for construction purposes (1.5.19)
    9. Construction operations at regulating tank 7 (1.5.7)
   10. Safety precautions for energized facilities (1.5.13)

F. Environmental Quality Protection

    1. Landscape preservation (1.6.1)
    2. Prevention of water pollution (1.6.2)
    3. Abatement of air pollution (1.6.3)
    4. Dust abatement (1.6.4)
    5. Noise abatement (1.6.5)
    6. Light abatement (1.6.6)
    7. Disturbance of bighorn sheep (1.6.7)
    8. Pesticides (1.6.8) and (3.6.4)
    9. Cleanup and disposal of waste material (1.6.9)

G. Construction

    1. Layout of work and quantity surveys (1.2.10)
    2. Explosives, drilling and blasting (1.5.14)
    3. Removing culvert and fence, Pumping Plant No. 7 (2.3.1, 2.4.1)
    4. Excavation
      a. General (3.2.1)
      b. Structure (3.2.4)
    5. Disposal of excavated material (3.2.5)
    6. Backfill (3.3.1, 3.3.2)
    7. Constructing pumping plant site embankments (3.4.1)
    8. Removing and relocating reinforced concrete pressure pipe (4.1.1)
    9. Concrete
      a. Placement schedule (5.1.2)
      b. Placement drawings (5.1.3)
   10. Concrete in switchyards (5.2.2)
   11. Cast-in-place concrete (5.3.1)
   12. Precast walls and panels (5.5.2)

13. Switchyard steel structures (8.4.1)
14. Mechanical (Section 13)
   a. Manifolds (13.2.1)
   b. Traveling cranes (13.3.1)
15. Painting, lining, and coating (Section 15)

A. INTRODUCTION AND ORGANIZATION

1. *Southern Nevada Construction Office:*

Mr. XX opened the conference with introductions of Water and Power Resources Service personnel in attendance, briefly explaining their individual responsibilities and functions on jobsites and in direct dealings with Jones/Smith. He also introduced representatives of the Division of Colorado River Resources and Southern Nevada Water System, State and operating agencies working with SNCO on this Project.

2. *Contractor:*

Mr. introduced members of his organization who will be working with SNCO on this Project. Mr. J. XX of T. C. Jones Inc., will monitor this Project for the contractor and will be the contact person in this joint venture.

3. *Contracting Officer and Authorized Representative:*

Mr. XX explained he is the Contracting Officer's authorized representative on this contract and has authority to act in details of administering the contract. XX is the Contract Specialist assigned the responsibility of general administration of the contract. Changes involving expenditures under $10,000 can be made at Project level provided they do not involve a time extension. Any changes of major consequence will be handled by the Contracting Officer, Mr. XX located in Denver. (Subsequent to this meeting, XX was appointed Acting Contracting Officer.)

B. GENERAL PROVISIONS

1. *Specifications and Drawings–Differences:*

The contractor was advised that they are obligated to notify SNCO as promptly as possible of any differences or errors they note in the drawings during the course of the work. It was stressed that prompt notification is of utmost importance in any matters of conflict or disagreement. If there is a conflict between the narrative portion of the specifications and drawings, the specifications narrative will govern. Should items appear on drawings that are not contained in the specifications, they are considered to be a part of both.

2. *Changes:*

It was again stressed that prompt notification to SNCO by the contractor is of utmost importance if there are contract changes or monetary adjustments of any kind.

3. *Differing Site Conditions:*

The emphasis again is on prompt notification to SNCO by the contractor should anything different be encountered than shown on the specifications, drawings, or subsurface data in specifications.

4. *Disputes:*

The contractor was advised of the procedure to be followed in resolving a difference in interpretation of specifications: the contractor is to notify SNCO of the difference, and SNCO will render an opinion. If the contractor disagrees, they may ask for findings of fact from the Contracting Officer. Should there still be a dispute, the contractor may appeal.

The contractor questioned how cost proposals should be handled. They were informed that an agreement should be reached prior to work being accomplished. It was pointed out that before initiating a change, it would be prudent for the contractor to discuss the matter with SNCO so they (the contractor) can be given insights to our requirements. The contractor inquired whether they would be notified of our decision by letter and was answered in the affirmative.

5. *Material and Workmanship:*

It was pointed out to the contractor that materials furnished are to be as specified, in good condition and operational. SNCO has the right to use defective materials until the contractor provides proper materials.

6. *Inspection and Acceptance:*

There will be SNCO inspectors at all sites of work. The contractor was advised of SNCO procedures to perform on-site inspections of all activities. Acceptance of any portion of the work will require that portion to be a completed unit of work.

7. *Superintendence by Contractor (G.P. 11 and 1.2.11):*

The contractor advised SNCO that they will have an office located at the jobsite in addition to offices in Denver and Novato, California. They require three copies of all correspondence, one copy to each office. Enclosures should be mailed only to the Project office. There is a new address for the Denver office (XX XX Street).

XX, Resident Engineer, will be the SNCO contact in the field. XX, Field Engineer, will be the ultimate field contact. The contractor will have supervisors on the jobsites who can speak for the contractor and resolve issues when problems arise without the necessity of a conference.

The contractor was told that the Water and Power Resources Service does not provide general space for a contractor, but right-of-way is available for temporary facilities. Arrangements can be made to use the existing facilities of the Southern Nevada Water System, other contractors, etc.

8. *Permits:*

It is the responsibility of the contractor to obtain all permits and licenses necessary to operate under the contract in the State of Nevada. The contractor stated that the City of North Las Vegas and Clark County now require use or building permits, and no temporary power can be obtained without these permits. The Southern Nevada Construction Office, however, does not feel a building permit is necessary.

9. *Other Contracts (G.P. 14 and 1.2.9):*

The contractor was informed that other contractors will be working at the same locations at approximately the same time, especially at the switchyard sites. A coordinated effort is expected between Smith/Jones and other contractors working in the area to accomplish the job.

10. *Equal Opportunity:*

This subject was bypassed for the present time. A separate discussion on this topic will be held in the future after submission of the Affirmative Action Plan by the contractor.

11. *Taxes:*

The contractor was advised that all taxes are their responsibility. Necessary taxes should be included in the bid.

12. *Supplements to General Provisions:*

Mr. XX pointed out that there is a supplement to the General Provisions, and the contractor acknowledged receipt of same.

C. CONTRACT ADMINISTRATION

1. *Labor and Wage Provisions:*

The contractor was advised of our payroll requirements. Anyone performing on-site work must appear on the payroll registers for either the contractor or a subcontractor except those classified as owner/operator (truck, backhoe, etc.).

The contractor has received a book setting forth labor standards provisions and the requirement to display safety posters. The contractor stated they need additional safety and EEO posters to be displayed at jobsites.

2. *Subcontracts:*

The contractor is responsible for all work performed by the subcontractor. The Southern Nevada Construction Office deals only with the primary contractor. Subcontract agreements are needed before firms can appear on the jobsite. Bonding is required on all subcontracts over $5,000. There are provisions for payment bond forms in accordance with paragraph 1.1.2. of the specifications.

The contractor was advised that subcontracts are needed only for subcontractors performing on-site work. If the contractor is just purchasing materials or equipment, only a purchase order need be submitted. A letter was mailed to the contractor on November 14, 1998, stating those bid items requiring purchase orders.

3. *Commencement, Prosecution and Completion of Work (1.2.3):*

The contractor will receive a formal letter from the Contracting Officer extending all completion dates under paragraph 1.2.3. of the specifications with the exception of Tabulation No. 1. The date of notice to proceed is October 11, 1998. A time extension of 179 days will be added to all completion dates, establishing a contract completion date of April 12, 2001.

4. *Liquidated Damages (1.2.4):*

Maximum liquidated damages will not exceed $600 per day.

5. *Construction Program (1.2.6):*

The contractor has submitted a construction program for approval. There was a question regarding the late starting date for construction of the switchyards. The contractor stated that work on other Government projects has been delayed because it is difficult to obtain materials from suppliers.

There is a dire need to have the switchyards completed earlier than stated in the contractor's schedule, especially Switchyards 2, 1A and Hacienda, as they are the biggest problems. Work on Switchyard No. 1 can start on November 1, 1999.

A letter will be sent to the contractor outlining revisions to the construction schedule. The contractor will have 30 days after receipt of the letter to make any necessary changes.

6. *Small Business and Minority Subcontracting (1.2.12 and 1.2.13):*

The contractor agrees to establish and conduct a small business subcontracting program which will enable small businesses and minorities to be considered as subcontractors and suppliers. The contractor will submit quarterly reports of subcontracting to small business concerns on Optional Form 61.

7. *Payments and Adjustments:*

a. *Funds Available for Earning (1.3.1):*

An allocation of 3.7 million dollars has been made for this contract for FY 98. A letter has been sent reserving 3.7 million dollars. If additional funds are needed, they must be requested. Should any allocated funds not be used, they will be allocated elsewhere.

b. *Progress Payments (1.3.2):*

The cut-off date for progress payments will be the 25th of the month to enable the Government to prepare the necessary documents and get them to the Finance Office by the 1st of the month so that a check can be issued the contractor by the 10th of the month. A representative from the SNCO field office will meet with the field representative of the contractor to agree on work accomplished during the month. The contractor was informed that the amount of work between the 25th and the end of the month is not estimated. A breakdown of the lump sum items must be submitted for use in determining payment. For payment for materials on inventory, the contractor was advised they must submit either a paid invoice or a letter indicating that title is vested in the contractaor and that the material is free of all liens and encumbrances from both the contractor and supplier.

Two forms will be transmitted to the contractor for completion and return with each month's progress estimate: (1) contractor's monthly summary of manhours and lost-time injuries report, and (2) certification of contractor conformity to labor standards. These forms can be obtained through SNCO.

The contractor will provide SNCO a listing of persons authorized to sign progress payment documents.

The contractor was advised to document all adjustments as soon as possible to that an equitable adjustment can be ascertained.

c. *Posting Securities in Lieu of Retained Percentages (1.3.3):*

The contractor has the option of depositing approved interest-bearing negotiable securities in escrow in lieu of retaining percentages. If the contractor elects to exercise this option, a request must be submitted to Denver in accordance with paragraph 1.3.3. of the specifications.

8. *Cost Reduction Incentive (1.3.14):*

This provision allows the contractor to share in any savings on a fifty-fifty basis. The contractor indicated his understanding of this provision.

9. *Correspondence:*

Mr. XX reviewed the procedure for forwarding correspondence to SNCO: all correspondence to SNCO should be directed to the Project Construction Engineer. The Project Construction Engineer should also receive an information copy of all correspondence directed to Denver for approval.

The contractor advised that correspondence should be directed to Smith/Jones Inc., XX XX Lane, XX, California, with information copies to Jones Construction, Inc., XX XX Street, Denver, Colorado. When the contractor has a local office, correspondence will be directed to that address with information copies to XX and Denver offices.

10. *Submittals:*

The contractor was advised that all submittals from their subcontractors or suppliers must be made through the prime contractor. The specifications state which submittals are to be transmitted to Denver and which are to be sent to this office. It was stressed by Mr. XX that care should be taken to ensure that submittals are made to the proper office, as at least five days will be lost if correspondence is directed to the wrong office. It was requested that SNCO receive copies of transmittal letters for any submittals to Denver.

D. SAFETY AND HEALTH

1. *General Requirement (1.1.4):*

The contractor has submitted his safety and health program which will be reviewed by this office and forwarded to Denver for approval. The contractor had expressed the desire to schedule the monthly safety meeting for the first Tuesday of the month. SNCO has another meeting scheduled for that time, so another date must be chosen. This will not affect approval of the contractor's safety program, however. A mutually acceptable date will be decided upon.

The contractor advised that he is aware of the OSHA regulations. Within the State of Nevada, the Nevada Industrial Commission has a safety division which carries out the OSHA regulations under a State plan. In answer to the contractor's question regarding frequency of visits by State inspectors, Mr. XX stated that State Industrial or Federal people may come onto the job at any time for a safety inspection. These inspections usually occur as the result of a worker reporting an unsafe condition.

There is a possibility the contractor may be allowed to use their own workers' compensation insurance carrier at some time during this contract. The contractor will check with the Nevada Industrial Relations Committee to determine if and when this may be done.

The contractor was given two copies of the booklet entitled "Safety and Health Regulations for Construction" and was informed they may have as many as they require. The contractor stated they have one copy at their office and will inform us if they require additional copies. This is the safety manual and contractor must abide by during construction activities. Approximately 80 percent of the book is taken directly from OSHA construction regulations. The last portion of the book is the Government supplement in which there are regulations governing areas OSHA does not cover. If the contractor wants to perform work outside these regulations and feels they have a plan that is as effective, and it is under the Government supplement, the SNCO Safety Manager can work with the contractor possibly to obtain a waiver directly from this office. There will be a safety meeting at a later date to cover these items in depth if necessary.

SNCO is to be advised of all accidents occurring on the job. The inspector is to be notified immediately of injuries. It is desirable that the inspector be informed of any accidents occurring near work sites so that this office will be aware of occurrences should any inquiries be made. Lost-time accident report forms are to be completed.

All equipment is to have a safety inspection prior to being placed in service. The contractor is bringing in a new crane which should have all the latest safety features. All off-highway earth handling equipment will be brake-tested. The contractor asked if each piece of equipment would be tested and was informed if equipment comes to the contractor directly from another of their jobs without going elsewhere it would not need to be tested. Equipment hired to haul to and from sites normally will not require testing if they are licensed by the County, but will be observed to ascertain that they are performing safely. Testing will be performed by the contractor and observed by Government inspectors prior to the equipment actually being used on the job.

Precautionary measures must be taken to prevent heavy equipment, particularly cranes, from going into powerlines. Various warning devices were discussed, including posting signs stating "CAUTION—POWERLINE OVERHEAD" in large, bold, red lettering or stringing large, orange balls along powerlines. Mr. XX commented this sometimes seems an insoluble problem, but recommended extraordinary measures be taken to ensure that equipment does not move into powerlines.

The contractor asked if the Bureau requires physical examinations for crane operators and was informed this requirement is contained in the safety manual. It was pointed out that crane operators and operators of hoisting equipment should have physicals to determine if there is any impairment of hearing or vision, color blindness, history of epilepsy, heart condition, or other serious health conditions. It was noted by SNCO that there had been recent problems in this regard, and they were pleased that the contractor recognizes and is concerned about this problem.

Paragraph 1.1.4. of the specifications sets forth the requirement that a weekly tool box safety meeting be held. The Government inspector in that area will probably sit in on these meetings. It was explained that the inspector on the jobsite will be in charge of ensuring that on-site safety rules are enforced by the contractor. In doing this, he will confine his discussions to the foreman level or higher unless there is imminent danger to employees. When the Safety Manager is on the jobsite, he will also confine his conversations to the superintendent or the foreman.

2. *Safety of the Public (1.5.10):*

All measures necessary to protect the public must be taken; primarily, anytime access to roads must be blocked or work creates a hazard.

3. *Rollover Protective Structures (1.5.11):*

Included in the contractor's safety program.

4. *Aerial Lifts (1.5.12):*

It was explained to the contractor that aerial lifts are considered to be any device used to transport personnel from one level to another. Contractor has determined aerial lifts will not be used.

5. *Protection of Existing Installations (1.5.15):*

The contractor is responsible for ascertaining where buried equipment is located and take necessary precautions to ensure it is not damaged.

The contractor's employees will be allowed to use sanitary facilities already at sites. The contractor will be required to maintain these facilities while in use.

6. *Security and Access to Work Sites (1.5.9):*

Security will be a big problem due to the high rate of vandalism in this area, particularly in the Twin Lakes area. Other areas, such as around Lake Mead, do not present such a problem. The contractor inquired about Hacienda and was assured there should be no problem at that site as there will be another contractor in the area and security can be handled jointly. There have been minimum problems at fenced facilities. The contractor suggested using guard dogs; there was no objection to this measure by SNCO representatives. Mr. XX suggested equipment be watched closely, as other contractors have had huge pieces of equipment stolen from jobsites.

7. *Electrical Power for Construction Purposes (1.5.18):*

Electrical power is available at several sites.

8. *Water for Construction Purposes (1.5.19):*

Water is available at all sites with the exception of Hacienda and Twin Lakes. Arrangements are now being made for water at these sites.

9. *Construction Operations at Pumping Plant 7 (1.5.7):*

The contractor was advised of the requirement to interface with other work already in existence. Connections should be coordinated through Southern Nevada Water System. Removal and relocation of the wasteway at Regulating Tank 7 cannot block the access road to Pumping Plant 7.

10. *Safety Precautions for Energized Facilities (1.5.13):*

The specifications contain the Bureau's electrical procedures for working on hot equipment. The contractor should coordinate with the power company to develop procedures for power outages. Shutdown schedules are more convenient during the winter months (October through April). During the summer months shutdown is inconvenient and at least one day's notice will be required.

F. ENVIRONMENTAL QUALITY PROTECTION

The contractor's attention was directed to those paragraphs in the specifications concerning this particular environment.

1. *Landscape Preservation (1.7.1):*

The contractor was cautioned to confine all work to rights-of-way as much as possible, as the desert environment does not "heal" as rapidly as other environments when disturbed. The work performed by this Project seems to be under constant scrutiny by a number of interested groups, and the contractor was assured that they would hear from one of these groups should any of the work operations appear to be detrimental to the environment. Landowners will have to be contacted for trespass rights in the course of the work.

Mr. XX pointed out that a portion of the work is in the Lake Mead Recreational Area administered by the National Park Service. They have been very cooperative in the past, but there is a tremendous amount of visitor traffic in this area and many people who will complain if they see anything extraordinary going on. The contractor was encouraged to coordinate any matters that might require attention by the National Park Service with that organization.

2. *Prevention of Water Pollution (1.6.2):*

The contractor was advised that anytime dewatering is performed as part of the work, water that could pollute Lake Mead should not be discharged into waterways. In areas where there may be large amounts of dewatering wastes off sites, particularly cement, it must be controlled and not dumped into drainage or sewer systems. Mr. XX stressed that no waste is to be washed into Lake Mead. The contractor commented their estimator had assured them they would encounter no water at any of the sites.

3. & 4. *Abatement of Air Pollution and Dust Abatement (1.6.3, 1.6.4):*

The contractor was advised that emission of dust into the air is scrutinized quite closely by the County dust control office. They were informed that methods using water to control the dust should be sufficient. The contractor inquired if a permit was required to disturb the topsoil and was informed that a permit was required.

5. *Noise Abatement (1.6.5):*

The contractor was informed that the County has established daylight operation hours in residential areas; there can be no activity before 6:00 a.m. or after 10:00 p.m.

6. *Light Abatement (1.6.6):*

Night work is permitted, if it is performed quietly, but no floodlights can be used that might disturb residents.

7. *Disturbance of Bighorn Sheep (1.6.7):*

The contractor was informed that there is a large herd of bighorn sheep in the area of operation, particularly around Pumping Plant 2 because of a watering hole. All of the Lake Mead Recreational Area controlled by the National Park Service is range for the bighorn sheep. The Park Service has requested that none of the Project activities be detrimental to these animals and that precautions be taken to assure that their water sources near jobsites are not polluted. The Park Service has requested they be notified at least 24 hours prior to any blasting operations so they may observe the effects of blasting on the sheep. The Park Service is quite sensitive about the treatment of the bighorn sheep, and it was stressed to the contractor that these animals should not be harrassed in any manner by the workers.

8. *Pesticides (1.6.8, 3.6.4):*

Application of pesticides must be performed in accordance with the manufacturer's regulations. None of the waste and none of the unused portions of mixing tubs are to be cleaned and dumped into any waterway that may eventually pollute any system.

9. *Cleanup and Disposal of Waste Material (1.6.9):*

The contractor will be responsible for hauling debris and waste material off the jobsites to a suitable disposal site. The contractor inquired if there would be a problem with burning waste materials, and was informed they should contact the County and the Fire Marshall for permission for burning. The contractor stated if there was any problem obtaining burning permits, they would haul the waste materials to disposal areas.

G. CONSTRUCTION

1. *Layout of Work and Quantity Surveys (1.2.10):*

Mr. XX, Field Engineer for SNCO, read the general provisions of paragraph 1.2.10 of the specifications. All survey work performed by the contractor will be subject to field and office review. SNCO will supply the contractor with survey books, and requested a copy of the contractor's survey notes be sent to this office for review for accuracy. The contractor asked when notes should be submitted and was informed the notes should be sent to this office as soon as completed so that there would not be a quantity of notes received for review at one time. Mr. XX clarified that SNCO is just interested in receiving copies of the sheets, not the book. The contractor was informed that the area is being staked now and should be finalized so that the contractor will have the data needed to begin construction when they reach the site.

2. *Explosives, Drilling and Blasting (1.5.14):*

Prior to any blasting operations, a blasting plan is required stating where, when, how, etc., the blasting is to be performed and the safety precautions to be employed during blasting. If any blasting is performed near homes, quite stringent methods must be employed to eliminate vibration.

3. *Removing Culvert and Fence, Pumping Plant No. 7 (2.3.1, 2.4.1):*

The contractor was informed that they would be required to remove some chain link fencing that would become the property of the Government upon removal. When the fencing is removed, SNCO will advise the contractor where to dispose of it.

4. *Excavation:*

   a. *General (3.2.1):*

   It was pointed out to the contractor that any overexcavation as a result of work operations not directed by the Government will be backfilled, compacted, etc., at the contractor's expense. Government expense is specifically for line and grade excavation.

   b. *Structure (3.2.4):*

   Blasting and line drilling operations will require controlled blasting operations.

5. *Disposal of Excavated Material (3.2.5):*

SNCO field engineers will meet with the contractor's field personnel before beginning field operations in Hacienda, Twin Lakes and switchyard areas, as the present specifications point out where excess material can be placed. There are specific areas for excavated material.

6. *Backfill (3.3.1, 3.3.2):*

No backfill will be placed against concrete walls until 28 days of cure has been obtained. Concrete strength will not be tested prior to the 28-day cure period. There are certain requirements in the specifications for compacted backfill, and this office will be conducting tests as material is compacted.

7. *Constructing Pumping Plant Site Embankments (3.4.1):*

Construction of forebay tank and flow control structures, including connection of intake manifold to the portion of intake manifold installed under these specifications, will be performed under another contract during the contract period covered by these specifications. To avoid interference with work of other contractors at these locations, the service yard embankments will not be constructed beyond the limits designated by the Contracting Officer until receipt of written notification that the work may begin as specified in paragraph 1.2.3. SNCO engineers will be working with the contractor in the field, as there are other contractors on site, and there are certain time periods when work can be carried out.

8. *Removing and Relocating Reinforced Concrete Pressure Pipe (4.1.1):*

The existing wasteway at Pumping Plant 7B will be removed and relocated to another facility if possible. The contractor was cautioned to use care when removing the pipe. If damage occurs, replacement will be at the contractor's expense. Existing pipe will be reused where possible, but damaged pieces will not be used.

9. *Concrete:*

   a. *Placement Schedule (5.1.2):*

   The contractor is required to submit a placement schedule 60 days after receipt of notice to proceed.

b. *Placement Drawings (5.1.3):*

Current requirement is that placement drawings be submitted for review and approval. Lift drawings are also required. All inserts should be shown on drawings. It was stressed that drawings should be submitted in a timely fashion, as typically they are submitted so late that there is not sufficient time for review prior to commencement of work. The contractor asked if drawings were to be submitted for everything, including Rebar. They were told this is the case and that Rebar was detailed separately. Rebar may be included on detail drawings. The contractor stated the only problem area they foresee will be the conduits. Mr. XX commented that paragraph 5.1.3 spells out what should be included on lift drawings.

10. *Concrete in Switchyards (5.2.2):*

It was noted there are a number of items included in paragraph 5.2.2 of the specifications that require additional information from the Contractor before the Government can design foundations. The contractor was requested to provide this information as soon as possible so that foundations may be designed more quickly. This item will be discussed more thoroughly at additional meetings in the next few weeks.

11. *Cast-In-Place Concrete (5.3.1):*

The contractor will use Bureau-mix designs for concrete. Concrete aggregates will require rescreening. The contractor stated that concrete buckets will be used for placing. The contractor also stated they are bringing in a big crane to be used in the placing operation.

12. *Precast Walls and Panels (5.5.2):*

The contractor will precast everything on site except T beams. XXX will construct the forms using fiberglass. The contractor was asked what kind of cure will be used and they indicated this would be discussed at a meeting to be held the following day. The mix design will be Type II cement consisting of 7.5 sacks per cubic yard with low sand content of 35-40% and no more than 3-inch slump. The concrete will be covered immediately after placement with Visqueen or a tarp. The schedule for the panels will begin in March 1980 with a completion time of 16 weeks. Mr. XX pointed out that the type of curing used on panels will be critical as related to temperatures, and that in this area heat is more a problem than cold. It was stressed that inadequate curing will prevent panels meeting the 5,000 pound stress requirement. The contractor was encouraged to look at Mr. XX's operations. The contractor indicated they would do more research on this subject.

13. *Switchyard Steel Structures (8.4.1):*

The Government will design one 69-kilovolt interrupter support after receipt of information from the contractor for the 69-kilovolt nondisconnect interrupter. SNCO requested this information be submitted as soon as possible in order to gain needed lead time for design.

14. *Mechanical:*

a. *Manifolds (13.2.1):*

The contractor was informed that the manifold alignment should be line-on-line because another contractor is involved. SNCO will work closely with the contractor, providing as much assistance as possible in fitting manifolds prior to placement. The contractor stated they will be using dummy sections between manifolds. This fact was very encouraging to SNCO engineers.

b. *Traveling Cranes (13.3.1):*

The contractor was informed that after cranes are assembled, they are obligated to operate the cranes for other contractors, if so needed, or operate them for the Government on an hourly rate based on actual cost of operation plus 10%. The cranes will be tested to 125% of capacity. After passing testing, the contractor is free to use the crane as they see fit.

It was noted at this point that qualified welders, as stipulated by specifications, will be required for welding performed on switchyard steel structures and on manifolds. Welds will be subject to nondestructive testing. Three types of tests will be performed: visual inspection, radiographic or ultrasonic testing, and magnetic particle and dye penetrant testing.

There will be a number of tests required after completion of construction: water supply will be tested for 125 pounds per square inch for 8 hours with no leaks; heating and air conditioning system shall be adjusted to deliver the specified cubic feet per minute under full-load conditions with acutal air quantity within 5% of quantity shown on drawings, and simulated dirty filter conditions will be tested; sewage system will have to be lead tested.

The contractor was questioned regarding what type of coating is to be used on manifolds. The contractor was informed that 1040 tapecoat only is not satisfactory. The contractor is uncertain at this time what type coating will be used, and their proposal will be submitted to this office for approval.

15. *Painting, Lining, and Coating (Section 15):*

Purchase orders for paint must be submitted to Denver for approval 45 days in advance of purchase. It was stressed that unapproved paint cannot be used.

Mr. XX announced that this completed the agenda items and that there was still time to discuss pertinent items if anyone had additional information of importance.

The contractor was asked when they anticipate being on the jobsite. They replied they will begin work the first week in December. Their office will be located at Twin Lakes with another (a trailer) at Pumping Plant 7B. The precast site will be located at Twin Lakes.

Mr. XXXX, of the Division of Colorado River Resources, stated that completion of Switchyard XX is critical to their construction schedule. A meeting should be scheduled between SNCO, Div. of Colorado River Resources, Southern Nevada Water System and Smith/Jones as soon as possible to coordinate these efforts plus the expansion of the XXXXXXXX Water Treatment Facility.

The conference adjourned at 11:30 AM.

Noted:_____   Date:_____

Project Construction Engineer

Concurrence: Smith/Jones, Inc.

_____

Title: _____

Date: _____

Appendix 4

A Task Force Report on the Impact of the Design/Build Process on Public Safety and the Professional Engineer's Obligations

I. Introduction

Purpose

The Design/Build Task Force of the Florida Institute of Consulting Engineers (FICE) studied the impact of the design/build process on public safety and the professional engineer's obligation to practice responsibly. Recognizing that design/build can be a viable procurement method, FICE charged this Task Force with reviewing how the engineering profession should practice under this methodology, and further, with developing recommendations for all parties involved in the process.

History of Design/Build

Design/build construction has been practiced in various forms for centuries. The earliest records of use of the concept reveal rulers who wanted large monuments and facilities constructed for their empires. The ruler contracted with a single individual to design and build a project and made that individual the single responsible party.

Over time this single party became the "master builder," who was usually both a designer and contractor for the envisioned projects. In many situations the plans for the project were very sketchy and required the designer's interpretation and guidance during the course of the actual construction. By necessity the designer was also the builder.

Over the ages more complex projects required greater expertise in design. The terms "architect" and "engineer" became more commonplace, and the traditional method of delivery as we know it today evolved. The designer's role became one of transforming a client's needs or goals to drawings from which a builder could construct. The contractor's role was not to interpret the client's wishes but rather to complete the project in accordance with the designer's plans and specifications for a mutually-agreed-upon price in a specific time frame.

While this traditional method of project delivery gave the client a system of checks and balances to guard against defects and deficiencies in the constructed facility, it did not give the owner single-source responsibility or accountability as the "master builder" concept had centuries earlier. This lack of accountability and the litigious nature of our society has once again made the single-source concept attractive to some clients and owners.

While the design/build, single-source concept is attractive to owners, it does, in most circumstances, require a more knowledgeable and responsive owner who can adequately select qualified design/build professionals, be

Reprinted from *Design/Build for Design Firms*, a publication of the American Consulting Engineers Council. For a copy or for further information, contact the Florida Institute of Consulting Engineers (FICE).

responsive to the design/build team, and be in a position to *negotiate* a contract with the design/build team that is fair to all parties with an adequate contingency to cover unknowns. This process has been used successfully in Europe and other parts of the world. In the U.S. it has been used successfully in the industrial sector.

Design/Build Teams

Design/build can be a viable procurement method in which a single entity, the design/build team, provides all the professional design and construction services necessary to build all or a portion of a facility for the client. This single-source responsibility distinguishes the design/build concept from other methods of project delivery.

Traditional Method

```
        ┌──────────┐
        │  OWNER   │
        └──────────┘
         ┌────┴────┐
    ┌────────┐  ┌────────────┐
    │  A/E   │  │  General   │
    │  Firm  │  │ Contractor │
    └────────┘  └────────────┘
                 │ Subcontractor
                 │ Subcontractor
                 │ Subcontractor
```

Historically, three approaches have been used by design/build contractors in the delivery of projects. The first approach utilizes complete in-house architectural, engineering and construction services. These design/build contractors rely solely on the capabilities of their own in-house staff. The divisions or disciplines of the design/build firm design, permit and construct the facility without the use of outside consultants.

The second form of design/build delivery is comprised of long-standing team relationships between separate corporations or firms on multiple projects. These teams are formulated either on a joint venture or contractor/subcontractor basis. In most instances, because of the financial and bonding responsibilities, the construction firm is the prime contractor or majority venture partner. However, architectural and engineering firms are more frequently taking the role of the prime or team leader.

Typically, these teams enter into joint marketing agreements and market a variety of project types. To complement a perceived or real weakness or to assist in local marketing, these teams may also associate with smaller local contractors or architectural and engineering consultants.

The third approach to design/build delivery is the partnership of convenience or "shotgun" approach. These teams are brought together for a specific project or venture. The make-up of these teams can be a combination of the two preceding approaches or the association of firms with little or no joint historical track record. Team members are selected for a variety of reasons, including technical expertise, financial backing or marketing ability.

No matter which team approach is chosen, careful selection of team members and partners, as well as proper definition of contractual relations between design professionals and construction entities are most important. Design/build is a process built upon trust. This report details the process from a professional perspective.

II. Recommendations

The FICE Design/Build Task Force recognizes that the concept of merging design and construction into a single entity has become attractive to many owners. Several reasons have brought certain sectors of the traditionally separate design and construction industries to the design/build format. A lack of confidence in the perceived ability of the architectural and engineering communities to control budgets, meet schedules and properly coordinate documents has led to this position. In addition, the litigious environment of our society has created adversarial posturing among owners, designers and contractors.

Design/build can, in certain situations, address these issues. However, the owner who cannot make decisions, insists on low fees, is not properly financed, and is not knowledgeable in the design and construction markets, will find that the design/build process does not solve his or her problem.

The FICE Design/Build Task Force recognized that design/build is a viable delivery process; that there is no significant professional or legal reason to prohibit the use of design/build; and that the marketplace, including governmental agencies and legislative bodies, regards design/build as a legitimate process. The Task Force studied the process over a 15-month period and makes the following recommendations regarding the professional use of design/build construction.

1. Owner/agent knowledge. The owner's experience in the design and construction process must be considered. Owners must understand:

- Facility requirements (space, usage, flexibility, etc.)
- The design process
- The construction process
- Historical cost data (design, construction and life cycle)
- Design/build contractual relationships (liability, bondability, insurability and terms/conditions)

2. Private-sector considerations. The private owner has significant flexibility to address design/build team selection. However, he or she should be knowledgeable and consider:

- Minimum scope requirements to allow design/build team flexibility in addressing quality, innovation and cost effectiveness
- Submittal response format regarding team requirements, method of selection and timing
- Invited proposers' list
- Qualifications-based selection addressing the design concept and detail, team qualifications, etc.

Design/Build

```
┌──────────┐
│  OWNER   │
└──────────┘
     │
     │
┌──────────────┐
│ Design/Build │
└──────────────┘
  ┌ A/E
  │ Firm
  │ Subcontractor
  │
  │ Subcontractor
  │
  └ Subcontractor
```

3. Public-sector considerations. The public owner usually has certain restrictions in the selection process. However, the owner should have some experience in the design/build process and consider:

- Agency selection process and criteria, including state statutes and local ordinances
- Scope requirements and design criteria, including agency experience in design/build team selection
- Scope requirements, including insurance and bonding requirements, time limitations, design/build team selection method and sample draft contract
- Advertising process and distribution of scope package
- Submittal review based on design/build team qualifications and experience, adherence to scope documents, and agency selection criteria

An agency with solid design/build experience should determine scope requirements to allow the design/build team flexibility in addressing quality, innovation and cost-effectiveness. On the other hand, an agency with nominal design/build experience should carefully develop detailed scope documents that address facility usage, space requirements, quality considerations, cost effectiveness and any other specific project requirements, in order to assure that the agency receives the desired facility.

The submittal process is costly for design/build team members.

4. Contract negotiations. The contractual language of the final agreement should address:

- Responsibilities and liabilities
- Insurance and bonding
- Incentives and penalties, if any
- Completion time
- Price of mutually agreed upon scope of work

Owners should be aware that the submittal process is costly for design/build team members. Consideration should be given to compensating short-listed or invited teams to assure quality responses. Owners should pay the design/build teams for the cost of proposal preparation. When the design entity is separate, and subordinate to the contractor, it should be reimbursed by the contractor for design/build proposal preparation.

The owner should also consider inspection services. Inspection of construction in a design/build project should be provided by the architect/engineer and included in the scope as part of the design/build process. Furthermore, the owner should be cognizant of the critical role that the architect/engineer plays in a design/build team and be assured that these parties are disclosed, properly selected and qualified. The low-bid mentality runs counter to the owner's interest and should be avoided.

III. Basis of Recommendations

Scope of Study

In preparing this report, the FICE Design/Build Task Force solicited input and opinions from all sectors that could potentially be involved with the design/build process. A general listing of these sources appears on page 17.

The Task Force has concluded that there is no clear recommendation concerning whether or not a given party should participate in a design/build project. Within a given sector, e.g., public owners, the Task Force found a difference of opinion as to the merits of design/build. The intent of this report is to bring to light the various facets, forms and intricacies of the design/build process. Having provided this information, it is left to readers to form their own conclusions as to the suitability of design/build based on the particular need or service.

Perceived Advantages

The following list of perceived advantages should be reviewed by prospective participants to judge conformance with their situations and needs. Items listed as an advantage for one party may be perceived as a disadvantage by another.

1. Single-Point Responsibility. By contracting with a single entity, the owner has only one primary contractual link to deal with. The owner has contracted for a completed project that meets the owner's needs. Under the conventional process, the owner contracts with an architect or engineer for the design and inspection, and then contracts separately for the construction. In this conventional process the owner may be responsible for coordination of design and construction, including any problems that may arise during construction. In contrast, under the design/build concept, coordination and liability become the responsibility of a single entity—the prime or design/build team leader—thereby freeing the owner from much of the traditional burden of administrative and liability management.

2. Reduced Time Frame. Coupling the design and construction may condense the total life of the process. This is particularly true for public projects with lengthy selection procedures for both designers and contractors. In addition, a unified design/build group has the ability to fast-track the process. The design/build process also has the ability to minimize the potential for lengthy litigation.

> *A unified design/build group has the ability to fast-track the process and to minimize the potential for lengthy litigation.*

*Given proper
qualifications-
based design/
build team
selection, the
designer can
provide quality
inspection
services.*

3. Design/Material Efficiency. Under design/build, the contractor and designer work together to maximize the strengths of the contractor. By having direct access to the designer, the contractor can propose materials and construction methods that will optimize the contractor's efficiency and potentially lead to a lower cost for the owner.

4. Innovative Design. When an owner provides a design/build team with his or her basic concepts and requirements, the team has the opportunity to develop a design that is based on innovation rather than a common design suited for general construction by all contractors. The strengths of both the designer and builder can be mobilized to the advantage of all parties.

Perceived Disadvantages

1. Single-Point Responsibility. Under the design/build contract, the contractor as prime will now assume liability for the design. This traditionally has not been the case. Design liability can have a lengthy statute of limitations, and the contractor therefore may assume lifetime responsibility for the design. This provision is generally not included in the contractor's bond, and its presence could have a severe impact on the ability of a contractor to secure a bond.

2. Increased Time Frame. For a building-type facility, a substantial amount of work may be required to develop scope documents for the proposal. This is particularly true when there are extensive equipment, finishes, and hardware needs on the project. This extra effort could actually result in a longer time frame than the conventional process. However, when an owner has standard detailed design criteria and construction methodologies, time frames can be reduced by design/build. An example of this would be a transportation project (roadway, bridge, etc.) where historically developed standards and specifications rigidly control the type and quality of facility to be built.

3. Material Quality. An owner who does not have a rigid set of standards and specifications in the design/build scope documents could leave the owner exposed to the use of substandard materials by the contractor in the quest to be low bidder, particularly if the selection is based predominantly on price.

4. Quality of Inspection. Questions concerning the propriety of the design/build team inspecting its own work were raised during the Task Force's interview process. There is a perception that the contractor would leverage the contractor's position over the designer to force the contractor to accept substandard construction or to overlook incorrect construction. There is also the concern that the contractor would not allow the participation of the designer in the inspection process. However, the Task Force concluded that, given proper qualifications-based design/build team selection, the designer can provide quality inspection services.

Selection Process

Although there is some common ground in the selection processes for private owners and governmental bodies, there are also many significant differences.

1. Private Owners. Experienced or repeat builder/owners have knowledge of construction costs, building materials, systems, construction time and reputable contractors. These owners select the one contractor considered best suited for the facility. The owner describes the building he or she has in mind, with its uses, quality, life span and approximate budget. The design/build team prepares a very detailed proposal, including some preliminary sketches and/or plans. The owner and the design/build contractor review the proposal and make any changes and/or additions that the owner may require. They then negotiate the price, and the proposal terms are incorporated into the contract. This is the purest and most direct method.

A private owner who takes proposals will require either the services of an in-house staff or, if not available, an architectural or engineering consultant to prepare a request for proposal similar to the requirements mentioned above, but including schematic drawings and outline specifications. The owner, with the help of the consultant, will prepare a list of well-known design/build contractors. A meeting may be scheduled with the contractors two to three weeks before proposals are due to answer any questions that may have arisen. When proposals are received they are reviewed by the owner and consultant, and one is selected that best meets or exceeds the owner's requirements within the budget. Cost-effective additions and/or deletions to the base contract may be considered. This procedure does not require that the lowest price be accepted.

2. Governmental Agencies. Governmental agencies are restricted by laws, rules and regulations. They do not always have the authority to prepare a selected proposers' list or take the proposer with the proposal that best meets the requirements of the proposed project, regardless of initial cost. These restrictions require that the request for proposal be much more detailed than those required for private work in order to put all proposers on the same basis. This process can prevent proposers from offering cost-effective and innovative variations to the request for proposal.

In Florida, design/build work for public agencies is procured on the basis of competitive negotiations (qualifications-based selection) rather than low bid, in accordance with the Consultants' Competitive Negotiation Act.

3. Request for Proposal (RFP). Requests for proposal may vary from one page, giving description, time and date to a single contractor, to full boilerplate specifications defining insurance, liabilities, construction time, bonding, etc., with outline specifications and descriptions including schematic drawings of the project.

Restrictions on governmental agencies can prevent proposers from offering cost-effective and innovative variations to the RFP.

A minimum RFP is used when an owner has selected one design/build team to negotiate for design and construction of a project. Such an RFP has a description of the basic requirements, time frame and approximate budget. The proposal submitted by the design/build team is detailed, so that the team can communicate to the owner exactly what the proposal covers. From this point negotiations between the two parties define the final scope and cost, and a contract is prepared.

A full RFP includes complete general conditions and supplementary general conditions, technical specifications, and drawings that have been taken to the design development stage. This RFP does not give the design/builder much freedom to propose other methods or materials.

The type of RFP format depends on the complexity of the project, the freedom that the owner wishes to retain, the flexibility the owner wishes to give to the design/build team, whether the owner is a private or governmental entity with restrictions on straight negotiations, and whether the owner is a knowledgeable entity or a novice.

Design/build construction does not lend itself to hard bidding.

4. Selection. Prior to issuing an RFP, the owner should have in mind how the selection is to be made. This process should be outlined for the design/build teams and then followed as closely as possible. Final award and negotiation of an appropriate contract should always address the issue of quality and qualifications.

Consultants' Competitive Negotiation Act (CCNA). State of Florida statutes, like those of most states, require that professional services be negotiated rather than bid. This is an effort on behalf of the public at large to assure that governmental agencies are selecting professional design firms based on appropriate qualifications for the project and, in this fashion, obtain design services that have the public's health and safety in proper focus. There has been some question whether design/build services, which provide a combination of contracting and design services, fall under CCNA. Several indications, including court rulings, pointed toward the concept that traditional design-related CCNA would not accommodate the design/build process.

Accordingly, the 1989 Florida Legislature addressed this issue and modified the CCNA to allow design/build selection for public work. This selection is not to be based upon price as in the traditional construction bid market. A process of scope document preparation and qualifications submittal must be considered in the selection of a design/build team. It is important to note that qualifications are considered an essential part of the selection process. A copy of the law appears in the appendix on page A-1.

Construction Bidding. The purchase of major capital facilities for state and local agencies has historically been required to be a low bid process. Design/build construction does not lend itself to hard bidding. As reflected in the CCNA modifications referenced above, even public-sector design/build selection has a qualifications-based selection process. Accordingly, many governmental entities and their personnel will be required to refocus their thought process for selection of construction services. In many respects this may yield the public more value for their dollar in that quality teams will be providing the construction and design for the facility, rather than a low-bid team looking for additional services and claims.

Conflicts of Interest

In recent years, building a project has become an adversarial process instead of a cooperative relationship. This is due in part to the limited money market and all parties bidding very tightly within the minimum construction budgets established by the owners.

1. **Economic Restraints.** When design/build teams are selected through the bidding process, economic restraints have been put in place. The price covers the RFP with no reserve for changes, variations or contingencies. The design group, as part of the design/build team, will design to satisfy the RFP, but since the design group is also in the bidding mode they may not have funds for study regarding alternate solutions. They must go with a solution that, by experience, they know will work.

2. **Quality of Construction.** The quality of construction is dependent on the price. The price will vary if the RFP requests a 10-, 20- or 40-year serviceable building or if the RFP outline specifications delineate the quality desired for the particular project. Quality construction means many things to many people. Knowledgeable owners always know what quality they require for each particular project. A novice owner may be led astray with catch phrases or implied warranties. The old saying is still true, "You get what you pay for."

3. **Inspection Personnel.** Conflicts of interest regarding inspection personnel rarely arise on negotiated contracts. On these contracts, the design/build contractor is going to correct any known inadequacies to avoid future problems and to keep a satisfied client/owner. On design/build bid contracts, where the situation seems to become more adversarial, the client/owner may feel the design professional's allegiance is to the design/build team—which it is. At the same time, the owner must understand that true design professionals would not jeopardize their reputation for basic design changes that are contrary to good practice. It would be a matter of opinion on quality changes, but should not be in code-related matters. In any event, many owners will retain the services of separate design professionals to perform inspections during construction. The Task Force does not necessarily endorse this practice. The outside party may not be knowledgeable of the project design history and, further, may introduce an element of second-guessing with resultant claims for changes and delays.

Conflicts of interest regarding inspection rarely arise on negotiated contracts.

Proposal Preparation

1. Qualifications and Experience. It is the Task Force's position that the qualifications, experience and interrelationships of the team should be the predominant factors in the selection of the team by the project sponsor. Thus, the design/build team should devote sufficient time in preparation of the proposal to adequately describe its qualifications and experience, specifically as related to the type of project. This will not only include qualifications and experience by individual team members on similar projects, but also as members of past design/build teams. References from previous projects for which the team members individually or collectively have served should be included. It is also important to prepare an organizational chart that describes not only the party responsible for the ultimate project execution, but also the contractual relationships within the design/build team. Makeup of the design team should be disclosed to the owner, and anticipated reporting lines of communication outlined.

2. Risk. In order to adequately demonstrate the team's design approach and project cost, limited development of plans and specifications will be required. The prime firm, be it either contractor or designer, must understand that there is an element of risk involved in executing a contract with a project sponsor for design/build services based on development of limited plans and specifications by the design entity of the team. Once fully developed documents are available, changes or additional details may be required in the project that were not anticipated during the early design and pricing phases. Additional costs may then be incurred by the design/build team. Thus, the prime entity must include a contingency within the contract that will allow for these almost certain, but undefined, costs. Obviously, the contingency amount will be in large measure a function of the detail of the plans and specifications at the time that the contract is executed.

3. Expense of Preparation. There are at least three approaches to allocating the expense of proposal preparation in the design/build method of project delivery. As in the traditional method of project delivery, where the design firm and construction contractor are separate, the cost of design/build proposal preparation can be considered as a cost of doing business and, thus, be built into the individual firm's overhead structure. However, it should be recognized that the costs involved on the part of the design entity of the design/build team will be much larger in the design/build method of delivery than the traditional approach. This additional design team cost is due to the requirement to produce a design sufficient for cost preparation by the construction entity. In the case of a design/build team consisting of multiple firms as opposed to a single firm possessing in-house design capabilities, the cost of preparation by the design entity may be high enough to preclude some firms from pursuing design/build work.

A second method of addressing the expense of preparation of the proposal is for the project sponsor to provide funds to the short-listed design/build teams to assist in the preparation of detailed proposals. This would be similar to project sponsors providing funds to architectural firms in a

The qualifications, experience and inter-relationships of the team should be the predominant factors in the selection of the team.

design competition. This may be an appropriate alternative since, in the design/build method, the project sponsor may receive some benefit from the design work performed by unsuccessful design/build teams.

It is the Task Force's recommendation that the majority of such reimbursable costs go to the design entity, since the contractor's portion of the proposal preparation process will not be significantly different in the design/build approach compared to that of the traditional method of project delivery.

A final means of allocating the expense of preparation of the proposal would be for the contractor to bear a portion of the expense of the preparation of the necessary preliminary design documents. There may be some elements of design carried out by subconsultants that are essentially completed during the proposal preparation phase. One such instance would be in the development of geotechnical engineering recommendations to assist the design team in the selection of an appropriate foundation alternative and preparation of preliminary foundation plans. In this instance, the majority if not all of the work of the geotechnical engineering consultant is performed during the proposal preparation process, and consideration should be given by the project sponsor to reimbursing this element of the work and other similar services.

No matter the approach, all parties should recognize that in many cases the design consultant's fee may be substantially higher than that for traditional design and construction. Risk, higher proposal preparation expense, contingencies, etc., all add to the fee level.

Bonding and Professional Liability Insurance

With the change in structure of design and construction also comes changes in liability. Certainly the design/build entity, whether all in-house, a joint venture, or a builder subcontracting with the design professional, is liable for negligent design or construction. The owner will have little interest in who is responsible and will certainly look to the design/build entity with whom he or she has signed the contract. The insurance industry is beginning to recognize the special circumstances of design/build contracts, but standardized processes have not yet been accepted.

It appears that if design and construction are handled on a joint venture basis between a construction firm and a design firm, both partners are fully liable for defects in either design or construction. For example, in *Kishwaukee Community Health Services Center v. Hospital Building and Equipment Co.*, 638 F. Supp. 1492 (N.D. Ill. 1986), the Court held that where a hospital owner sued the contractor and two design professionals who had been retained for the design of the project, the defendants were hired "as one cohesive group, with each liable under the contract." Thus the architect was held liable not only for design errors, but also for construction errors. *See* generally, *Block,* "As the Walls Came Tumbling

The project sponsor may receive some benefit from the design work performed by unsuccessful design/build teams.

Down, Architects' Expanding Liability Under Design/Build Construction Contracting," 17 John Marshall L. Rev. 1/(1981).

Until policies are updated, these new roles can give rise to insurance coverage disputes. In *United States Fidelity & Guaranty Co. v Continental Casualty Co.*, 153 Ill. App. 3d 185, 505 N.E. 2d 1072 (1987), a workman was injured at a job site at which an architectural firm was allegedly in charge of the construction work. U.S.F.&G. provided multi-peril insurance coverage, which excluded coverage for personal injury or property damage arising from providing "professional services." Continental's insurance covered liabilities for errors and omissions or negligent acts resulting from the firm's performance of "professional services." Continental argued that the architectural firm's activities were more in the nature of a "design/build architect" rather than a "traditional architect" and that its professional liability insurance policy covered only "traditional architectural" services, not "design/build" architectural services. The court held that the Continental insurance policy was not specific enough to exclude claims arising from job site activities as a design/build architect. Clearly this is an invitation to insurance companies to update the language of their policies.

The architect was held liable not only for design errors, but also for construction errors.

1. Architects and Engineers Professional Errors and Omissions (E&O). For design professionals, the purchase of E&O coverage has historically been necessitated by a special exclusion written into the general commercial liability policy barring claims for design errors. The typical E&O policy agrees to indemnify the insured for negligent acts, errors and omissions arising out of the rendering of a described professional service. It is important that design professionals who are contemplating or involved with design/build work confirm that they have appropriate coverage with their insurance carrier. All design/build parties should note that professional E&O insurance specifically excludes any coverage associated with commercial general liability.

2. Commercial General Liability. This is the standard litigation coverage purchased by architects, engineers and contractors. It provides insurance for bodily injury, property damage and legal defense arising out of the insured's premises, operations, products and completed operations. For contractors, limits are available in excess of $10 million per occurrence. The problem is that this policy specifically excludes any coverage for design services or pollution incident claims at the job site.

3. Asbestos Abatement. Since every asbestos abatement project (removal or encapsulation) carries risk and responsibility, contractors involved in this work should also be wary of insufficient general liability coverage. Most policies provide coverage for asbestos abatement operations by explicitly amending the pollution exclusion on a general liability coverage form.

4. Engineers' Pollution Liability. Almost all professional E&O policies contain an exclusion for claims arising out of a pollution incident, although some policies delete the exclusion for the design of drinking water systems or sewage treatment plants. To fill the gap, this policy provides coverage for pollution claims derived from the rendering of a negligent professional act, error or omission for described operations. This coverage is written on either a blanket/specified sites or project-specified basis.

5. Asbestos Consultants E&O. This liability insurance policy eliminates the pollution exclusion to provide coverage for errors, acts or omissions arising out of the design of asbestos abatement projects. It has the potential to be a redundant coverage, so consultants should check to see if this design work is covered in the engineer's pollution liability policy. Asbestos and E&O policies usually have limits up to $1 million in coverage.

6. Commercial Automobile Liability. General liability and professional liability insurance policies exclude claims arising out of the maintenance, use or operation of motor vehicles. New policy forms exclude pollution claims. This can have a significant impact if a contractor transports materials that could lead to an environmental damage claim. In fact, it has been estimated that 70 percent of hazardous waste transporters do not have pollution coverage. It is available, but it must be added by specific endorsement and can be expensive.

7. Workers' Compensation. This policy pays for claims arising out of the injury, death or disability of employees on a no-fault basis, and does not contain a pollution exclusion. However, insurance companies are reluctant to write workers' compensation coverage on the higher risk classes of remedial action contractors, because they cannot charge a rate commensurate with the increased hazards. This coverage is usually available from state-assigned risk pools that are obligated to provide the insurance.

8. Payment and Performance Bonds. These bonds insure the performance of the contractor and payment to all subcontractors. The number of sureties providing bonding coverage for design/build contractors is limited. Bid bonds should be required in order to prevent unqualified bidders who cannot provide a bond from bidding. The sureties specifically exclude design coverage in the performance bond but may include design coverage for payment. RFPs should confirm these clauses.

It is important to note that all design/build team members, whether contractors or consultants, should confirm the adequacy of their insurance coverages. If their traditional role is modified (such as a contractor becoming a sub to an architect or engineer or a consultant working for a contractor in a design/build environment), then appropriate insurance coverage should be arranged. Project-specific coverage for E&O may be appropriate. Job-site safety responsibility should be analyzed with appropriate contractual language and insurance coverages addressed by all parties.

Contractors and owners who are not familiar with design profession E&O coverages should recognize that professional liability insurance does not cover general liability and vice versa. They should also recognize that E&O insurance is typically written on a "claims made" basis, and is only effective while the policy is in force.

All design team members, whether contractors or consultants, should confirm the adequacy of their insurance coverages.

Legal Status and Maturity

The rapid growth of the design/build construction process in the early 1980s threatened to outpace the ability of legislatures, the courts, the professional societies, and the insurance industry to keep up. However, aided by recognition within the construction industry that design/build is a reality, the legal structures are attempting to become current.

1. Historical Perspective. In the early 19th century, "packaged dealers" practiced a form of design/build, offering both design and construction services. Architects sought to distinguish themselves from packaged dealers, and adopted ethical principles which required them to put the owners' interests above their own and forbade architects from acting as packaged dealers. These prohibitions against packaged dealing and design/build carried over into the American Institute of Architects (AIA) Code of Ethics and state regulatory language for over 100 years.

Legislatures and the courts have repeatedly faced the problem of determining appropriate treatment for design/build contracts.

Despite these limitations, in the 20th century the design/build concept has been used with considerable success, principally in connection with complex industrial facilities such as power plants and chemical plants. At the other end of the spectrum, it has been used with success in simple buildings or prefabricated facilities.

In the 1970s, design/build grew in popularity in intermediate facilities, which have traditionally been performed under the "golden triangle" of construction, consisting of owner, architect/engineer, and contractor. These include office buildings, hospitals, libraries, waste treatment projects and schools.

In 1978, the AIA Board of Directors authorized a three-year experiment permitting architects to participate in design/build. This experiment came in response to a call for an end to the ethical prohibition against architects engaging in design/build. By 1980, the AIA board dropped the ethical prohibitions, canceled the experiment and authorized the drafting of AIA design/build contract documents. The law has been struggling to catch up ever since.

2. Traditional Legal Approach. Under the traditional legal approach that permeates statutes, regulations and contract documents, public owners often engage in a competitive negotiation process to select a design professional. The long-standing position of professional societies was that price should not be the primary factor in the selection procedure, and should be considered only after the best-qualified firm has been selected. The federal A/E procurement law (the Brooks Act) mandates a non-bidding system for Federal A/E contracts. In Florida, the Consultants' Competitive Negotiation Act, *Fla. Stat.*, S287.055 (1988), provides that the agency first determine the three most qualified firms and then attempt to negotiate a price.

Language expressly hostile to the design/build concept was written into the Federal Acquisition Regulations, which provides: "No contract for the construction of a project shall be awarded to the firm that designed the project or its subsidiaries or affiliates, except with the approval of the head of the agency or authorized representative (S36.209)."

State licensing laws also enforced separate roles. In New York, a state that still prohibits the corporate practice of architectural and engineering services, the New York State Education Department issued an opinion as recently as 1983 that it would be unlawful for an A/E to contract with a construction company to provide design services, since this would involve the licensed professional in illegal fee splitting.

3. The Legislative and Judicial Response to Design/Build Contracts.
With the change in position by the AIA and the perceived cost and time advantages associated with design/build contracts, the legislatures and the courts have repeatedly faced the problem of determining appropriate treatment for design/build contracts. The net result is an era of rapid change to accommodate the new-found legal creature. This era is by no means over, but a trend toward acceptance of the design/build process has emerged.

To remove any questions of the authority of public agencies to use the design/build concept, legislative bodies have sometimes expressly authorized the approach. For example, in Florida the Legislature has on various occasions provided express statutory authority for design/build activity for the construction of public and private projects.

4. The Design/Build Contract.
Given the authority to proceed, the industry generally relies upon the AIA design/build documents that were published in December 1985. The A191 document sets forth an Owner-Design/Builder Agreement; the A491 document sets forth a Design/Builder-Contractor Agreement; and B901 sets forth a Design/Builder-Architect Agreement.

Federal agencies are beginning to issue their own design/build agreements. GSA first published a design/build contract in March 1987. The Associated General Contractors of America also issued standards for design/build agreements for lump-sum projects and guaranteed maximum price projects.

All three AIA Design/Build documents are divided into two parts, each of which is a contract unto itself. Part I determines the owner's program, alternative design and construction approaches, and early budget estimates. Little, if any, significant construction occurs. At or near the end of this phase, the design/builder submits to the owner a proposed contract sum for the second phase (final design and construction). If the owner accepts the proposal, then the parties execute the second part of the contract, which stipulates a fixed price for the remaining work. Often, time and budget constraints are known up front, and the solicitation may require offerors to stipulate to a guaranteed maximum price including both phases. Under the AIA document, the design/builder's compensation for final design and construction becomes fixed without competition from other firms.

Prudent owners should define their needs with the greatest possible precision at the outset and choose a design/builder with a strong track record of providing reliable service.

In most states, if an engineer or architect signs and seals documents that have the corporation's title block or name on the document, then that corporation must be certified to practice engineering or architecture as a firm.

If the owner elects not to proceed to Phase II of the contract, he or she may be required to obtain new design work.

5. Owner Obligations. In the design/build format, prudent owners should define their needs with the greatest possible precision at the outset and choose a design/builder with a strong track record of providing reliable service. Due to the very structure of the design/build concept, owners must do more than what would be required under either a construction competitive bidding analysis alone or a professional designer's competitive negotiation analysis alone.

Design/build is a sophisticated project delivery system, and requires greater coordination between owner and contractor. The speed of design/build puts pressure on owners to comply with requirements for timely delivery of owner-furnished equipment and materials. Similarly, the processing of shop drawings needs to be reliable. Owners need to recognize that under this process they must live with decisions made in the early stages. Otherwise, the process is slowed down and the benefits are not achieved.

Licensing and Permitting

Much has been written about the selection of professional design and construction services and the statutory requirements for public project selection. As in any matter regarding statutory requirements, each member of a design/build team and the owner should seek appropriate legal counsel or guidance. However, three important matters need special reference:

1. Appropriate individual licenses and corporate certifications must be in place to avoid any potential fines and statutory non-conformance. These licenses include individual architectural, contracting and engineering licenses, as well as appropriate Certificates of Authorization to practice the various professions. There has been a misconception on the part of many contractors that if they have in-house engineers who are themselves individually licensed, the contracting entity can practice engineering or architecture. In general, that is not true. In most states, if an engineer or architect signs and seals documents that have the corporation's title block or name on the document, then that corporation must be certified to practice engineering or architecture as a firm. This can be onerous to some contractors, but if a contractor is going to lead a design/build entity, the contractor must recognize that it has professional design responsibilities.

Florida Statute 287.055, reproduced in the appendix on page A–1, now clearly allows a contractor to offer design/build services, without certificates of authorization to practice architecture or engineering, provided such professional services are sub-contracted to properly licensed and certified individuals or firms. However, if the contracting entity intends to practice engineering and/or architecture in-house, then the contracting entity must have the appropriate certificates of authorization. Similarly, Board of

Professional Engineers Rules such as the Statements on Responsibilities of Professional Engineers for structural design must be followed.

2. Sequencing of permit applications with appropriate support documentation as required for submittal to various permitting agencies must be studied. Most agencies prefer complete document submittal prior to start of construction. However, a design/build contract usually has certain fast-track elements included, which may require phased submittal of permit documents. An understanding should be reached with the permitting authority in order to avoid delays and potential legal action later.

3. Industrial licensing exemptions should be carefully studied prior to assuming that an exemption exists. Such exemptions are often confusing to owners, professionals and contractors, and may require special attention.

Sources

The FICE Design/Build Task Force solicited input and opinions from all sectors that could potentially be involved with the design/build process. A general listing of these sources includes:

- **Owners:** Public (city, county, state, federal); quasi-governmental authorities; private (industry, institutions, developers)

- **Contractors:** Design/build; general; subcontractor/supplier

- **Designers:** Architectural; engineering; architectural/engineering (A/E and E/A)

- **Institutes/Associations:** American Consulting Engineers Council (ACEC); National Society of Professional Engineers (NSPE); American Institute of Architects (AIA); Associated General Contractors (AGC); Florida Association of Building Officials

- **Regulatory and Governmental Agencies:** Department of Professional Regulation; Board of Professional Engineers; Board of Architecture; Construction Industry Licensing Board; Department of Community Affairs; Department of General Services; Board of Regents; Department of Transportation

- **Other:** Attorneys; professional liability insurance companies; bonding/surety companies

Under the design/build process the owner must live with decisions made in the early stages.

Credits

This Florida Institute of Consulting Engineers (FICE) Design/Build Report is a result of a 15-month effort on the part of the Task Force. Task Force members were:

Raymond F. Messer, P.E.
Chairman, FICE Task Force
Walter P. Moore and Associates, Inc.
Tampa, Florida

Joseph P. LoBuono, P.E.
LoBuono, Armstrong & Associates
Tallahassee, Florida

Joseph J. Motta, P.E.
Wade-Trim, Inc.
Tampa, Florida

Emmanuel N. Nicolaides, P.E.
Hufsey-Nicolaides
Miami, Florida

Rafael M. Couret, P.E.
DSA Group, Inc.
Tampa, Florida

Myron L. Hayden, P.E.
C-E Environmental, Inc.
Tallahassee, Florida

Ernest A. Cox, P.E.
Westinghouse Environmental & Geotechnical
Services, Inc.
Altamonte Springs, Florida

Peter D. Brown, E.I.
Peter Brown Construction
Clearwater, Florida

The principal authors of the report were Raymond F. Messer, Joseph P. LoBuono, Emmanuel N. Nicolaides and Peter D. Brown. Editing and final preparation of the report were completed by Frances A. Conaway.

Many people, too numerous to list, made themselves available to the Task Force to discuss design/build. The Task Force owes a deep debt of gratitude to these participants, without whom this report could not have been prepared.

Appendix 5

Avoiding and Resolving Disputes During Construction

Successful Practices and Guidelines

Prepared by the Technical Committee on Contracting Practices
of the Underground Technology Research Council

Sponsored by the
American Society of Civil Engineers
and the
American Institute of Mining, Metallurgical
and Petroleum Engineers

An updated and revised edition of
Avoiding and Resolving Disputes in Underground Construction (1989)

Published by the
American Society of Civil Engineers
1801 Alexander Bell Drive
Reston, VA 20191

The Technical Committee on Contracting Practices
of
The Underground Technology Research Council
1991

Chairman:

P.E. (Joe) Sperry
Tunnel Construction Consultant
636 Paloma Drive
Boulder City, NV 89005

Frank Carr, Esquire
Chief Trial Attorney
U.S. Army Corps of Engineers
20 Massachusetts Avenue, N.W.
Washington, D.C. 20314

John D. Coffee
Area Engineer
Federal Highway Administration
711 South Capitol Way, #501
Olympia, WA 98501

Oliver T. Harding
Washington State Dept of
 Transportation
1501 25th Avenue South
Seattle, WA 98144

Ronald E. Heuer
Geotechnical Consultant
3317 West Ringwood Road
McHenry, IL 60050

Martin N. Kelley
Vice President, Engineering
(Retired)
Kiewit Construction Group, Inc.
11835 Frances Street (Residence)
Omaha, NB 68144

Vladimir Khazak
Director, Technical Services
Municipality of Metro Seattle
821 Second Avenue
Seattle, WA 98104

John F. MacDonald
Project Manager
Guy F. Atkinson Company
448 South Hills St., #206
Los Angeles, CA 90013

Al Mathews
Al Mathews Corporation
Post Office Box 4039
Federal Way, WA 98063

Norman A. Nadel
Chairman
Nadel Associates, Inc.
420 Clock Tower Commons
Brewster, NY 10509

John E. Reeves
Chief, Office of Highway
 Construction
California Department of
 Transportation
Post Office Box 942873
Sacramento, CA 94272

Robert A. Rubin, Esquire
Postner & Rubin
17 Battery Place
New York, NY 10004

Robert J. Smith, Esquire
Wickwire Gavin, P.C.
Post Office Box 1683
Madison, WI 53701

B-2.   Disputes Review Board Specification

(This specification should be included in the Special Provisions and noted in the Instructions to Bidders.)

1.   Introduction

A Disputes Review Board will be established to assist in the resolution of disputes, claims and other controversies arising out of the work of this project.

This specification describes the purpose, procedure, function and key features of the Disputes Review Board. Appended to this specification is a Three Party Agreement formalizing the creation of the Board.

The Board will assist in and facilitate the timely and equitable resolution of disputes between the Owner and the Contractor, in an effort to avoid construction delay and litigation.

It is not intended for the Owner or the Contractor to default on their normal responsibility to amicably and fairly settle their differences by indiscriminately assigning them to the Board. It is intended that the Board encourage the Owner and Contractor to resolve potential disputes without resorting to this appeal procedure.

Either the Owner or the Contractor may appeal a dispute to the Board. Appeal to the Board should be initiated as soon as it appears that the normal Owner-Contractor dispute resolution effort is not succeeding, and prior to enacting other dispute resolution procedures or filing of litigation by either party.

The Board shall fairly and impartially consider disputes referred to it, and shall provide written recommendations, to the Owner and the Contractor, to assist in the resolution of these disputes.

Although the recommendations of the Disputes Review Board should carry great weight for both the Owner and the Contractor, they are not binding on either party.

2.   Continuance of Work During Dispute

At all times during the course of the dispute resolution process, the Contractor shall continue with the work as directed, in a diligent manner and without delay, or shall conform to the Owner's decision or order, and shall be governed by all applicable provisions of the Contract. Records of the work shall be kept in sufficient detail to enable payment in accordance with applicable provisions in the Contract, if this should become necessary.

3.   *Membership*

The Disputes Review Board will consist of one member selected by the Owner and approved by the Contractor, one member selected by the Contractor and approved by the Owner, and a third member selected by the first two members and approved by both the Owner and the Contractor. Normally, the third member will act as Chairman for all Board activities.

It is desirable that all Disputes Review Board members be experienced with the type of construction involved in this project, and interpretation of contract documents. The goal in selecting the third member is to complement the construction experience of the first two and to provide leadership for the Board's activities.

It is imperative that Board members show no partiality to either the Contractor or the Owner, or have any conflict of interest.

The criteria and limitations for membership are:

a.   *No member shall have an ownership interest in any party to the contract, or a financial interest in the contract, except for payment for services on the Disputes Review Board.*

b.   *Except for fee-based consulting services on other projects, no member shall have ever been previously employed by, or have had financial ties to, any party to the contract.*

c.   *No member shall have had a recent, close, professional or personal, relationship with any key member of any party to the contract.*

d.   *No member shall have had substantial prior involvement in the project, of a nature which could compromise his ability to impartially participate in the Board's activities.*

e.   *During his tenure as a Disputes Review Board member, no member shall be employed by any party to the contract.*

f.   *During his tenure as a Disputes Review Board member, no member shall engage in a discussion or make an agreement, with any party to the contract, regarding employment after the contract is completed.*

Before their appointments are final, the first two prospective members shall submit complete disclosure statements for the approval of both the Owner and Contractor. Each statement shall include a resume of experience, together with a declaration describing all past, present and anticipated or planned future relationships to this project and with all parties involved in the construction contract. Disclosure of recent, close, professional or personal, relationships with all key members of all parties to the contract shall be included. The third Board member shall supply such a statement to the first two Board members and to the Owner and Contractor before his appointment is final.

The Owner and the Contractor shall each select and negotiate a working agreement with their respective member within six weeks after award of the contract. Immediately after approval, the Owner and Contractor will notify their members to begin selection of the third member. The first two members shall ensure that the third member meets all of the criteria listed above. The third member shall be selected within four weeks after the first two members are notified to proceed with his selection. In the event of an impasse in selection of the third member, that member shall be selected by mutual agreement of the Owner and the Contractor. In so doing, they may, but are not required to, consider the nominees offered by the first two members.

The Owner, the Contractor, and all three members of the Disputes Review Board shall execute the Disputes Review Board Three Party Agreement within four weeks after the selection of the third member.

4.   Operation

The Disputes Review Board shall formulate its own rules of operation. It is not desirable to adopt hard and fast rules for the functioning of the Board. The entire procedure shall be kept flexible to adapt to changing situations. The Board shall initiate, with the Owner's and Contractor's concurrence, new rules or modifications to old ones, whenever this is deemed appropriate.

In order to keep abreast of construction developments and progress, the members will be promptly informed of construction activity by means of regular written progress reports and other relevant data from the Owner. The Board shall visit the project and meet with representatives of the Owner and the Contractor at regular intervals and at times of critical construction events. The frequency of these visits shall be as agreed among the Owner, the Contractor and the Board, depending on the progress of the work.

The regular meetings shall be held at the job site. Each meeting shall consist of an informal round table discussion followed by a field inspection of the work. The round table discussion shall be attended by selected personnel from the Owner and the Contractor. The Agenda shall generally include the following:

a.   Meeting convened by the Chairman of the Disputes Review Board.

b.   Opening Remarks by the Owner's representative.

c.   A description by the Contractor of:
        work accomplished since the last meeting,
        current status of the work schedule,
        schedule for future work,
        potential disputes, claims and other controversies,
        proposed solutions for these problems.

d.   Discussion by the Owner's representative of:
        the work schedule as he views it,
        Potential disputes, claims and other controversies,
        status of past disputes,
        claims and other controversies

e.   Set tentative date for next meeting.

If it is considered necessary by the parties, the Owner will prepare minutes of regular meetings and circulate them for comments, revisions and/or approval of all concerned.

The field inspection shall cover all active segments of the work. The Board shall be accompanied by representatives of both the Owner and Contractor.
Seeking any Board member's advice or consultation during the meetings or at any other time is expressly prohibited.

5.   *Procedure and Schedule for Dispute Resolution*

Disputes shall be considered as quickly as possible, taking into consideration the particular circumstances and the time required to prepare detailed documentation. Steps may be omitted as agreed to by both parties, and the time periods stated below may be shortened in order to hasten resolution.

a.   If the Contractor objects to any decision, action or order of the Owner, the Contractor may file a written protest

*with the Owner, stating clearly, and in detail, the basis
for the objection, within one week after the event.*

b.   *The Owner will consider the written protest and make its
decision on the basis of the pertinent contract provisions,
together with the facts and circumstances involved in the
dispute. The decision will be furnished in writing to the
Contractor, within two weeks after receipt of the
Contractor's written protest.*

c.   *This decision shall be final and conclusive on the
subject, unless a written appeal to the Owner is filed by
the Contractor, within one week of receiving the decision.
Both parties are encouraged to pursue the matter further
to attempt to settle the dispute. When it appears that the
dispute cannot be resolved on the job, either party may
appeal to the Disputes Review Board.*

d.   *In addition to the above procedure, by written Notice of
Appeal, either the Owner or Contractor may appeal any
dispute, claim or other controversy to the Board.
Simultaneous with submittal to the Board, a copy of the
Notice of Appeal shall be provided to the other party.
The Notice of Appeal shall state clearly and in full detail
the specific issues of the dispute to be considered by the
Board.*

e.   *When a dispute is appealed to the Board, it shall first be
decided when to conduct the hearing. If the matter is
not urgent, it may heard at the next regular Board
meeting. For an urgent matter, the Board shall meet at
its earliest convenience.*

f.   *During the hearing, the Contractor and the Owner shall
each have ample opportunity to be heard and to offer
evidence. Detailed procedures are given in Section 6.
The Board's recommendations for resolution of the
dispute will be given in writing, to both the Owner and
the Contractor, within two weeks of completion of the
hearings. In exceptionally difficult cases, this time may
be extended by mutual agreement of all parties.*

g.   *If requested by either party, the Board shall meet with
the Owner and Contractor to provide additional
clarification of its recommendation.*

h.   *Within two weeks of receiving the Board's
recommendations, or such other time specified by the
Board, both the Owner and the Contractor shall respond
to the other and to the Board in writing, signifying either
acceptance or rejection of the Board's recommendations.*

*The failure of either party to respond within the specified
period shall be deemed an acceptance of the Board's
recommendations. If, with the aid of the Board's
recommendations, the Owner and the Contractor are
able to resolve their dispute the Owner will promptly
process any required contract changes.*

i.   *Should the dispute remain unresolved, either party may
appeal the decision back to the Board, or may resort to
other methods of settlement.*

*Although both the Owner and the Contractor should place great
weight on the Disputes Review Board recommendations, they
are not binding. If the Board's recommendations do not resolve
the dispute, the written recommendations, including any
minority report, will be admissible as evidence in any subsequent
dispute resolution proceeding.*

6.   **Conduct of Hearing**

*The Board may request that written documentation and
arguments from both parties be submitted to each member
before the hearing begins.*

*Normally the hearing will be conducted at the job site.
However, any location that would be more convenient and still
provide all required facilities and access to necessary
documentation is satisfactory. Private sessions of the Board
may be held at any convenient location.*

*The third member of the Board will act as Chairman of the
hearing, or he may appoint one of the other members.
Normally each member keeps his own notes, and a formal
transcript is not prepared. In special cases, when requested by
either party, the Board may allow preparation of a transcript by
a Court Reporter. Audio or video recordings should not be
permitted.*

*The Owner and the Contractor shall have representatives at all
hearings. The Contractor will first discuss the dispute, followed
by the Owner. Each party will then be allowed successive
rebuttals until all aspects are fully covered. The Board members
may ask questions, request clarification, or ask for additional
data. In large or complex cases, additional hearings may be
necessary in order to consider and fully understand all the
evidence presented by both parties. Both the Owner and*

Contractor shall be provided full and adequate opportunity to present all of their evidence, documentation and testimony regarding all issues before the Board.

During the hearings, no Board member shall express any opinion concerning the merit of any facet of the case.

After the hearings are concluded, the Board shall meet to formulate its recommendations. All Board deliberations shall be conducted in private, with all individual views kept strictly confidential. The Board's recommendations, together with an explanation of its reasonings, shall be submitted as a written report to both parties. The recommendations shall be based on the pertinent contract provisions, applicable laws and regulations, and the facts and circumstances involved in the dispute.

The Board shall make every effort to reach a unanimous recommendation. If this proves impossible, the dissenting member may prepare a minority report.

7.   Compensation

Fees and expenses of all three members of the Board shall be shared equally by the Owner and the Contractor. The Owner will prepare and mail minutes and progress reports, will provide administrative services, such as conference facilities and secretarial services, and will bear the cost of these services. If the Board desires special services, such as legal consultation, accounting, data research, and the like, both parties must agree, and the costs will be shared by them as mutually agreed.

The Contractor shall pay the invoices of all Board members after approval by both parties. The Contractor will then bill the Owner for 50% of such invoices.

Appendix 6

Quality in the Constructed Project

Manuals and Reports on Engineering Practice No. 73

QUALITY
in the
CONSTRUCTED
PROJECT

A Guide for Owners, Designers and Constructors

Volume 1

Published by the
American Society of Civil Engineers
1801 Alexander Bell Drive
Reston, VA 20191

First Edition (Revised version of the preliminary manual published in 1988)

Matrix of Suggested Responsibilities for Owners, Contractors, and Design Professionals

TASK	O	DP	C	PRINCIPAL REFERENCE
Assemble owner's advisers	P			Ch. 2
Establish project requirements	P	A*	A*	Ch. 2
Arrange for project financing	P			Ch. 2
Define structure and organization of project team	P			Ch. 3
Plan coordination and communication process	P	A*	A*	Ch. 4
Establish policy and procedure for selection of design professional	P			Ch. 5
Select design professional	P			Ch. 5
Negotiate and sign owner/design professional agreement	P+	P+		Ch. 6
Perform alternative design studies	R	P	A*	Ch. 7
Evaluate project impacts	R	P		Ch. 7
Select preferred alternatives	P	A		Ch. 7
Complete project planning and set design criteria	R	P	A*	Ch. 8
Assemble and manage a qualified, multidiscipline design team	R	P		Ch. 8
Coordinate design disciplines and perform final design	R	P		Ch. 9
Coordinate planning activities with appropriate regulatory agencies	P	A		Ch. 10
Obtain necessary approvals and permits from regulatory agencies	P	A		Ch. 12
Set organization of field construction team	P	A	A*	Ch. 12
Establish policy and procedure for selection of constructor	P	A		Ch. 13
Select constructor by competitive bidding or other means	P	A	A*	Ch. 13
Prepare plans, specifications and other construction contract documents	R+	P+		Ch. 14
Sign construction contract	P+		P+	Ch. 14
Organize for construction	R	A	P	Ch. 15
Initiate job safety and first aid program	R		P	Ch. 15
Submit plans for temporary construction	R	R	P	Ch. 16
Present required contract submittals, including shop drawings		A	P	Ch. 16
Administrative review of contract submittals	P	A	A	Ch. 16
Technical review of contract submittals	R	P	A	Ch. 16
Construct project facilities as specified by contract documents	R	R	P	Ch. 17
Administer constructor, design professional, and other contracts	P	A		Ch. 17
Make appropriate progress payments	P			Ch. 17
Prepare and negotiate contract change orders	P+	A	P+	Ch. 17
Plan and staff for project start-up, operation, and maintenance	P	A	A	Ch. 18
Administer QA/QC programs for design activities	R	P		Ch. 19
Administer QA/QC programs for construction activities	R	R	P	Ch. 19
Consider, evaluate, and use computers as appropriate for all phases of project	P	P	P	Ch. 20
Specify use of peer review as necessary	P	A	A	Ch. 21
Perform competently and on schedule under contract terms	P	P	P	
Seek to avoid conflicts; or resolve them short of litigation	P+	P+	P+	Ch. 22

O:	Owner	P:	Primary Responsibility
DP:	Design Professional	A:	Advising or Assisting
C:	Constructor	R:	Reviewing

*If design professional or constructor are not yet under contract, these services are supplied by qualified advisers.
+Review by legal counsel is indicated.

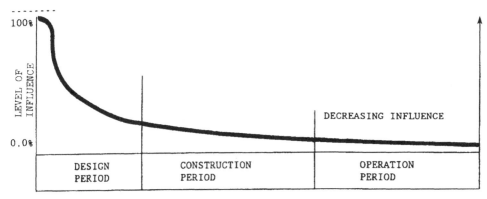

FIG. 1-1. Opportunity to Influence Project Characteristics

TABLE 4-1.   Roles of Team Members

Participant	Project Beginning	Design Phase	Construction Phase	Project Completion
Owner	Forming and informing the group. Leading in outlining project requirements for the design professional.	Contributing to decisions in support of design. Participating in design reviews. Communicating changes when necessary.	Providing for qualified inspection and testing as required by contract documents and regulatory agencies. Administering contracts.	Maintaining group coordination and getting the group's attention on follow-up or completion items.
Design Professional	Assisting with project objectives and program requirements. Leading the development of process for coordination among team members.	Leading the design effort. Involving the owner and others at appropriate times. Preparing necessary design plans and specifications.	Technical support for required interpretations, changes, shop drawing reviews, or field problems, in a timely way. Field observation.	Assisting with follow-up work, completing required manuals and documents, assisting with start-up.
Constructor	Being an early participant. Contributing to alternative studies and scheduling.*	Assisting in vendor selection and constructability reviews.*	Performing the construction effort. Involving others at appropriate times, such as shop drawings, inspections, tests, etc. Field observation.	Leading the follow-up. Guiding vendor and subcontractor follow-up work.

*Often the constructor is not yet selected for the project. If early selection is possible, the constructor can make these contributions.

Appendix 7

Federal Acquisition Regulations (FAR) Provisions

33.210   Contracting officer's authority.

Except as provided in this section, contracting officers are authorized, within any specific limitations of their warrants, to decide or settle all claims arising under or relating to a contract subject to the Act. This authorization does not extend to—

(a) A claim or dispute for penalties or forfeitures prescribed by statute or regulation that another Federal agency is specifically authorized to administer, settle, or determine; or

(b) The settlement, compromise, payment or adjustment of any claim involving fraud.

[48 FR 42349, Sept. 19, 1983. Redesignated and amended at 50 FR 2270, Jan. 15, 1985; 51 FR 36972, Oct. 16, 1986]

33.211   Contracting officer's decision.

(a) When a claim by or against a contractor cannot be satisfied or settled by mutual agreement and a decision on the claim is necessary, the contracting officer shall—

(1) Review the facts pertinent to the claim;

(2) Secure assistance from legal and other advisors;

(3) Coordinate with the contract administration office or contracting office, as appropriate; and

(4) Prepare a written decision that shall include a—

(i) Description of the claim or dispute;

(ii) Reference to the pertinent contract terms;

(iii) Statement of the factual areas of agreement and disagreement;

(iv) Statement of the contracting officer's decision, with supporting rationale;

(v) Paragraph substantially as follows:

This is the final decision of the Contracting Officer. You may appeal this decision to the Board of Contract Appeals. If you decide to appeal, you must, within 90 days from the date you receive this decision, mail or otherwise furnish written notice to the Board of Contract Appeals and provide a copy to the Contracting Officer from whose decision the appeal is taken. The notice shall indicate that an appeal is intended, reference this decision, and identify the contract by number. Instead of appealing to the Board of Contract Appeals, you may bring an action directly in the U.S. Claims Court (except as provided in the Contract Disputes Act of 1978, 41 U.S.C. 603, regarding Maritime Contracts) within 12 months of the date you receive this decision. If you appeal to the Board of Contract Appeals, you may, solely at your election, proceed under the Board's small claims procedure for claims of $10,000 or less or its accelerated procedure for claims of $50,000 or less; and

(vi) Demand for payment prepared in accordance with 32.610(b) in all cases where the decision results in a finding that the contractor is indebted to the Government.

(b) The contracting officer shall furnish a copy of the decision to the contractor by certified mail, return receipt requested, or by any other method that provides evidence of receipt. This requirement shall apply to decisions on claims initiated by or against the contractor.

(c) The contracting officer shall issue the decision within the following statutory time limitations:

(1) For claims of $50,000 or less, 60 days after receiving a written request from the contractor that a decision be rendered within that period, or within a reasonable time after receipt of the claim if the contractor does not make such a request.

(2) For claims over $50,000, 60 days after receiving a certified claim; *provided, however*, that if a decision will not be issued within 60 days, the contracting officer shall notify the contractor, within that period, of the time within which a decision will be issued.

(d) The contracting officer shall issue a decision within a reasonable time, taking into account—

(1) The size and complexity of the claim;

(2) The adequacy of the contractor's supporting data; and

(3) Any other relevant factors.

(e) In the event of undue delay by the contracting officer in rendering a decision on a claim, the contractor may request the agency BCA to direct the contracting officer to issue a decision in a specified time period determined by the BCA.

(f) Any failure of the contracting officer to issue a decision within the required time periods will be deemed a decision by the contracting officer denying the claim and will authorize the contractor to file an appeal or suit on the claim.

(g) The amount determined payable under the decision, less any portion already paid, should be paid, if otherwise proper, without awaiting contractor action concerning appeal. Such payment shall be without prejudice to the rights of either party.

[48 FR 42349, Sept. 19, 1983. Redesignated at 50 FR 2270, Jan. 15, 1985, and amended at 54 FR 34755, Aug. 21, 1989]

52.202-1 Definitions.

As prescribed in subpart 2.2, insert the following clause in solicitations and contracts except when (a) a fixed-price research and development contract that is expected to be $2,500 or less is contemplated or (b) a purchase order is contemplated. Additional definitions may be included; *provided*, they are consistent with this clause and the Federal Acquisition Regulation.

DEFINITIONS (APR 1984)

(a) *Head of the agency* (also called *agency head*) or *Secretary* means the Secretary (or Attorney General, Administrator, Governor, Chairperson, or other chief official, as appropriate) of the agency, including any deputy or assistant chief official of the agency, and, in the Department of Defense, the Under Secretary and any Assistant Secretary of the Departments of the Army, Navy, and Air Force and the Director and Deputy Director of Defense agencies; and the term *authorized representative* means any person, persons, or board (other than the Contracting Officer) authorized to act for the head of the agency or Secretary.

(b) *Contracting Officer* means a person with the authority to enter into, administer, and/or terminate contracts and make related determinations and findings. The term includes certain authorized representatives of the Contracting Officer acting within the limits of their authority as delegated by the Contracting Officer.

(c) Except as otherwise provided in this contract, the term *subcontracts* includes, but is not limited to, purchase orders and changes and modifications to purchase orders under this contract.

52.233-1 Disputes.

As prescribed in 33.214, insert the following clause:

DISPUTES (APR 1984)

(a) This contract is subject to the Contract Disputes Act of 1978 (41 U.S.C. 601-613) (the Act).

(b) Except as provided in the Act, all disputes arising under or relating to this contract shall be resolved under this clause.

(c) *Claim*, as used in this clause, means a written demand or written assertion by one of the contracting parties seeking, as a matter of right, the payment of money in a sum certain, the adjustment or interpretation of contract terms, or other relief arising under or relating to this contract. A claim arising under a contract, unlike a claim relating to that contract, is a claim that can be resolved under a contract clause that provides for the relief sought by the claimant. However, a written demand or written assertion by the Contractor seeking the payment of money exceeding $50,000 is not a claim under the Act until certified as required by subparagraph (d)(2) below. A voucher, invoice, or other routine request for payment that is not in dispute when submitted is not a claim under the Act. The submission may be converted to a claim under the Act, by complying with the submission and certification requirements of this clause, if it is disputed either as to liability or amount or is not acted upon in a reasonable time.

(d) (1) A claim by the Contractor shall be made in writing and submitted to the Contracting Officer for a written decision. A claim by the Government against the Contractor shall be subject to a written decision by the Contracting Officer.

(2) For Contractor claims exceeding $50,000, the Contractor shall submit with the claim a certification that—

(i) The claim is made in good faith;

(ii) Supporting data are accurate and complete to the best of the Contractor's knowledge and belief; and

(iii) The amount requested accurately reflects the contract adjustment for which the Contractor believes the Government is liable.

(3) (i) If the Contractor is an individual, the certification shall be executed by that individual.

(ii) If the Contractor is not an individual, the certification shall be executed by—

(A) A senior company official in charge at the Contractor's plant or location involved; or

(B) An officer or general partner of the Contractor having overall responsibility for the conduct of the Contractor's affairs.

(e) For Contractor claims of $50,000 or less, the Contracting Officer must, if requested in writing by the Contractor, render a decision within 60 days of the request. For Contractor-certified claims over $50,000, the Contracting Officer must, within 60 days, decide the claim or notify the Contractor of the date by which the decision will be made.

(f) The Contracting Officer's decision shall be final unless the Contractor appeals or files a suit as provided in the Act.

(g) The Government shall pay interest on the amount found due and unpaid from (1) the date the Contracting Officer receives the claim (properly certified if required), or (2) the date payment otherwise would be due, if that date is later, until the date of payment. Simple interest on claims shall be paid at the rate, fixed by the Secretary of the Treasury as provided in the Act, which is applicable to the period during which the Contracting Officer receives the claim and then at the rate applicable for each 6-month period as fixed by the Treasury Secretary during the pendency of the claim.

(h) The Contractor shall proceed diligently with performance of this contract, pending final resolution of any request for relief, claim, appeal, or action arising under the contract, and comply with any decision of the Contracting Officer.

52.212–11 Variation in Estimated Quantity.

As prescribed in 12.403(c), insert the following clause in solicitations and contracts when a fixed-price construction contract is contemplated that authorizes a variation in the estimated quantity of unit-priced items:

VARIATION IN ESTIMATED QUANTITY (APR 1984)

If the quantity of a unit-priced item in this contract is an estimated quantity and the actual quantity of the unit-priced item varies more than 15 percent above or below the estimated quantity, an equitable adjustment in the contract price shall be made upon demand of either party. The equitable adjustment shall be based upon any increase or decrease in costs due solely to the variation above 115 percent or below 85 percent of the estimated quantity. If the quantity variation is such as to cause an increase in the time necessary for completion, the Contractor may request, in writing, an extension of time, to be received by the Contracting Officer within 10 days from the beginning of the delay, or within such further period as may be granted by the Contracting Officer before the date of final settlement of the contract. Upon the receipt of a written request for an extension, the Contracting Officer shall ascertain the facts and make an adjustment for extending the completion date as, in the judgement of the Contracting Officer, is justified.

52.212–12 Suspension of Work.

As prescribed in 12.505(a), insert the following clause in solicitations and contracts when a fixed-price construc-

tion or architect-engineer contract is contemplated:

SUSPENSION OF WORK (APR 1984)

(a) The Contracting Officer may order the Contractor, in writing, to suspend, delay, or interrupt all or any part of the work of this contract for the period of time that the Contracting Officer determines appropriate for the convenience of the Government.

(b) If the performance of all or any part of the work is, for an unreasonable period of time, suspended, delayed, or interrupted (1) by an act of the Contracting Officer in the administration of this contract, or (2) by the Contracting Officer's failure to act within the time specified in this contract (or within a reasonable time if not specified), an adjustment shall be made for any increase in the cost of performance of this contract (excluding profit) necessarily caused by the unreasonable suspension, delay, or interruption, and the contract modified in writing accordingly. However, no adjustment shall be made under this clause for any suspension, delay, or interruption to the extent that performance would have been so suspended, delayed, or interrupted by any other cause, including the fault or negligence of the Contractor, or for which an equitable adjustment is provided for or excluded under any other term or condition of this contract.

(c) A claim under this clause shall not be allowed (1) for any costs incurred more than 20 days before the Contractor shall have notified the Contracting Officer in writing of the act or failure to act involved (but this requirement shall not apply to a claim resulting from a suspension order), and (2) unless the claim, in an amount stated, is asserted in writing as soon as practicable after the termination of the suspension, delay, or interruption, but not later than the date of final payment under the contract.

52.236–2 Differing Site Conditions.

As prescribed in 36.502, insert the following clause in solicitations and contracts when a fixed-price construction contract or a fixed-price dismantling, demolition, or removal of improvements contract is contemplated and the contract amount is expected to exceed the small purchase limitation. The contracting officer may insert the clause in solicitations and contracts when a fixed-price construction or a fixed-price contract for dismantling, demolition, or removal of improvements is contemplated and the contract amount is expected to be within the small purchase limitation.

DIFFERING SITE CONDITIONS (APR 1984)

(a) The Contractor shall promptly, and before the conditions are disturbed, give a written notice to the Contracting Officer of (1) subsurface or latent physical conditions at the site which differ materially from those indicated in this contract, or (2) unknown physical conditions at the site, of an unusual nature, which differ materially from those ordinarily encountered and generally recognized as inhering in work of the character provided for in the contract.

(b) The Contracting Officer shall investigate the site conditions promptly after receiving the notice. If the conditions do materially so differ and cause an increase or decrease in the Contractor's cost of, or the time required for, performing any part of the work under this contract, whether or not changed as a result of the conditions, an equitable adjustment shall be made under this clause and the contract modified in writing accordingly.

(c) No request by the Contractor for an equitable adjustment to the contract under this clause shall be allowed, unless the Contractor has given the written notice required; *provided*, that the time prescribed in (a) above for giving written notice may be extended by the Contracting Officer.

(d) No request by the Contractor for an equitable adjustment to the contract for differing site conditions shall be allowed if made after final payment under this contract.

(End of clause)

(R 7-602.4 1968 FEB)

(R 1-7.602-4)

52.236-3 Site Investigation and Conditions Affecting the Work.

As prescribed in 36.503, insert the following clause in solicitations and contracts when a fixed-price construction contract or a fixed-price dismantling, demolition, or removal of improvements contract is contemplated and the contract amount is expected to exceed the small purchase limitation. The contracting officer may insert the clause in solicitations and contracts when a fixed-price construction or a fixed-price contract for dismantling, demolition, or removal of improvements is contemplated and the contract amount is expected to be within the small purchase limitation.

SITE INVESTIGATION AND CONDITIONS AFFECTING THE WORK (APR 1984)

(a) The Contractor acknowledges that it has taken steps reasonably necessary to ascertain the nature and location of the work, and that it has investigated and satisfied itself as to the general and local conditions which can affect the work or its cost, including but not limited to (1) conditions bearing upon transportation, disposal, handling, and storage of materials; (2) the availability of labor, water, electric power, and roads; (3) uncertainties of weather, river stages, tides, or similar physical conditions at the site; (4) the conformation and conditions of the

ground; and (5) the character of equipment and facilities needed preliminary to and during work performance. The Contractor also acknowledges that it has satisfied itself as to the character, quality, and quantity of surface and subsurface materials or obstacles to be encountered insofar as this information is reasonably ascertainable from an inspection of the site, including all exploratory work done by the Government, as well as from the drawings and specifications made a part of this contract. Any failure of the Contractor to take the actions described and acknowledged in this paragraph will not relieve the Contractor from responsibility for estimating properly the difficulty and cost of successfully performing the work, or for proceeding to successfully perform the work without additional expense to the Government.

(b) The Government assumes no responsibility for any conclusions or interpretations made by the Contractor based on the information made available by the Government. Nor does the Government assume responsibility for any understanding reached or representation made concerning conditions which can affect the work by any of its officers or agents before the execution of this contract, unless that understanding or representation is expressly stated in this contract.

52.243-1 Changes—Fixed-Price.

As described in 43.205(a)(1), insert the following clause. The 30-day period may be varied according to agency procedures.

CHANGES—FIXED-PRICE (AUG 1987)

(a) The Contracting Officer may at any time, by written order, and without notice to the sureties, if any, make changes within the general scope of this contract in any one or more of the following:

(2) Drawings, designs, or specifications when the supplies to be furnished are to be specially manufactured for the Government in accordance with the drawings, designs, or specifications.

(2) Method of shipment or packing.

(3) Place of delivery.

(b) If any such change causes an increase or decrease in the cost of, or the time required for, performance of any part of the work under this contract, whether or not changed by the order, the Contracting Officer shall make an equitable adjustment in the contract price, the delivery schedule, or both, and shall modify the contract.

(c) The Contractor must assert its right to an adjustment under this clause within 30 days from the date of receipt of the written order. However, if the Contracting Officer decides that the facts justify it, the Contracting Officer may receive and act upon a proposal submitted before final payment of the contract.

(d) If the Contractor's proposal includes the cost of property made obsolete or excess by the change, the Contracting Officer shall have the right to prescribe the manner of the disposition of the property.

(e) Failure to agree to any adjustment shall be a dispute under the Disputes clause. However, nothing in this clause shall excuse the Contractor from proceeding with the contract as changed.

52.236-21 Specifications and Drawings for Construction.

As prescribed in 36.520, insert the following clause in solicitations and contracts when a fixed-price construction contract or a fixed-price dismantling, demolition, or removal of improvements contract is contemplated and the contract amount is expected to exceed the small purchase limitation. The contracting officer may insert the clause in solicitations and contracts when a fixed-price construction or a fixed-price contract for dismantling, demolition or removal of improvements is contemplated and the contract amount is expected to be within the small purchase limitation.

SPECIFICATIONS AND DRAWINGS FOR CONSTRUCTION (APR 1984)

(a) The Contractor shall keep on the work site a copy of the drawings and specifications and shall at all times give the Contracting Officer access thereto. Anything mentioned in the specifications and not shown on the drawings, or shown on the drawings and not mentioned in the specifications, shall be of like effect as if shown or mentioned in both. In case of difference between drawings and specifications, the specifications shall govern. In case of discrepancy in the figures, in the drawings, or in the specifications, the matter shall be promptly submitted to the Contracting Officer, who shall promptly make a determination in writing. Any adjustment by the Contractor without such a determination shall be at its own risk and expense. The Contracting Officer shall furnish from time to time such detailed drawings and other information as considered necessary, unless otherwise provided.

(b) Wherever in the specifications or upon the drawings the words *directed, required, ordered, designated, prescribed,* or words of like import are used, it shall be understood that the *direction, requirement, order, designation,* or *prescription,* of the Contracting Officer is intended and similarly the words *approved, acceptable, satisfactory,* or words of like import shall mean *approved by,* or *acceptable to,* or *satisfactory to* the Contracting Officer, unless otherwise expressly stated.

(c) Where *as shown, as indicated, as detailed,* or words of similar import are used, it shall be understood that the reference is made to the drawings accompanying this contract unless stated otherwise. The word *provided* as used herein shall be understood to mean *provide complete in place,* that is *furnished and installed.*

(d) Shop drawings means drawings, submitted to the Government by the Contractor, subcontractor, any lower tier subcontractor pursuant to a construction contract, showing in detail (1) the proposed fabrication and assembly of structural elements and (2) the installation (i.e., form, fit, and attachment details) of materials or equipment. It includes drawings, diagrams, layouts, schematics, descriptive literature, illustrations, schedules, performance and test data, and similar materials furnished by the contractor to explain in detail specific portions of the work required by the contract. The Government may duplicate, use, and disclose in any manner and for any purpose shop drawings delivered under this contract.

(e) If this contract requires shop drawings, the Contractor shall coordinate all such drawings, and review them for accuracy, completeness, and compliance with contract requirements and shall indicate its approval thereon as evidence of such coordination and review. Shop drawings submitted to the Contracting Officer without evidence of the Contractor's approval may be returned for resubmission. The Contracting Officer will indicate an approval or disapproval of the shop drawings and if not approved as submitted shall indicate the Government's reasons therefor. Any work done before such approval shall be at the Contractor's risk. Approval by the Contracting Officer shall not relieve the Contractor from responsibility for any errors or omissions in such drawings, nor from responsibility for complying with the requirements of this contract, except with respect to variations described and approved in accordance with (f) below.

(f) If shop drawings show variations from the contract requirements, the Contractor shall describe such variations in writing, separate from the drawings, at the time of submission. If the Contracting Officer approves any such variation, the Contracting Officer shall issue an appropriate contract modification, except that, if the variation is minor or does not involve a change in price or in time of performance, a modification need not be issued.

(g) The Contractor shall submit to the Contracting Officer for approval four copies (unless otherwise indicated) of all shop drawings as called for under the various headings of these specifications. Three sets (unless otherwise indicated) of all shop drawings, will be retained by the Contracting Officer and one set will be returned to the Contractor.

(h) This clause shall be included in all subcontracts at any tier.

52.249-10 Default (Fixed-Price Construction).

As prescribed in 49.504(c)(1), insert the following clause in solicitations and contracts for construction when a fixed-price contract is contemplated and the contract amount is expected to exceed the small purchase limitation. The clause may also be used when the contract amount is not expected to exceed the small purchase limitation, if appropriate (e.g., if completion dates are essential).

DEFAULT (FIXED-PRICE CONSTRUCTION) (APR 1984)

(a) If the Contractor refuses or fails to prosecute the work or any separable part, with the diligence that will insure its completion within the time specified in this contract including any extension, or fails to complete the work within this time, the Government may, by written notice to the Contractor, terminate the right to proceed with the work (or the separable part of the work) that has been delayed. In this event, the Government may take over the work and complete it by contract or otherwise, and may take possession of and use any materials, appliances, and plant on the work site necessary for completing the work. The Contractor and its sureties shall be liable for any damage to the Government resulting from the Contractor's refusal or failure to complete the work within the specified time, whether or not the Contractor's right to proceed with the work is terminated. This liability includes any increased costs incurred by the Government in completing the work.

(b) The Contractor's right to proceed shall not be terminated nor the Contractor charged with damages under this clause, if—

(1) The delay in completing the work arises from unforeseeable causes beyond the control and without the fault or negligence of the Contractor. Examples of such causes include (i) acts of God or of the public enemy, (ii) acts of the Government in either its sovereign or contractual capacity, (iii) acts of another Contractor in the performance of a contract with the Government, (iv) fires, (v) floods, (vi) epidemics, (vii) quarantine restrictions, (viii) strikes, (ix) freight embargoes, (x) unusually severe weather, or (xi) delays of subcontractors or suppliers at any tier arising from unforeseeable causes beyond the control and without the fault or negligence of both the Contractor and the subcontractors or suppliers; and

(2) The Contractor, within 10 days from the beginning of any delay (unless extended by the Contracting Officer), notifies the Contracting Officer in writing of the causes of delay. The Contracting Officer shall ascertain the facts and the extent of delay. If, in the judgment of the Contracting Officer, the findings of fact warrant such action, the time for completing the work shall be extended. The findings of the Contracting Officer shall be final and conclusive on the parties, but subject to appeal under the Disputes clause.

(c) If, after termination of the Contractor's right to proceed, it is determined that the Contractor was not in default, or that the delay was excusable, the rights and obligations of the parties will be the same as if the termination had been issued for the convenience of the Government.

(d) The rights and remedies of the Government in this clause are in addition to any other rights and remedies provided by law or under this contract.

52.212-5 Liquidated Damages—Construction.

As prescribed in 12.204(b), the contracting officer may insert the following clause in solicitations and contracts for construction, except contracts on a cost-plus-fixed-fee basis (see 12.202):

LIQUIDATED DAMAGES— CONSTRUCTION (APR 1984)

(a) If the Contractor fails to complete the work within the time specified in the contract, or any extension, the Contractor shall pay to the Government as liquidated damages, the sum of ——— [Contracting Officer insert amount] for each day of delay.

(b) If the Government terminates the Contractor's right to proceed, the resulting damage will consist of liquidated damages until such reasonable time as may be required for final completion of the work together with any increased costs occasioned the Government in completing the work.

(c) If the Government does not terminate the Contractor's right to proceed, the resulting damage will consist of liquidated damages until the work is completed or accepted.

Appendix 8

Charting the Course: The 1994 Construction Industry Survey on Dispute Avoidance and Resolution—Part I

Construction Lawyer

November, 1996

Construction Dispute Resolution

*5 CHARTING THE COURSE: THE 1994 CONSTRUCTION INDUSTRY SURVEY ON DISPUTE AVOIDANCE AND RESOLUTION—PART I

Thomas J. Stipanowich
Leslie King O'Neal

WESTLAW LAWPRAC INDEX
AMS — Arbitration/Mediation/Settlement/Other Forms of ADR

Introduction

Never in history has so much effort been directed at improving the procedures by which controversies are resolved. The "quiet revolution" affecting dispute resolution reflects widespread concerns with the limitations of formal adjudicatory systems and a renewed emphasis on negotiated solutions. A landmark 1994 survey cosponsored by the Forum on the Construction Industry provides an unprecedented look at current industry trends, and reflects significant changes in the way disputes are resolved and, increasingly, avoided altogether. The results, which will be summarized beginning in this issue, bear important implications for Forum members and their clients. Excerpts follow.

TABLE B

PERCEIVED EFFECTIVENESS OF VARIOUS APPROACHES (FORUM ATTORNEYS)
(1 = Very Ineffective; 5 = Very Effective)

	Binding Arbitration	Dispute Review Board	Early Neutral Evaluation	Mediation	Mini-trial	Nonbinding Arbitration
Reducing Cost of Resolving Dispute	3.04	3.19	3.69	**3.95**	2.76	2.53
Reducing Time to Resolve Dispute	3.25	3.30	3.54	**3.93**	2.86	2.60
Enhancing Understanding of Dispute	3.36	3.24	3.65	**3.86**	3.71	3.35
Minimizing Future Disputes	3.06	3.10	3.16	**3.48**	2.64	2.45
Opening Channels of Communication	2.49	3.24	3.76	**3.89**	2.70	2.71
Preserving Job Relationships	2.08	3.13	**3.66**	3.53	2.30	2.29
Meeting Job Budget	2.09	2.92	**3.49**	3.07	2.08	2.09

TABLE C

PERCEIVED EFFECTIVENESS OF VARIOUS APPROACHES (DESIGN PROFESSIONALS)

(1 = Very Ineffective; 5 = Very Effective)

	Binding Arbitration	Dispute Review Board	Early Neutral Evaluation	Mediation	Mini-trial	Nonbinding Arbitration	Partnering
Reducing Cost of Resolving Dispute	2.74	3.14	3.72	3.55	2.47	2.58	**3.87**
Reducing Time to Resolve Dispute	2.59	2.95	3.55	3.38	2.26	2.35	**3.90**
Enhancing Understanding of Dispute	2.84	3.44	**3.62**	3.36	3.11	2.89	3.01
Minimizing Future Disputes	2.60	2.77	3.21	2.90	2.29	2.32	**3.94**
Opening Channels of Communication	2.07	2.89	3.53	3.23	2.03	2.20	**4.44**
Preserving Job Relationships	1.78	2.68	3.35	2.61	1.77	1.79	**4.32**
Meeting Job Budget	1.60	2.28	2.76	2.13	1.58	1.60	**3.80**
Meeting Job Schedule	1.55	2.40	2.99	2.19	1.57	1.62	**3.99**

TABLE D
PERCEIVED EFFECTIVENESS OF VARIOUS APPROACHES (CONSTRUCTION CONTRACTORS)
(1 = Very Ineffective; 5 = Very Effective)

	Binding Arbitration	Dispute Review Board	Early Neutral Evaluation	Mediation	Mini-trial	Nonbinding Arbitration	Partnering
Reducing Cost of Resolving Dispute	3.07	2.89	3.29	3.25	2.45	2.95	4.05
Reducing Time to Resolve Dispute	3.08	2.81	3.07	3.14	2.18	2.84	4.06
Enhancing Understanding of Dispute	3.36	3.05	3.16	3.40	3.02	3.40	3.41
Minimizing Future Disputes	2.71	2.51	2.96	2.68	2.13	2.52	3.92
Opening Channels of Communication	2.34	2.37	2.87	2.72	1.79	2.37	4.45
Preserving Job Relationships	2.07	2.13	2.63	2.24	1.79	1.95	4.36
Meeting Job Budget	1.90	1.84	2.59	1.77	1.44	1.87	3.81

Appendix 9

Data-Collection Survey: Claim Susceptibility for Construction Projects*

Directions

The following questions survey construction projects for sensitivity to disputes. The purpose of the survey is to find a correlation between project variables and project-dispute susceptibility. This information will help predict susceptibility for future projects.

Please complete a separate packet for each project to be surveyed. Projects should be completed and have no outstanding disputes. Each packet should take approximately 30 minutes to answer. The survey contains 37 questions and is divided into four parts:

- People—organization and individuals of the major contractual parties

- Project—project information and variables

- Process—preconstruction planning and contract information

- Performance review—severity of disputes on the completed project

The first three sections refer to project environment variables, and the last section rates the effectiveness of the project relative to disputes. The survey also requests some background information on the project.

Each of the questions has a separate scale to which respondents rate their particular project. The scale ranges from 1 (worst) to 6 (best) for all questions.

People

Questions 1-7 refer to the owner and contractor of the project. Each question must be answered twice—first for the owner, then the contractor. An owner is the organization managing the construction for whomever will own it. The contractor is the organization responsible for the project's design *and* construction or may be the general contractor only responsible for the overall construction.

---

*From "Are Contract Disputes Predictable?" published by the American Society of Civil Engineers' *Journal of Construction Engineering and Management*, Vol. 121, No. 4, 1995.

1. Capable management. Consider the owner/contractor's organization and level of skill of the upper management. Upper management are those people at a home office or corporate level responsible for the overall success of the project. Their responsibility stretches beyond "the contract" to include long-term business concerns and customer satisfaction objectives.

1a. Was the *owner's* upper management heavily involved in the overall management of the project?
(No-Yes)   1   2   3   4   5   6

1b. Was the *contractor's* upper management heavily involved in the over-all management of the project?
(No-Yes)   1   2   3   4   5   6

1c. Were previous contractual parties satisfied with the *owner's* upper managerial support and responses?
(No-Yes)   1   2   3   4   5   6

1d. Were previous contractual parties satisfied with the *contractor's* upper managerial support and responses?
(No-Yes)   1   2   3   4   5   6

2. Effectiveness of responsibility structures. If the owner/contractor's responsibilities within the organization are clear and if the people are allowed to make decisions on matters within their control, the responsibility structure is effective.

2a. Was the *owner's* responsibility structure effective?
(No-Yes)   1   2   3   4   5   6

2b. Was the *contractor's* responsibility structure effective?
(No-Yes)   1   2   3   4   5   6

3. Organization's experience with this type of project. This deals with the experience of the owner/contractor's organization as a whole and not with the individuals.

3a. Did the *owner's* organization have experience with this type of project?
(None-a lot)   1   2   3   4   5   6

3b. Did the *contractor's* organization have experience with this type of project?
(None-a lot)   1   2   3   4   5   6

4. Success of past projects. A company may be deemed successful and reputable if previous projects have performed well with regard to schedule, budget, and minimal disputes.

4a. Would the *owner's* organization be considered "successful" based on its history before this project?
(No-Yes) 1 2 3 4 5 6

4b. Would the *contractor's* organization be considered "successful" based on its history before this project?
(No-Yes) 1 2 3 4 5 6

5. Individuals' experience/competence. The owner/contractor's individuals responsible for construction and management are project managers, project engineers, superintendents, etc.

5a. What was the experience and competence level of the *owner's* project individuals?
(Low-High) 1 2 3 4 5 6

5b. What was the experience and competence level of the *contractor's* project individuals?
(Low-High) 1 2 3 4 5 6

6. Individuals' motivation (reward structures). Individuals' motivation may be a result of the organization's goals or the way in which employees are compensated.

6a. Did the *owner's* individuals have direct, tangible, personal incentives to avoid or resolve disputes?
(No-Yes) 1 2 3 4 5 6

6b. Did the *contractor's* individuals have direct, tangible, personal incentives to avoid or resolve disputes?
(No-Yes) 1 2 3 4 5 6

7. Interpersonal skills. A high level of interpersonal skills may be inherent or a result of experience or training for these skills.

7a. How would you classify the *owner's* individuals' interpersonal skills?
(Low-High) 1 2 3 4 5 6

7b. How would you classify the *contractor's* individuals' interpersonal skills?
(Low-High) 1 2 3 4 5 6

8. Team building. Team building, or "partnering," is considered to be a commitment between the organizations for the purpose of achieving specific business objectives for the length of the project. The relationship is based upon trust, dedication to common goals, and understanding of each other's individual expectations and values. Was a formal "Team Building" approach honestly carried out before the project began?
(No-Yes) 1 2 3 4 5 6

9. History together. A history between the owner and contractor on past projects may affect the current project. No experience together has a neutral rating. How would the history of the owner and contractor together on previous projects rate? (If no prior history, mark 4).
   (Poor-Good)   1   2   3   4   5   6

10. Uneven power balance. A stronger company may have an advantage in settling disputes, due to its financial, experiential, or technical levels. Disagreements between parties may be difficult to solve when one feels threatened. Did the owner and contractor have equivalent abilities and resources in order to absorb costs associated with disputes?
   (No-Yes)   1   2   3   4   5   6

11. Expectations of further work. While a project is under way, organizations may have expectations of work together on future projects. These expectations can affect how diplomatically disputes are handled. To what extent did the success of this project affect the possibility of work together on future projects?
   (Low-High)   1   2   3   4   5   6

Project

12. Environmental issues. This category considers the natural or physical environment in which the project was constructed. Was the project considered to be environmentally sensitive?
   (Yes-No)   1   2   3   4   5   6

13. Public interference. Problems can arise when construction projects conflict with the public's prerogative for comfort and safety. Traffic interference may occur or a high-profile project, such as a hazardous-waste incinerator, may cause local discontent. What was the probability and intensity of public interference for this project?
   (High-Low)   1   2   3   4   5   6

14. Site limitations. Project site limitations include, but are not limited to, storage and access for staging and setup. How did the project rate in terms of site limitations and demand for space?
   (Poor-Good)   1   2   3   4   5   6

15. Remoteness. Was the project located in an area where materials and technical expertise were locally available *and* adequate?
   (No-Yes)   1   2   3   4   5   6

16. Availability of capable craftsmen/subs. What was the availability of skilled workers and subcontractors for successful completion of this project?
    (Low-High) 1 2 3 4 5 6

17. Pioneer projects. A "pioneer" project includes aspects such as new technology, which have never been constructed or used before. To what extent was this a "pioneer" project?
    (High-Low) 1 2 3 4 5 6

18. Design complexity. This entails the complexity of the design, not innovation, of the project. A nuclear-power plant is not a "pioneer" project, but does have a complex design. What was the level of design complexity for this project?
    (High-Low) 1 2 3 4 5 6

19. Construction complexity. What was the level of construction complexity and innovation needed for this project?
    (High-Low) 1 2 3 4 5 6

20. Size. Large-scale projects increase the possibility of disputes. Was the project considered to be unusually large?
    (Yes-No) 1 2 3 4 5 6

Process

21. Input from all groups involved. Preconstruction planning includes potential for construction and value engineering studies, in which information is shared among parties. To what extent was quality input shared during this phase of preconstruction?
    (Low-High) 1 2 3 4 5 6

22. Financial planning. The quality of financial planning is sometimes dependent on the size of the project, the current economic situation, and the possibility of changes or additions to the contract. What was the level of experience and effort of the financial planners, and what was the adequacy of the financial plan?
    (Low-High) 1 2 3 4 5 6

23. Permits and regulations. Were the regulatory requirements, such as building permits and environmental impact studies, identified and completed in a timely manner before construction?
    (No-Yes) 1 2 3 4 5 6

24. Scope definition. The scope defines what work is included in the project. Regarding the type of project and contract used, was the scope appropriately defined?
    (No-Yes) 1 2 3 4 5 6

25. Realistic obligations. Contractual obligations, such as scheduling, budget, and quality of work, need to be practical to avoid disputes. Were the contractual obligations considered to be realistic by both parties before construction began?
    (No-Yes) 1 2 3 4 5 6

26. Risk identification/allocation. Organizations involved in a construction project should identify relevant risks associated with the project and allocate risk liability to the project participants appropriately by means of the contract. Proper allocation ensures assignment to the party with the best ability to deal with the risk. Methods of proper allocation are good contractual language or use of the appropriate contract (turnkey, design/build, general contractor, etc.) How well were risks identified and properly allocated?
    (Poor-Good) 1 2 3 4 5 6

27. Adequacy of technical plans/specifications. Adequacy of technical plans/specs include development and review, completeness, clarity, and organization of the bid documents. How did the adequacy of technical plans/ specs rate?
    (Poor-Good) 1 2 3 4 5 6

28. Formal dispute resolution process. Formal dispute-resolution processes are helpful in keeping disputes from becoming claims. Such processes include, but are not limited to, many ADR methods: DRB, rapid response teams, minitrials, mediations, negotiation. How well did the contract spell out such a formal dispute resolution process?
    (Poor-Good) 1 2 3 4 5 6

29. Operating procedures. Such procedures include payment procedures, schedule-updating procedures and requirements, submittal methods, and meeting and communication procedures. To what extent were these procedures spelled out and reasonable?
    (Low-High) 1 2 3 4 5 6

Performance

30. What was the type of project for which this survey was completed? (heavy highway-civil, industrial, commercial, etc.)

31. What role did the "contractor" of this project play? (general contractor, design/builder, construction manager)

32. What was the general payment method used by this contract? (lump sum, cost plus, unit price)

33. What was the original estimated dollar amount for this project?

34. How would you rate the frequency of disputes which arose at the field level?
    (High-Low)   1   2   3   4   5   6

35. How would you rate the severity of the largest dispute/claim that arose on your project?
    (High-Low)   1   2   3   4   5   6

36. What is the estimated number of disputes that were settled beyond the field level?

37. What was the final dollar amount of these disputes that were settled beyond the field level?

Appendix 10

Trends and Evolving Risks in Design/Build, BOT and BOOT Projects*

The Construction Industry (CII) in 1990 published the results of a broadly based "Assessment of Construction Industry Project Management Practices and Performance," the University of Texas, April 1990, page 25. The Assessment presented the Owner, Architect/Engineer, and the Contractor perceptions of success in achieving project goals:

Table 8 Comparison of Owner, Architect/Engineer, and Contractor Performance			
Performance Measure	**Average Percent Frequency of Meeting or Performing Better**		
	Owner	**A/E**	**Contractor**
Cost	61	61	67
Schedule	66	61	71
Technical	80	78	
Quality			83
Safety			85
Profit		65	75

*Kris R. Neilsen, ASCE/ICE Triennial Conference, Session VI: Pitfalls in International Engineering and Construction: What to Watch For, Philadelphia, Pennsylvania, September 1996.

Table 2 Cultural Traits Which Increase Risk in Multi-national Consortia			
Cultural Trait	**Europe**	**USA**	**Japan**
Objectives for business entity	Continuity and social values	Profitability	Permanent existence
Basic business principles	Fair competition	Fair competition	Impartial (fair) sharing
Characteristical features	Reliability	Self-assertion	Harmony
Business style	Client first policy	Short-term competitive relationships with long term focus	Long-term credible relationships
Working condition	Individual	Individual often within a team	Teamwork
Employment form	Long-term employment; Improvement of position by changing jobs	Improvement of position by changing jobs	Lifetime employment
Employment attitude	Employing individual	Employing the individual's skills	Employing individual
Principles of behavior	Participate, create, and work skills	Participate in education and manifestation of skill	Attend, learn, and labor
Wage system	Ability, achievement, and rank	Ability and achievement	Seniority and achievement

Table 2			
Cultural Traits Which Increase Risk in Multi-national Consortia			
Cultural Trait	**Europe**	**USA**	**Japan**
Measure of business achievement	Short-range profit	Short-range profit	Contracts awarded and long-range profits
Changes to business entity	Slow	Rapid	Slow
Decision-making process	Discussions between superiors and subordinates	Moving from top-down to flat team	Bottom-up and mutual-agreement
Working environment	Individual Offices	Individual Spaces	Large, shared offices
Loyalty to organization	Medium	Little	Great
Competition with organization	Avoid	Broad	Avoid
Relations between colleagues	Friendship	Individual with movement to task teams	Sense of commonness
Perception of work	Responsibility	Responsibility	Life Employment
Decision criteria tendency	Results-oriented	Ideas, philosophy, and processes	Results-oriented
Human resources	Long-range assets	Floating assets	Fixed assets
Reward	Big	Big	Small (bonuses, promotions, and salary)
Punishment	Relocation or dismissal	Dismissal	Relocation
In-house education	Permanent	Considered little, self-enlightenment promoted	Systematic and seriously taken
Salary difference	Medium	Big	Small

Table 3 Agreements with BOT/BOOT Consortium	
Party	**Contract Subject**
Government	Authorizations, approvals, and assurances — possible equity participant; party to agreement for power produced; plant purchase option
Shareholders/Equity Owners	Equity subscription agreement — base commitment; overrun commitment
Lenders	Term loan agreement by banks; insurance companies; others
Design-Build Consortium	Design and construction contract cost schedule, and performance guarantees
Plant Operation	Operating Agreement
Insurers	Insurance — Builder's Risk; efficacy; force majeure, etc.
Creditworthy Power Purchaser/Consumers	Power purchase agreement
Fuel Supplier	Fuel supply agreement

Appendix 11

AAA Demand for Arbitration and Answer

American Arbitration Association
CONSTRUCTION INDUSTRY ARBITRATION RULES
DEMAND FOR ARBITRATION

MEDIATION If you want the AAA to contact the other party and attempt to arrange a mediation, please check this box.	☐

TO: Name of Respondent			Name of Representative (if known)		
Address			Address		
City	State	Zip Code	City	State	Zip Code
Phone No.	Fax No.		Phone No.	Fax No.	

THE NAMED CLAIMANT, A PARTY TO A WRITTEN AGREEMENT PROVIDING FOR ARBITRATION UNDER THE CONSTRUCTION INDUSTRY ARBITRATION RULES, HEREBY DEMANDS ARBITRATION THEREUNDER. (ATTACH THE ARBITRATION CLAUSE.)

NATURE OF DISPUTE (Please give enough details to enable the AAA to select arbitrators with appropriate experience.):

DOLLAR AMOUNT OF CLAIM:
$

OTHER RELIEF SOUGHT:

PLEASE DESCRIBE APPROPRIATE QUALIFICATIONS FOR ARBITRATOR(S) TO BE APPOINTED TO HEAR THIS DISPUTE:

CLAIMANT IS:
☐ Owner ☐ Design Professional (specify_____) ☐ Contractor
☐ Subcontractor (specify_____) ☐ Other (specify_____)

RESPONDENT IS:
☐ Owner ☐ Design Professional (specify_____) ☐ Contractor
☐ Subcontractor (specify_____) ☐ Other (specify_____)

ESTIMATED TIME NEEDED FOR HEARINGS OVERALL: _____ hours _____ days

Copies of this demand are being filed with the American Arbitration Association at its _____ office. Claimant requests that the AAA commence the administration of the arbitration. Under the rules, you may file an answering statement within ten days after notice from the AAA.

CLAIMANT REQUESTS THAT ARBITRATION HEARINGS BE HELD AT THE FOLLOWING LOCALE:

Signature (may be signed by a representative)		Title	Date

Name of Claimant			Name of Representative		
Address			Address		
City	State	Zip Code	City	State	Zip Code
Phone No.	Fax No.		Phone No.	Fax No.	

TO INSTITUTE PROCEEDINGS, PLEASE SEND THREE COPIES OF THIS DEMAND AND THE ARBITRATION AGREEMENT, WITH THE FILING FEE, AS PROVIDED FOR IN THE RULES, TO THE AAA. SEND THE ORIGINAL DEMAND TO THE RESPONDENT.

Source: Construction ADR Task Force Report, October 26, 1995, American Arbitration Association.

American Arbitration Association
CONSTRUCTION INDUSTRY ARBITRATION RULES
ANSWERING STATEMENT

MEDIATION If you want the AAA to contact the other party and attempt to arrange a mediation, please check this box. ☐

TO: Name of Claimant			Name of Representative (if known)		
Address			Address		
City	State	Zip Code	City	State	Zip Code
Phone No.	Fax No.		Phone No.	Fax No.	

RESPONDENT ANSWERS CLAIMANT'S DEMAND FOR ARBITRATION AS FOLLOWS:
(Please describe the dispute and any counter-claim in sufficient detail so the AAA may select an arbitrator with appropriate qualifications and experience.) AAA Case # (if known)_____.

DOLLAR AMOUNT OF COUNTER-CLAIM:	OTHER RELIEF SOUGHT:
$	

PLEASE DESCRIBE APPROPRIATE QUALIFICATIONS FOR ARBITRATOR(S) TO BE APPOINTED TO HEAR THIS DISPUTE:

ESTIMATED TIME NEEDED FOR HEARINGS OVERALL: _____ hours _____ days

RESPONDENT REQUESTS THAT ARBITRATION HEARINGS BE HELD AT THE FOLLOWING LOCALE:

Signature (may be signed by a representative)	Title	Date

Name of Respondent			Name of Representative		
Address			Address		
City	State	Zip Code	City	State	Zip Code
Phone No.	Fax No.		Phone No.	Fax No.	

PLEASE SEND THREE COPIES OF THIS ANSWERING STATEMENT, WITH THE FILING FEE FOR ANY COUNTER-CLAIM, AS PROVIDED FOR IN THE RULES, TO THE AAA. SEND THE ORIGINAL ANSWERING STATEMENT TO THE CLAIMANT.

Index